INORGANIC SYNTHESES

Volume 32

●●●●●●●

Editor-in-Chief
MARCETTA YORK DARENSBOURG
Texas A&M University

●●●●●●●●●●●●●●●●●●●●●●●●●●●●●

INORGANIC SYNTHESES

Volume 32

A Wiley-Interscience Publication
JOHN WILEY & SONS, INC.

New York   Chichester   Weinheim   Brisbane   Singapore   Toronto

Library of Congress Catalog Number: 39–23015

ISBN 0–471–24921–1

Printed in the United States of America

10 9 8 7 6 5 4 3 2 1

This volume is dedicated to Sir Geoffrey Wilkinson (1921–1996), who was for many years one of the International Associates of Inorganic Syntheses. His monumental contributions to inorganic and organometallic chemistry, and in particular his emphasis on the importance of synthetic chemistry, will long be remembered.

PREFACE

The importance of continuing advances in ligand design for the role of coordination chemistry in research and practical applications can hardly be overstated. Thus a major focus of this volume of *Inorganic Syntheses* is on ligand synthesis. To assure that the associated organic synthesis does not overwhelm our inorganic tradition, in almost all cases examples of the synthesis of transition metal complexes using these ligands are also included. As indicated above, the term "continuing advances" signifies that future volumes could be similarly dedicated. Currently developing hot areas, such as the ligand–transition metal complex templating of the peptide bundle assembly, or ligands designed for stereoselective oxidation reactions, are not discussed herein. However, the mindful reader will undoubtedly find inspiration within this volume, from those ligands designed for other research purposes. Rather it is the bulk quantity, gram to multigram lots, of water-solubilizing phosphine ligands for organometallics already proven to be of importance to product and catalyst separation in industry that form the basis of Chapter 1. Included are phosphines made water soluble by sulfonation (i.e., ionic phosphines) and nonionic phosphines rendered water soluble by H-bonding interactions.

During the past two decades, a remarkable synergism has developed in protein crystallography, enzymology, biophysical chemistry, and inorganic synthesis. The ligands and coordination complexes in Chapter 2 provide examples of attempts to model the function and form of the coordination environment of the active sites in metalloenzymes. The biomimetic ligand syntheses herein focus on models for the common metal-binding amino-acid residues histidine (imidazoles), cysteine (thiolates), and methionine (thioethers). Derivatives of the highly useful pyrazolylborates find form in syntheses producing neutral bi- and tridentate pyrazolyls, as well as several tripodal ligands. The requirement of isolated metal sites, which the protein catalyst so easily manages, has inspired the synthetic design of sterically bulky ligands, which has lead to abiological ligands and industrially important fundamental studies of small molecule activation.

The initial entry of Chapter 3 proves that sophisticated ligands are not a requirement for useful coordination complexes. An alternative synthesis is offered for *cisplatin*, the widely used anticancer chemotherapeutic agent which contains the ultimate neutral ligand, ammonia. The coordination complexes of Chapter 3 traverse the periodic table and provide a number of labile ligand complexes which might be used as precursors for ligand exchange with those of Chapters 2 and 3. Chapters 4 and 5 are contributed syntheses of cluster compounds and of hydrides. Although short, the hydride chapter presents pearls, epitomizing the spirit of *Inorganic Syntheses* for cleverly improved and useful syntheses.

I was gratified to hear, during a workshop for new graduate students, the admonition to be skeptical of the literature, for it was only the collections of *Inorganic Syntheses* and *Organic Syntheses* that one could really trust to have been checked. Nevertheless, even with the rigorous reviewing and checking procedures of *Inorganic Syntheses*, there are occasional errors in interpretation, and we acknowledge one such in Chapter 6. We are grateful to our readers for bringing this to our attention and to them as well as the original author for returning to the problem and sorting it out.

Although many members of the editorial board of *Inorganic Syntheses* served as active reviewers of the submissions of this volume, two stand out for special recognition—Du Shriver and John Shapley. Du's careful editing of all manuscripts that crossed his desk was the starting point for my own editing, and helped me considerably; John's offer to check cluster syntheses went beyond the call of duty. In this regard, I must mention that the impact of *Inorganic Syntheses* on entire areas of development in inorganic chemistry depends on the enthusiasm or willingness (or lack thereof) of the checkers. To all contributors and checkers, I acknowledge my deepest gratitude. Finally, I must thank two undergraduates for keeping chaos at bay—Matthew Miller, a summer researcher from Grove City College, who responded to my call for help during the final assembly of the manuscript and Curtis Franke, a TAMU chemistry major/student worker, for his tremendous organizational skills while being the exemplar "Good Ag."

MARCETTA YORK DARENSBOURG

College Station, Texas

NOTICE TO CONTRIBUTORS
AND CHECKERS

The *Inorganic Syntheses* series is published to provide all users of inorganic substances with detailed and foolproof procedures for the preparation of important and timely compounds. Thus the series is the concern of the entire scientific community. The Editorial Board hopes that all chemists will share in the responsibility of producing *Inorganic Syntheses* by offering their advice and assistance in both the formulation and the laboratory evaluation of outstanding syntheses. Help of this kind will be invaluable in achieving excellence and pertinence to current scientific interests.

There is no rigid definition of what constitutes a suitable synthesis. The major criterion by which syntheses are judged is the potential value to the scientific community. An ideal synthesis is one that presents a new or revised experimental procedure applicable to a variety of related compounds, at least one of which is critically important in current research. However, syntheses of individual compounds that are of interest or importance are also acceptable. Syntheses of compounds that are readily available commercially at reasonable prices are not acceptable. Corrections and improvements of syntheses already appearing in *Inorganic Syntheses* are suitable for inclusion.

The Editorial Board lists the following criteria of content for submitted manuscripts. Style should conform with that of previous volumes of *Inorganic Syntheses*. The introductory section should include a concise and critical summary of the available procedures for synthesis of the product in question. It should also include an estimate of the time required for the synthesis, an indication of the importance and utility of the product, and an admonition if any potential hazards are associated with the procedure. The Procedure should present detailed and unambiguous laboratory directions and be written so that it anticipates possible mistakes and misunderstandings on the part of the person who attempts to duplicate the procedure. Any unusual equipment or procedure should be clearly described. Line drawings should be included when they can be helpful. All safety measures should be stated clearly. Sources of unusual starting materials must be given, and, if possible, minimal standards of purity of reagents and solvents should be stated. The scale should be reasonable for normal laboratory operation, and any problems involved in scaling the procedure either up or down should be discussed. The criteria for judging the purity of the final product should be delineated clearly. The Properties section should supply and discuss those physical and chemical characteristics that are relevant to judging the purity of the product and to permitting its handling and use in an intelligent manner. Under References, all pertinent literature citations should be listed in order. A style sheet is available from the Secretary of the Editorial Board.

The Editorial Board determines whether sumitted syntheses meet the general specifications outlined above. Every procedure will be checked in an independent laboratory, and publication is contingent upon satisfactory duplication of the syntheses.

Each manuscript should be submitted in duplicate to the Secretary of the Editorial Board, Professor Jay H. Worrell, Department of Chemistry, University of South Florida, Tampa, FL 33620. The manuscript should be typewritten in English. Nomenclature should be consistent and should follow the recommendations presented in *Nomenclature of Inorganic Chemistry*, 2nd ed., Butterworths & Co, London, 1970 and in *Pure and Applied Chemistry*, Volume 28, No. 1 (1971). Abbreviations should conform to those used in publications of the American Chemical Society, particularly *Inorganic Chemistry*.

Chemists willing to check syntheses should contact the editor of a future volume or make this information known to Professor Worrell.

TOXIC SUBSTANCES AND LABORATORY HAZARDS

Chemicals and chemistry are by their very nature hazardous. Chemical reactivity implies that reagents have the ability to combine. This process can be sufficiently vigorous as to cause flame, an explosion, or, often less immediately obvious, a toxic reaction.

The obvious hazards in the syntheses reported in this volume are delineated, where appropriate, in the experimental procedure. It is impossible, however, to foresee every eventuality, such as a new biological effect of a common laboratory reagent. As a consequence, *all* chemicals used and *all* reactions described in this volume should be viewed as potentially hazardous. Care should be taken to avoid inhalation or other physical contact with all reagents and solvents used in this volume. In addition, particular attention should be paid to avoiding sparks, open flames, or other potential sources which could set fire to combustible vapors or gases.

A list of 400 toxic substances may be found in the *Federal Register*, Volume 40, No. 23072, May 28, 1975. An abbreviated list may be obtained from *Inorganic Syntheses*, Vol. 18, p. xv, 1978. A current assessment of the hazards associated with a particular chemical is available in the most recent edition of *Threshold Limit Values for Chemical Substances and Physical Agents in the Workroom Environment* published by the American Conference of Governmental Industrial Hygienists.

The drying of impure ethers can produce a violent explosion. Further information about this hazard may be found in *Inorganic Syntheses*, Volume 12, p. 317.

CONTENTS

Chapter Three **TRANSITION METAL COMPLEXES
AND PRECURSORS**

**Chapter Five   MAIN GROUP AND TRANSITION METAL
HYDRIDES**

Chapter Six   TITANIUM(III) CHLORIDE

(A Correction to Inorganic Syntheses, Vol. 24, pp. 181 (1986)

INORGANIC SYNTHESES

Volume 32

Chapter One

LIGANDS FOR WATER-SOLUBILIZING ORGANOMETALLIC COMPOUNDS

1. (META-SULFONATOPHENYL)DIPHENYLPHOSPHINE, SODIUM SALT AND ITS COMPLEXES WITH RHODIUM(I), RUTHENIUM(II), IRIDIUM(I)

Submitted by FERENC JOÓ,* JÓZSEF KOVÁCS,* ÁGNES KATHÓ,* ATTILA C. BÉNYEI,* TARA DECUIR,[†] and DONALD J. DARENSBOURG[†]
Checked by ALEX MIEDANER and DANIEL L. DUBOIS*[‡]

The preparation of (meta-sulfonatophenyl)diphenylphosphine, sodium salt in 1958 initiated the era of water-soluble transition metal complexes with tertiary phosphines as ligands.[1] Water-soluble analogs of several prominent PPh$_3$-containing complexes were synthesized[2-4] with this ligand (commonly abbreviated as TPPMS) of which the preparations of RhCl(TPPMS)$_3$, [RuCl$_2$(TPPMS)$_2$]$_2$ and trans-IrCl(CO)(TPPMS)$_2$, are given in this section.[§]

* Institute of Physical Chemistry, Lajos Kossuth University, Debrecen, H-4010 Hungary.
[†] Department of Chemistry, Texas A&M University, College Station, TX, 77843.
[‡] National Renewable Energy Laboratory, 1617 Cole Boulevard, Golden, CO 80401.
[§] Commercial starting materials were used: PPh$_3$ (Aldrich), RhCl$_3 \cdot$ 3H$_2$O (Fluka), RuCl$_3 \cdot$ 3H$_2$O (Johnson Matthey), RuCl$_2$(PPh$_3$)$_3$ (Fluka) trans-IrCl(CO)(PPh$_3$)$_2$ (Strem Chemicals), and 20% fuming sulfuric acid (Aldrich).

1

A. (*META*-SULFONATOPHENYL)DIPHENYLPHOSPHINE, SODIUM SALT (MONO-SULFONATED TRIPHENYLPHOSPHINE, TPPMS)

$$P(C_6H_5)_3 + SO_3 \rightarrow P(C_6H_5)_2(C_6H_4\text{-}m\text{-}SO_3H)$$

$$P(C_6H_5)_2(C_6H_4\text{-}m\text{-}SO_3H) + NaOH + H_2O \rightarrow$$

$$P(C_6H_5)_2(C_6H_4\text{-}m\text{-}SO_3Na \cdot 2H_2O)$$

Procedure

This procedure is a slight modification of that described in reference 1. The essential difference is that in order to avoid multiple sulfonation, *incomplete monosulfonation* of PPh$_3$ is employed and the micellar aggregates of TPPMS/ PPh$_3$ formed upon neutralization are broken with prolonged heating of the aqueous emulsion under argon.

■ **Caution.** *The reaction must be carried out in a well-ventilated hood! Since 20% fuming sulfuric acid and 50% NaOH are strongly caustic, appropriate rubber gloves and safety glasses must be used throughout the preparation!*

In a 100-mL Erlenmeyer flask chilled in ice water, 20 g (76.3 mmol) of finely ground PPh$_3$ is dissolved in 50 mL of 20% fuming sulfuric acid (contains 230 mmol free SO$_3$). The phosphine is added in small portions, then the flask is covered with a watchglass and is gently shaken to dissolve most of PPh$_3$ before the next addition. Complete dissolution of PPh$_3$ requires approximately 30 min. The solution is then left to warm to room temperature and placed on top of a boiling water (steam) bath where it is kept for 75 min with occasional swirling. After cooling to room temperature, the solution is carefully poured onto 400 g of crushed ice in a 1-L beaker.

■ **Caution.** *Pouring the sulfonation mixture into ice generates a high amount of heat and should be done carefully to avoid splashing and local boiling!*

A milky emulsion is obtained with some gummy material which sticks to the wall of the beaker and to the glass stirring rod.* This emulsion is further cooled in ice water and carefully neutralized by slow addition of approximately 140 mL 50% aqueous NaOH solution until pH 3–4, yielding a white precipitate.

---

* At this point the checkers observed a clear solution.

■ **Caution.** *Neutralization of the strongly acidic sulfonation mixture with strong NaOH solution generates high amounts of heat and should be done carefully to avoid splashing and local boiling!*

The resulting solution is allowed to stand at room temperature for 30 min after which the precipitate is filtered with *gentle* suction on a Büchner funnel and, finally, as much of the remaining moisture is pressed out with a suitable flat glass stopper. This raw product contains TPPMS, nonsulfonated starting material, and some sodium sulfate. The crude product is dissolved in 800 mL hot water, placed in an Erlenmeyer flask on top of a boiling water bath and flushed with a stream of argon. During this time the milky emulsion slowly clears,* and in 1–3 h a clear, or only very slightly cloudy, solution is obtained while the nonsulfonated PPh_3 collects on the bottom of the flask.† This solution is left to cool until the PPh_3 melt solidifies and then is quickly filtered through a Celite (Hyflo Super-Cell) pad. Typically, 4 g (15 mmol) of crystalline PPh_3 is recovered, which can be used in subsequent preparations. The filtrate is allowed to cool under argon and then placed in a refrigerator overnight. The product crystallizes as small shiny white flakes and is dried over P_4O_{10}. Yield: 12.5 g, 51% (based on reacted PPh_3). This material contains 2–5% PPh_3, is virtually free (HPLC) from phosphine oxide (OTPPMS) and multiply sulfonated byproducts, and is suitable for subsequent preparation of water-soluble metal complexes via ligand exchange of PPh_3-containing starting materials. Analytically pure TPPMS is obtained by recrystallization of 10 g of this compound from 100 mL water under argon. Yield: 5.7 g, 57% (overall yield 29%).

Anal. Calcd. for $C_{18}H_{18}O_5PSNa$: C, 53.99; H, 4.53: Found: C, 54.33; H, 4.60.

Properties

The solubility of the sodium salt of (*meta*-sulfonatophenyl)diphenylphosphine, TPPMS in water is approximately 12 g/L at room temperature.[5] It dissolves slightly in cold ethanol, but is soluble at elevated temperatures. It is virtually insoluble in acetone and aliphatic, aromatic, or chlorinated hydrocarbons, but is soluble at room temperature in tetrahydrofuran. The compound crystallizes with two waters of crystallizations; however, the anhydrous form can also be obtained.[3] Dry TPPMS is stable to air but is oxidized rapidly when wet, especially in basic aqueous solutions. It is a highly surfactant compound and forms aggregates and micelles in neutral aqueous

---

* The checker observed a clear solution.
† In two separate preparations, the checkers did not observe separation of PPh_3.

solutions.[5] Purity of the product can be checked by RP HPLC. The column utilized consisted of Nucleosil 5 C_{18}, 250 mm; eluants A: 10 v/v% CH_3OH + 90 v/v% 0.5 w/v% aqueous $(NH_4)_2CO_3$; B: 90 v/v% CH_3OH + 10 v/v% 0.5 w/v% aqueous $(NH_4)_2CO_3$; gradient: 0–2 min 90% A + 10% B, 15–35 min 10% A + 90% B, 38–40 min 90% A + 10% B, continuous change; detection at $\lambda = 260$ nm. The samples (0.2 mg/mL) were prepared in eluant B, injection 20 μL. The retention times were: TPPMS oxide, 11.0 min; TPPMS, 16.1 min; PPh$_3$ oxide, 18.3 min; PPh$_3$, 24.3 min. Spectral properties: UV-vis (water) λ_{max} 260 nm (ε 6.7×10^2 M^{-1} cm^{-1}); IR (KBr): $\nu(SO_3)$ 1196 cm^{-1}; NMR (δ; D_2O, 200 MHz) ^1H: 6.25 ppm (broad singlet) and 6.98 ppm (doublet); ^{31}P: -3.88 ppm (singlet).

TPPMS serves as ligand in a variety of catalysts for hydrogenation,[6] hydroformylation, and C–C bond formation.[7] In aqueous solutions, it reacts with activated olefins,[8] alkynes,[9] and aliphatic as well as aromatic aldehydes,[10] giving the corresponding substituted alkylphosphonium salts.

B. CHLOROTRIS[(*META*-SULFONATOPHENYL)
DIPHENYLPHOSPHINE]-RHODIUM(I), SODIUM SALT,
RhCl(TPPMS)₃ (WATER-SOLUBLE ANALOG OF
WILKINSON'S CATALYST)

$$RhCl_3 \cdot 3H_2O + 4TPPMS \rightarrow RhCl(TPPMS)_3 + TPPMSO + 2HCl + 2H_2O$$

Procedure

The procedure is analogous to the preparation of RhCl(PPh$_3$)$_3$.[11] It is important to note that exchange of PPh$_3$ by TPPMS gives an isolable product only in *wet* THF.[3] In a 100-mL Schlenk flask equipped with a reflux condenser, an exit bubbler and a magnetic stirring bar, 30 mL 96% ethanol (4% water) is deoxygenated by refluxing in an argon or N_2 stream for 15 min. 1.4 g (3.5 mmol) finely ground TPPMS is dissolved in the hot ethanol. 150 mg RhCl$_3$ · 3H$_2$O ($\sim$ 40% Rh; 0.6 mmol Rh) is dissolved in 5 mL deoxygenated, hot 96% ethanol and added to the hot solution of the phosphine under an argon or N_2 stream. Residual rhodium trichloride is washed into the phosphine solution using another 1 mL portion of alcohol. The deep cherry red solution quickly turns brown and a fine, orange-yellow precipitate appears. The solution is refluxed while stirring for 20 min and then allowed to cool to room temperature. Using Schlenk techniques, the precipitate is filtered, washed with 5 mL of 96% deoxygenated alcohol and dried first in an argon or N_2 stream and then in vacuo over P_4O_{10} overnight. Yield of the hydrated complex: 563 mg, 70% for Rh.

Anal. Calcd. for $C_{54}H_{54}O_{13}ClP_3S_3Na_3Rh$ (tetrahydrate): C, 48.4; H, 4.06. Found: C, 49.3; H, 3.43.

Properties

RhCl(TPPMS⁻Na⁺)₃ is an orange-yellow crystalline material, highly soluble in water, practically insoluble in anhydrous THF, ethanol, acetone or diethyl ether. As a solid, it is moderately stable to air oxidation and can be kept for several months in a dry, cool place with no decomposition. In aqueous solutions, RhCl(TPPMS)₃ is rapidly oxidized by O_2. Moreover, it undergoes internal redox processes, even in anaerobic conditions, yielding Rh(III)-containing products and phosphine oxide. Such decomposition is accelerated in basic solutions and by heating. Spectral properties: UV (0.1 M HCl) λ_{max} 356 nm $(\varepsilon\ 3.2 \times 10^3\ M^{-1}\ cm^{-1})$, 444 nm $(\varepsilon\ 8.0 \times 10^2\ M^{-1}\ cm^{-1})$; IR (KBr): $v(SO_3)$ 1198 cm^{-1}; NMR $(\delta;\ D_2O,\ 200\ MHz)$ ^1H gives a complex pattern of broad peaks in the range of 6.38 to 7.42 ppm, ^{31}P exhibits a doublet of doublets centered at 34.2 ppm and a doublet centered at 53.2 ppm, at ratio of 2:1, respectively. These resonances are broad, which obscures the P–P coupling. RhCl(TPPMS)₃ is an active catalyst for hydrogenation of a variety of alkenes and alkynes both in homogeneous aqueous solutions and in aqueous/organic biphasic systems[6] and is also used for the hydrogenation of biomembranes.[12] In general, its properties are analogous to those of RhCl(PPh₃)₃.

C. DICHLOROBIS[(*META*-SULFONATOPHENYL) DIPHENYLPHOSPHINE]RUTHENIUM(II), SODIUM SALT

$$2\,RuCl_2(PPh_3)_3 + 4\,TPPMS \rightarrow [RuCl_2(TPPMS)_2]_2 + 6\,PPh_3$$

The complex can be prepared by the reaction of $RuCl_3 \cdot 3H_2O$ and TPPMS in refluxing 96% ethanol.[2] The procedure described here gives a purer product and more consistent yields.

Procedure

In a 100-mL Schlenk flask equipped with a reflux condensor, an exit bubbler, and a magnetic stirring bar, 24 mL dry THF is deoxygenated by refluxing in an Ar or a N_2 stream for 10 min. In the hot THF, 384 mg (0.4 mmol) $RuCl_2(PPh_3)_3$ is dissolved completely, followed by the addition of 384 mg (0.96 mmol) TPPMS, dissolved in 6 mL hot THF, under a constant flow of Ar or N_2. The deep brown solution yields a light brown precipitate in approximately 1 min. The mixture is refluxed under Ar or N_2 for 60 min, chilled in ice water for 5 min, and then stirred at room temperature for 30 min. The precipitate is collected under an inert atmosphere in a Schlenk filter, washed with cold, deoxygenated THF, and dried first in an argon or nitrogen stream and then in vacuo over P_4O_{10} overnight. Yield of the anhydrous complex: 331 mg, 92% based on $RuCl_2(PPh_3)_3$.

Anal. Calcd. for $C_{36}H_{28}Cl_2O_6P_2S_2Na_2Ru$ (anhydrous): C, 48.01; H, 3.14. Found: C, 47.65; H, 3.47.

Properties

$[RuCl_2(TPPMS)_2]_2$ is formed as a light brown powder, which is soluble in water, slightly soluble in THF and practically insoluble in anhydrous ethanol, acetone, and diethyl ether. Its dimeric nature is supported by NMR and is analogous to $[RuCl_2(PPh_3)_2]_2$.[13] In general, its chemical properties closely resemble those of $RuCl_2(PPh_3)_3$.[14] In the solid state, it is moderately stable to air oxidation and can be kept for several months in a dry, cool place with no decomposition. However, in aqueous solutions $[RuCl_2(TPPMS)_2]_2$ is rapidly oxidized by O_2, yielding green solutions containing a Ru(III) species. It readily absorbs H_2 in the presence of excess TPPMS, yielding $HRuCl(TPPMS)_3$.[2] Spectral properties: UV-vis (0.1 M HCl) λ_{max} 478 nm $(\varepsilon\, 3.16 \times 10^2\ M^{-1}\,cm^{-1})$; IR (KBr): $\nu(SO_3)$ 1194 cm^{-1}; NMR (δ; D_2O, 200 MHz). The ^1H NMR consists of a complex pattern of broad peaks in the range of 6.91 to 8.08 ppm. The ^{31}P NMR consists of two broad singlets at 55.3 and 54.8 ppm. The product was shown to be contaminated with TPPMS oxide, which appears as a singlet at 37.1 ppm.* In aqueous solutions at moderate temperatures ($> 50°C$), $[RuCl_2(TPPMS)_2]_2$ catalyzes the hydrogenation of a variety of alkenes, alkynes, and oxo-compounds (e.g., aldehydes).[6]

D. *trans*-CHLOROCARBONYLBIS[(*META*-SULFONATOPHENYL) DIPHENYL-PHOSPHINE]IRIDIUM(I), SODIUM SALT (WATER-SOLUBLE ANALOG OF VASKA'S COMPLEX)

$$trans\text{-}IrCl(CO)(PPh_3)_2 + 2\,TPPMS \rightarrow trans\text{-}IrCl(CO)(TPPMS)_2 + 2\,PPh_3$$

---

* A singlet at 37.1 ppm was not observed by the checkers. However, a resonance was observed for free PPh_3 which decreased when an aqueous solution was extracted with toluene.

Procedure

In a 50-mL Schlenk flask equipped with a reflux condensor, an exit bubbler, and a magnetic stirring bar, 25 mL dry THF is deoxygenated by refluxing in an Ar or N_2 stream for 20 min. In the hot THF, 387 mg (0.5 mmol) *trans*-IrCl(CO)(PPh$_3$)$_2$, Vaska's complex, is dissolved, and 400 mg (1.0 mmol) TPPMS, dissolved in 2 mL hot THF, is added under a constant stream of argon. The clear golden yellow solution becomes turbid after approximately 5 min. The mixture is refluxed under Ar for 30 min, then allowed to cool to room temperature. The solution is stirred at room temperature for 30 min. The bright yellow precipitate is collected on a Schlenk fritted-glass filter, washed with cold, deoxygenated THF, and dried first in a stream of inert gas and then in vacuo over P_4O_{10} overnight. Yield of the anhydrous complex: 350 mg, 71% based on *trans*-IrCl(CO)(PPh$_3$)$_2$.

Anal. Calcd. for $C_{37}H_{36}O_7ClP_2S_2Na_2Ir$ (anhydrous): C, 45.12; H, 2.87. Found: C, 45.10; H, 3.30.

Properties

trans-IrCl(CO)(TPPMS)$_2$ is a bright yellow crystalline compound, soluble in water and 2-methoxyethanol, slightly soluble in dry THF, and practically insoluble in anhydrous ethanol, acetone, and diethyl ether. In general, its chemical properties closely resemble those of *trans*-IrCl(CO)(PPh$_3$)$_2$.[15] As a solid, it is stable to air oxidation. In solution it readily reacts with O_2 and H_2, and in aqueous solutions these reactions are irreversible. Spectral properties: UV-vis (water) λ_{max} 328 nm (ε 8.5 × 10^2 M^{-1} cm^{-1}), 376 nm (ε 7.5 × 10^2 M^{-1} cm^{-1}), 428 nm (ε 2.2 × 10^2 M^{-1} cm^{-1}); IR (KBr): ν(CO) 1966 cm^{-1}, ν(SO$_3$) 1200 cm^{-1}; IR(H$_2$O): ν(CO) 1978 cm^{-1}; NMR (δ; d$_6$-DMSO, 200 MHz) The ^{1}H NMR consists of a complex pattern between 6.61 and 7.15 ppm. The ^{31}P NMR has a single peak at 25.1 ppm. It is necessary to record the ^{31}P NMR spectra in DMSO, rather than D$_2$O, because the formation of what is believed to be an hydroxide species, Ir(CO)(*OH*)(TPPMS)$_2$, upon dissolution in water. The hydroxide product can be identified as one of two peaks in the ^{31}P NMR, 27.8 ppm (OH species) and 25.4 ppm (Cl species). (O$_2$)IrCl(CO)(TPPMS)$_2$ can be obtained from oxygenated solutions of *trans*-Ir(CO)Cl(TPPMS)$_2$ in 2-methoxyethanol, ν(OO) 855 cm^{-1}, ν(CO) 2005 cm^{-1}. In aqueous solutions at moderate temperatures (> 50°C), *trans*-IrCl(CO)(TPPMS)$_2$ slowly catalyzes the hydrogenation of water soluble alkenes, but this process is accompanied by extensive isomerization.

Acknowledgments

The authors would like to thank the National Science Foundation (Grant INT-9313951) for their continued support. Gratitude is in order for the support given by the Hungarian National Research Foundation (OTKA, Grants F4021, T7527, and T4022, respectively). We also thank Johnson Matthey P.l.c. for a loan of $RuCl_3 \cdot 3H_2O$.

References

1. S. Ahrland, J. Chatt, N. R. Davies, and A. A. Williams, *J. Chem. Soc.*, **1958**, 264.
2. Z. Tóth, F. Joó, and M. T. Beck, *Inorg. Chim. Acta*, **42**, 153 (1980).
3. A. F. Borowski, D. J. Cole-Hamilton, and G. Wilkinson, *Nouv. J. Chem.*, **2**, 137 (1978).
4. W. A. Herrmann, J. A. Kulpe, W. Konkol, and H. Bahrmann, *J. Organometal. Chem.*, **389**, 85 (1990).
5. B. Salvesen and J. Bjerrum, *Acta Chem. Scand.*, **16**, 735 (1962).
6. P. A. Chaloner, M. A. Esteruelas, F. Joó, and L. A. Oro, *Homogeneous Hydrogenation*, Kluwer Academic Publishers, Dordrecht, The Netherlands, 1994, p.183.
7. P. Kalck and F. Monteil, *Adv. Organometal. Chem.*, **34**, 219 (1992).
8. C. Larpent and H. Patin, *Tetrahedron*, **44**, 6107 (1988).
9. C. Larpent, G. Meignan, and H. Patin, *Tetrahedron*, **46**, 6381 (1990).
10. D. J. Darensbourg, F. Joó, Á. Kathó, J. N. W. Stafford, A. Bényei, and J. H. Reibenspies, *Inorg. Chem.*, **33**, 175 (1994).
11. J. A. Osborn and G. Wilkinson, *Inorg. Synth.*, **28**, 77 (1990).
12. L. Vígh, F. Joó, P. R. van Hasselt, and P. J. C. Kuiper, *J. Mol. Catal.*, **22**, 15 (1983).
13. P. R. Hoffmann and K. G. Caulton, *J. Am. Chem. Soc.*, **97**, 4221 (1975).
14. F. H. Jardine, *Progr. Inorg. Chem.*, **31**, 265 (1984).
15. J. P. Collman, C. T. Sears, Jr., and M. Kubota, *Inorg. Synth.*, **28**, 92 (1990).

2. SYNTHESES OF WATER-SOLUBLE PHOSPHINES AND THEIR TRANSITION METAL COMPLEXES

Submitted by WOLFGANG A. HERRMANN* and
CHRISTIAN W. KOHLPAINTNER[†]
Checked by BRIAN E. HANSON[‡] and XIANXING KANG[‡]

Water-soluble phosphine ligands and metal complexes[1] have elicited considerable interest since the commercialization of a two-phase hydroformylation process by Ruhrchemie in 1984.[2] A water-soluble rhodium

* Technische Universität München, Lichtenbergstrasse 4, D-85747 Garching, Germany.
[†] Hoechst AG, Werk Ruhrchemie, Otto-Roelen-Strasse 3, D-46128 Oberhausen, Germany.
[‡] Department of Chemistry, Virginia Tech, Blacksburg, VA 24061.

complex based on tppts, the trisulfonated derivative of triphenylphosphine (tpp), is utilized to produce 400,000 metric tons of butyraldehyde annually. The modifying ligand tppts is prepared by reaction of tpp with sulfur trioxide in concentrated sulfuric acid.[3] The raw material after hydrolysis is a mixture of mono- (tppms), di- (tppds), and trisulfonated triphenylphosphine (tppts) as well as the corresponding oxides and sulfides.[4] The workup procedure[3] by neutralization and repeated precipitation from methanol/water mixtures is inconvenient and produces considerable amounts of sodium sulfate. The mixture obtained, consisting of tppds and tppts as well as small amounts of phosphine oxides, is sufficiently pure for industrial purposes but precludes the synthesis of analytically pure and well-defined metal complexes. Several purification methods for water-soluble phosphines have been tried in the past without success, or were only applicable for analytical purposes.[5,6] The poor availability of an exactly defined ligand complicates the preparation of the corresponding metal complexes. Additional difficulties are caused by slight oxidation of the phosphine by atmospheric oxygen in the aqueous solution. Analytically pure metal complexes of tppts could therefore not be obtained by conventional methods of purification.[7−9] In 1990, the technique of gel permeation chromatography was introduced to this area by Herrmann et al.[10] With help of this versatile tool, more than 40 water-soluble metal complexes of tppts could be purified and characterized even by elemental analyses.[11] Recently, a newly developed sulfonation technique using boric acid as a protective agent, provides the access to well-defined analytically pure sulfonated phosphines without further purification by chromatography.[12] Water-soluble metal complexes of sulfonated 2,2'-bipyridine have also been reported.[13] This particular ligand has been recommended for rhodium recovery from hydroformylation product streams.[14] The following preparation methods are examples of the syntheses of some important representatives of easily accessible water-soluble ligands and transition metal complexes thereof.[12,15−16]

General Remarks

All manipulations are performed under nitrogen atmosphere using standard Schlenk techniques.[17] Water is purged with argon or nitrogen for 15 min, heated at reflux for 12 h, and then allowed to cool under argon. Organic solvents are heated at reflux for 12 h and stored under inert gas atmosphere. Sephadex G-15, G-25, G-50* and Fractogel TSK[†] are degassed in a vacuum

---

* Pharmacia BioTech GmbH, Munzinger Strasse 9, D-79111 Freiburg, Germany.
[†] Merck KGaA, Frankfurter Strasse 250, D-64293 Darmstadt, Germany.

(10^{-2} Torr) without stirring. Reagents were used as provided by the suppliers without further purification.

■ **Caution.**   *Oleum[‡] (sulfur trioxide dissolved in concentrated sulfuric acid, so-called "fuming sulfuric acid") is a heavy, fuming, yellow liquid with a sharp penetrating odor. It reacts violently with water. Oleum is a severe eye, mucous membrane, and skin irritant. Wearing appropriate protective gloves and clothing to prevent skin contact is mandatory. All preparations must be carried out in an efficiently operating fume hood.*

Gel Permeation Chromatography

1. *Apparatus*   As shown in Fig. 1, the experimental setup starts with a storage vessel (**A**) for degassed water (2L two-necked flask connected to an argon inlet). The discharge tube (**B**) is introduced via a rubber septum into the flask and contains a metal frit for solvent filtration to prevent the column head from contamination with particles. If properly used and if blocking is avoided, the column can be used for more than a year without change. The second neck of the flask allows refilling of the eluent water and is fitted with a glass stopper during chromatography. The end of the tube is connected to a three-way valve (**C**) by which the probe application can be varied. This is advantageous over application by probe loops since larger volumes can be easily measured out. The three-way valve is followed by a peristaltic tube-pump (**D**) which allows exact addition of small amounts and enables the exclusion of oxygen. With the peristaltic tube-pump "Minipuls 3" (Abimed-Gilson Co.)[§] the rotation speed of the pumphead can be regulated in steps of 0.01 rpm (digital display). This device also contains an interface for automated computer control. After 10 h of operation (due to initial tube expansion), it is useful to draw a calibration curve by given tube diameter and measurement of the solvent stream at different speeds of rotation. The pressure on the tube can be regulated via a screw on the pumphead. This is done by screwing it on until the solvent flow caused by inert-gas pressure comes to an end and then further tightening by a third of a rotation.

The column (**E**) is located immediately behind the pump, providing the shortest possible tube connection. The geometry of the column is as important as the quality of the gel packing. The column should be made of polymethylmethacrylate with an internal diameter of 24 mm and length of 1.5 m (Fig. 2), and with PVC flanges on both ends. With the inner threads of the flanges, the punch and the filling funnel are fixed to the column. The filling

---

[‡] Source of oleum in different concentrations: Merck KGaA.
[§] Gilson Co., 72 Rue Gambetta/B.P. 45, F-95400 Villiers Le Bel, France.

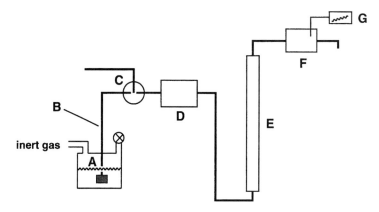

Figure 1. Experimental design of gel permeation chromatography setup; see text.

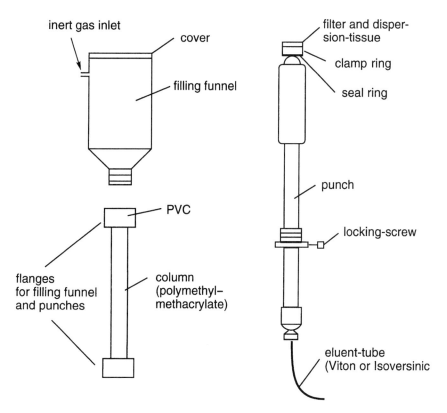

Figure 2. Column and packing apparatus; see text.

funnel is applied with an argon inlet and a flange cover. The punch can be moved upward to avoid dead volumes between gel packing and punch. The upper part of the punch consists of two conical Teflon parts with seal rings, which are pressed together when the locking screw is tightened. The punch ends in a filter tissue with smaller mesh width than the gel particle size. The tissue in turn is fixed with a clamp ring and is followed by another (coarse-meshed) filter tissue for solvent dispersion over the whole column cross-section.

The device **F** is a detector as shown in Fig. 1. Since solutions of tppts and other aryl-group containing water-soluble phosphines absorb UV light strongly, it is better to use a refractometer than usual UV detectors. A refractometer facilitates the detection of inorganic salts such as sodium chloride and sodium sulfate. Here refractometers from Büchi were used and operated with regular x/t-recorders (**G**).

■ **Caution.** *Vacuum degassing of Sephadex gel may produce violent evolution of gas. A safety shield should surround the evacuated flask.*

2. *Column packing*   To achieve sufficient separation, the column must be packed extremely homogenously.[18] For the exclusion of oxygen, a Sephadex gel is degassed in a round-bottomed flask with side arm (10^{-3} torr for several hours). As a consequence of remaining moisture, a violent evolution of gas is observed in the first 15 min. When the gel rests completely smooth even after swirling of the flask, inert gas is purged through the flask and Argon-saturated water (about 1 L per 100 g of dry gel) is added. Swelling starts immediately and is finished after 3 h at room temperature for Sephadex supports up to G-50. Within this period the gel must be swirled several times and, to assure complete swelling, it is allowed to stand overnight. The gel must not be magnetically stirred, otherwise its structure is destroyed. Before the column is filled it must be stood up and clamped exactly vertical; during filling exposure to any form of heat and air must be avoided. In the next step, the column punch is filled with water over the discharge tube by means of syringes and introduced into the column from below. Afterward the funnel is set on top and the whole column is flushed with argon for 5 min by introducing the gas from the upper end of the column. Water is added through the funnel until a level of about 6 cm on the punch is reached. The pump is connected to the upper end of the column and the water level is reduced to about 3 cm so that the tubes are freed from bubbles. The tube is then connected to the filling funnel and the amount of water over the swollen gel is adjusted to a ratio of gel/water from 2/3 up to 1/3. The gel is now cautiously suspended and transferred in one step into the column. When about 10 cm of gel are sedimented on the punch, the pump is started with a flow of

2 mL/(cm^2·h). After the whole gel has dropped down, residual water in the funnel is pipetted off without perturbing the gel. Now the second punch is filled with water and set on the column immediately after the filling funnel has been removed. This is done by introducing a small piece of tube into the column and tightening the locking-screw until the seal ring contacts the column wall. The punch is moved down to the gel surface and the surplus of water escapes through the inlet tube, which is now closed with a tube clamp as well as the tube at the lower end of the column. The pump is connected to the storage vessel and the column is equilibrated with a high flow of about 20–30 mL/(cm^2·h). Thus the gel-packing settles down and, as a consequence, the punch must be pushed down once again onto the gel surface. The column is now ready to use.

A solution of Blue Dextran 2000* (1 mg/mL) may be applied to check the quality of the packing and to test the exclusion limits. The blue dye is fixed to a dextrane of high molecular weight ($2 \cdot 10^6$) and thus should migrate as a compact zone without interaction with the gel. If the dextran section is diffuse, the column must be packed once again. Particularly the beginning of the column must be packed homogeneously since inhomogenities amplify during further separation. Therefore, it is often better to turn the column around (upside-down method) since the packing of the lower part is in most cases better.

A. DISODIUM SALT OF 3,3'-PHENYLPHOSPHINEDIYL-BENZENESULFONIC ACID (TPPDS)

$$P(C_6H_5)_3 + 2H_3SO_4^+ \rightarrow (C_6H_5)P(C_6H_4SO_3H)_2 + 2H_3O^+$$

$$(C_6H_5)P(C_6H_4SO_3H)_2 + 2N(C_8H_{17})_3 \rightarrow$$

$$[HN(C_8H_{17})_3]_2[(C_6H_5)P(C_6H_4SO_3)_2]$$

$$[HN(C_8H_{17})_3]_2[(C_6H_5)P(C_6H_4SO_3)_2] + 2NaOH \rightarrow$$

$$(C_6H_5)P(C_6H_4SO_3Na)_2 + 2N(C_8H_{17})_3 + 2H_2O$$

Procedure

In a 100-mL, two-necked flask, 4.80 g (77.8 mmol) boric acid, H_3BO_3, is dissolved in 20 mL conc. H_2SO_4 (96 wt %); 20 mL of oleum (65 wt % SO_3) is added dropwise. The reaction mixture is heated to 60°C and kept 45 min under a vacuum (10^{-2} Torr) to remove excessive sulfur trioxide. 3.0 g

* Pharmacia BioTech GmbH, Munzinger Strasse 9, D-79111 Freiburg, Germany.

(11.4 mmol) triphenylphosphine is added and the mixture is stirred for four days at 60°C. After cooling to room temperature, 50 mL of degassed water is added slowly. 16 mL (36.9 mmol) triisooctylamine dissolved in 50 mL toluene is added. After vigorous stirring for 10 min, the reaction mixture is transferred to a separation funnel. The lower water phase is removed and the organic phase washed three times with 20 mL of water each time to remove the boric acid. The water phases are collected and discarded. Under stirring a solution of sodium hydroxide in water (7.5 molar) is added slowly to the mixture until a pH of 11.8 is reached. The pH is measured by a standard glass electrode. The lower water phase is separated and neutralized with sulfuric acid (3.0 molar). The water is removed in a vacuum (10^{-2} Torr) until a precipitate forms. Then 80 mL of methanol is added. The mixture is filtered and after evaporation of the solvent a white solid remains. Yield: 4.69 g (94%).

Anal. Calcd. for $C_{18}H_{17}O_8S_2PNa_2$, 502.4; C, 43.03; H, 3.41; O, 25.48; S, 12.76; Na, 9.15; P, 6.17. Found: C, 42.71; H, 3.54; O, 25.68; S, 12.69; Na, 9.0; P, 6.38.

Properties

The white microcrystalline solid is slightly air sensitive and exhibits in the ^{31}P NMR (161.8 MHz, D_2O) a singlet at -3.38 ppm.

B. TRISODIUM SALT OF 3,3′,3″-PHOSPHINETRIYLBENZENE-SULFONIC ACID (TPPTS)

$$P(C_6H_5)_3 + 3SO_3 \xrightarrow{H_2SO_4} P(C_6H_4SO_3H)_3$$

$$P(C_6H_4SO_3H)_3 + 3N(C_8H_{17})_3 \rightarrow [HN(C_8H_{17})_3]_3[P(C_6H_4SO_3)_3]$$

$$[HN(C_8H_{17})_3]_3[P(C_6H_4SO_3)_2] + 3NaOH \rightarrow$$

$$P(C_6H_4SO_3Na)_3 + 3N(C_8H_{17})_3 + 3H_2O$$

Method A.[4] 96 g of 30 wt % oleum is placed in a 250-mL flask equipped with a stirrer, thermometer, and dropping funnel, and cooled in an external ice bath to an internal temperature of 15°C. Over a period of 1 h, 10.5 g (40 mmol) of triphenylphosphine and a further 32 g of 30 wt % oleum is added with stirring. The reaction temperature is kept between 15 and 20°C. After the addition of oleum and triphenylphosphine has been completed, the mixture is stirred for 24 h at 20°C. Subsequently, the content of the flask is added to a 1-L flask containing 300 g of water having a temperature of about

10°C. During the addition the internal temperature is kept between 20 and 40°C by intensive external cooling.

The homogeneous sulfonation mixture is placed in a 1-L flask and stirred with a mixture of 47.7 mL (110 mmol) triisooctyl amine and 180 mL toluene. After the addition has been completed, the reaction mixture is stirred for a further 30 min and left to separate for 30 min. The lower phase is separated and discarded.

Subsequently, to the extraction product a 5% aqueous sodium hydroxide solution is added while stirring. At a pH of 5.50 the addition of sodium hydroxide is interrupted. The water phase is separated and discarded. The addition of base is then continued until the mixture reaches a pH of 6.0–6.3. The aqueous solution which contains mainly tppts, minor amounts of tppds, and their oxides is separated. By concentration of the aqueous solution until the beginning of crystallization, followed by filtration, washing with methanol, and drying, the sodium salt of the 3,3',3''-phosphinetriylbenzenesulfonic acid can be obtained as a white solid with a purity of at least 95%. For highest purities the solid is dissolved in water and transferred to a Sephadex G-15 column for chromatography.

Method B. [12c,d] In a 100-mL two-necked, round-bottomed flask equipped with a magnetic stirrer and dropping funnel, 14.1 g (228 mmol) boric acid, H_3BO_3, is dissolved in conc. H_2SO_4 (96 wt %) in a vacuum (10^{-2} Torr) while vigorously stirring. 4.0 g (15.25 mmol) of triphenylphosphine is added, and after complete dissolution 67 mL of oleum (65 wt % SO_3) is slowly dropped over a period of 30 min into the solution. The temperature of the reaction mixture should not exceed 20°C. The mixture is stirred for 2 h at room temperature and afterwards hydrolyzed by means of a degassed water/ice mixture (240 g). 54 g (152 mmol) triisooctylamine, dissolved in 153 mL of toluene, is added to the hydrolysis mixture. After vigorous stirring for 15 min, the previously colorless organic phase is slightly yellow and the aqueous phase is colorless. In a separation funnel the lower water phase is removed and the organic phase is washed three times with 50 mL of water. The water phases are collected and discarded. While stirring, a solution of sodium hydroxide in water (7.5 molar) is added slowly to the mixture until a pH of 7.0, as measured by a standard glass electrode, is reached. The aqueous phase is separated and worked up as described in procedure A. The resulting solid contains 10 wt % TPPDS and TPPMS without oxides. For higher purities, the solid is dissolved in water and transferred to a column for chromatography on Sephadex G-15.

Anal. Calcd. for $C_{18}H_{18}O_{12}S_3PNa_3$, 622.47: C, 34.73; H, 2.91; O, 30.84; S, 15.45; Na, 11.08; P, 4.98. Found: C, 34.84; H, 2.80; O, 30.62; S, 15.73; Na, 11.24; P, 4.75.

Properties

Tppts is a white solid which is slightly air sensitive. Deaerated solutions of tppts are stable if stored under inert gas atmosphere. The ^{31}P NMR (161.8 MHz, D_2O) exhibits a singlet at $\delta - 5.1$ ppm. Detailed NMR data can be found in reference 19. Tppts is widely used as a ligand for metal-catalyzed reactions in water (e.g., hydrogenation, hydroformylation, carbonylation, and Heck reactions).[1]

C. OCTASODIUM SALT OF 2,2′-BIS-[METHYL(SULFOPHENYL)-PHOSPHINEDIYL]-1,1′-BINAPHTHALINE-4,4′,7,7′-OCTASULFONIC ACID (BINAS-8)

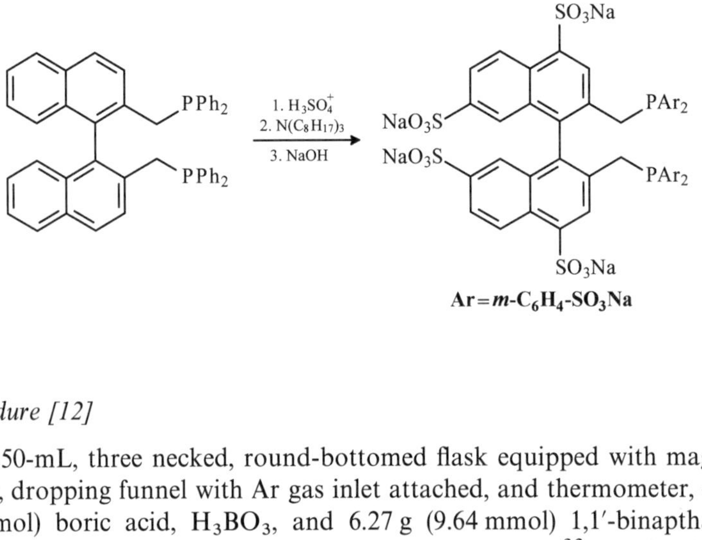

$Ar = m\text{-}C_6H_4\text{-}SO_3Na$

Procedure [12]

In a 250-mL, three necked, round-bottomed flask equipped with magnetic stirrer, dropping funnel with Ar gas inlet attached, and thermometer, 4.77 g (77 mmol) boric acid, H_3BO_3, and 6.27 g (9.64 mmol) 1,1′-binapthalene-2,2′-diylbis(methylene)bis(diphenylphosphine), NAPHOS,[33] are dissolved in 20 mL conc. H_2SO_4 (96 wt %). For 30 min the reaction mixture is held in a vacuum (10^{-2} Torr). The vacuum is released by Ar and 63 mL of oleum (65 wt % SO_3) are slowly added while not exceeding temperatures of 20°C. The reaction mixture is stirred for 100 h and then carefully hydrolyzed on crushed and degassed ice. 51.1 g (144.5 mmol) of triisooctylamine, dissolved in 230 mL toluene, is added to the hydrolysis product. After vigorous stirring for 15 min, the previously colorless organic phase is slightly yellow and the aqueous phase is decolored. In a separation funnel the lower water phase is removed and the organic phase washed three times with 50 mL water. The water phases are collected and discarded. While stirring, an aqueous solution of sodium hydroxide (7.5 molar) is added slowly to the mixture to a pH of

11.8. The water phase is separated and neutralized with an aqueous solution of sulfuric acid (3.0 molar). The solvent is removed in a vacuum (10^{-2} torr) and the residue is dissolved in 150 mL of methanol. The mixture is filtered under Ar atmosphere, and after evaporation of the solvent a white solid remains, consisting of the eightfold sulfonated derivative (BINAS-8). Yield: 18.44 g (93%).

Anal. Calcd. for $C_{46}H_{46}O_{33}S_8P_2Na_8$, 1629.24: C, 33.91; H, 2.85; O, 32.41; S, 15.75; Na, 11.29; P, 3.80. Found: C, 34.01; H, 2.90; O, 32.50; S, 15.59; Na, 11.18; P, 3.81.

Properties

The slightly yellow, moderately air-sensitive compound exhibits in the ^{31}P NMR (161.8 MHz, D_2O) a singlet at $\delta = -12.55$ ppm. The ^{1}H–^{1}H-COSY NMR shows the following pattern: $\delta = 3.62$ [d, ^{2}J(HH) = 13.8, 2H, CH_AH_B], 3.66 [d, ^{2}J(HH) = 13.8, 2H, CH_AH_B], 7.21 [dd, ^{3}J(HH) = 6.9, ^{3}J(PH) = 6.8, 2H, $H_{16'}$], 7.52 [tr, ^{3}J(HH) = 7.7, 2H, $H_{15'}$], 7.66 [tr, ^{3}J(HH) = 7.6, 2H, H_{15}], 7.72 [dd, ^{3}J(HH) = 7.5, ^{3}J(PH) = 7.2, 2H, H_{16}], 7.77 [d, ^{3}J(HH) = 9.0, 2H, H_5], 7.90 [d, ^{3}J(PH) = 5.4, 2H, $H_{12'}$], 7.97 (s, 2H, H_3), 7.98 [d, ^{3}J(HH) = 7.5, 2H, $H_{14'}$], 8.05 [d, ^{3}J(PH) = 5.8, 2H, H_{12}], 8.1 [d, ^{3}J(HH) = 7.7, 2H, H_{14}], 8.7 (s, 2H, H_8), 9.0 [d, ^{3}J(HH) = 9.0, 2H, H_6]. The Rh(I)-complex of this ligand represents the most active water-soluble catalyst in the hydroformylation of propene.[32] Palladium-BINAS catalysts are highly active in the carbonylation of benzylic chlorides to give phenyl acetic acids.[34]

D. DISODIUM SALT OF 5*H*-1-(3′-SULFOPHENYL)-DIBENZOPHOSPHOL-4-SULFONIC ACID

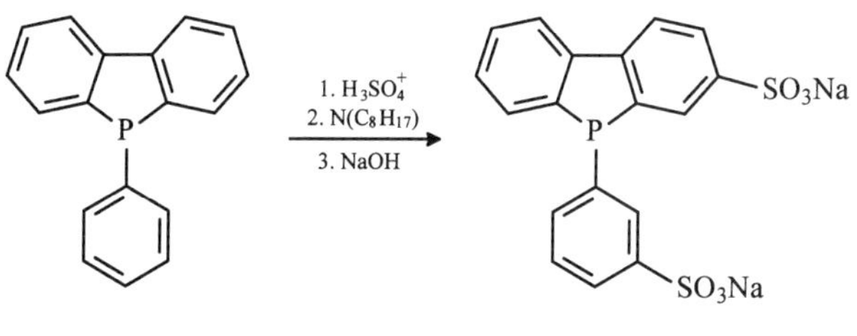

Procedure

All steps are to be carried out under Ar atmosphere. In a 100-mL, two-necked flask, 2.40 g (38.3 mmol) of boric acid, H_3BO_3, is dissolved in 8 mL conc.

H_2SO_4 (96 wt %). 10 mL of 65 wt % oleum is slowly added with help of a dropping funnel. The funnel is removed and replaced with a stop cock and the mixture is heated to 60°C. Excess of sulfur trioxide is removed in a vacuum (10^{-2} torr) over a period of 45 min. The vacuum is released by Ar and after cooling to 25°C, 500 mg (1.9 mmol) 5-*H*-phenyl-dibenzophosphol[20] is added while stirring. After complete dissolution, the reaction mixture is heated to 145°C for 15 h. After completed reaction, the mixture is cooled to room temperature and 20 mL of water is slowly added. Then 4 mL (9.2 mmol) of triisooctylamine is dissolved in 30 mL of toluene and added to the reaction mixture. After vigorous stirring for 10 min, the mixture is anaerobically transferred to a separation funnel. The lower water phase is removed and the organic phase is washed with water (3×20 mL). The water phases are collected and discarded. With stirring a solution of sodium hydroxide in water (7.5 molar) is then added slowly to the mixture until a pH of 11.8 is reached. The water phase is separated (toluene phase discarded) and neutralized with an aqueous solution of sulfuric acid (3.0 molar). The volume of the water phase is reduced in vacuo (10^{-2} torr) until a white precipitate forms; 80 mL of methanol are added. The mixture is filtered and after evaporation of the solvent a white solid remains. Yield: 852 mg (89%).

Properties

The compound is a white solid which is slightly air-sensitive. The ^{31}P NMR spectrum (161.8 MHz, D_2O) exhibits a singlet at $\delta 17.42$. The substitution pattern has been resolved by applying ^{1}H–^{1}H-COSY NMR: $\delta = 6.90$ [dt, $^3J(HH) = 7.3$ Hz, $^4J(HH) = 1.9$ Hz, 2H, H_8, H_9], 7.01 [tt, $^3J(HH) = 7.7$, $^3J(PH) = 12.0$, $^4J(HH) = 1.2$, 1H, H_{16}], 7.06 [td, $^3J(HH) = 7.7$, $^4J(PH) = 3.2$, 1H, H_{15}], 7.20 [dd, $^3J(HH) = 8.0$, $^4J(PH) = 4.4$, 1H, H_1], 7.32 [dt, $^3J(HH) = 7.3$, $^4J(HH) = 1.9$, 2H, H_6, H_7], 7.43 [dt, $^3J(PH) = 14.0$, $^4J(HH) = 1.2$, 1H, H_{12}], 7.55 [ddd, $^3J(HH) = 7.7$, $^4J(HH_{16} = 1.8$, $^4J(HH_{15})$ = 1.3, 1H, H_{14}], 7.81 [dd, $^3J(HH) = 8.0$, $^4J(PH) = 2.0$, 1H, H_2], 8.41 [dd, $^3J(PH) = 14.0$, $^4J(HH) = 2.0$, 1H, H_4].

E. HEXASODIUM-HEXACARBONYL-BIS[3,3',3''-PHOSPHINETRIYLBENZENESULFONATO]DICOBALTATE-HEXAHYDRATE, $Co_2(CO)_6(TPPTS)_2 \cdot 6H_2O$

$Co_2(CO)_8 + 2P(C_6H_4SO_3Na)_3 \cdot 3H_2O \rightarrow$

$$Na_6[Co_2(CO)_6(P(C_6H_4SO_3)_3)_2] \cdot 6H_2O + 2CO$$

Procedure

In a 50-mL, two-necked flask, 100 mg (0.3 mmol) of octacarbonyl dicobalt, $Co_2(CO)_8$, is dissolved in 10 mL of toluene. Then 400 mg (0.7 mmol) tppts is dissolved in 10 mL water and added to the solution. The reaction mixture is stirred for 3 h at room temperature. During the reaction the organic phase is decolorized, whereas the water phase turns brown. The mixture is transferred to a separation funnel while maintaining exclusion of oxygen. After the phases have separated, the organic phase is washed twice, in each case with 5 mL of water, and the combined aqueous phases are washed twice, in each case with 5 mL of toluene. The organic phases are discarded. The water phases are collected and the solvent is removed in a vacuum. The crude product is purified by column chromatography on Sephadex G-25. Yield: 370 mg (81%); brown powder.

Anal. Calcd. for $C_{42}H_{36}Co_2Na_6P_2O_{30}S_6$; 1530.84: C, 32.95; H, 2.37; O, 31.35; P, 4.05; S, 12.57. Found: C, 32.44; H, 2.37; O, 31.25; P, 3.97; S, 12.13.

Properties

$Co_2(CO)_6(tppts)_2$ is a brown-colored solid which is moderately stable in air, but is best handled and stored under an inert gas atmosphere. The compound is very soluble in water and insoluble in organic solvents like toluene or hexane. It exhibits in the ^{31}P NMR (109.3 MHz, D_2O, 5°C) a singlet at δ 68.8 ppm. The IR displays a strong carbonyl stretching vibration at 1954 cm^{-1}. The significant *SO*-absorptions are detectable at 1224 (sh, vst), 1200 (vst), 1039 (vst), and 623 (vst) cm^{-1}. The compound has been used for carbonylation of phenyl ethyl bromide[21] and on supported aqueous phase catalysts for the hydroformylation of olefins.[22]

F. NONASODIUM-CARBONYL-HYDRIDO-TRIS(3,3′,3″-PHOSPHINETRIYLBENZENESULFONATO)COBALTATE-NONAHYDRATE, HCo(CO)(TPPTS)$_3$·9H$_2$O

$$CoCl_2 + 3P(C_6H_4SO_3Na)_3 \cdot 3H_2O + CO \xrightarrow{NaBH_4}$$

$$HCo(CO)[P(C_6H_4SO_3Na)_3]_3 \cdot 9H_2O + 2NaCl$$

Under Ar or N_2 atmosphere, in a 50-mL, two-necked flask equipped with a gas inlet and dropping funnel, 120 mg (0.52 mmol) of $CoCl_2 \cdot 5H_2O$ is dissolved in 10 mL of water. Then 1.64 g (3.0 mmol) of tppts is added and the reaction mixture is cooled to 5°C. After the solids have dissolved with stirring, a solution of 32 mg (0.9 mmol) of $NaBH_4$ in 20 mL of water is added dropwise over a period of 1 h. At the same time carbon monoxide (CO) is introduced from a lecture bottle through two safety bottles and the gas inlet into the flask. The gas flow is adjusted to one bubble per second as measured in the reaction solution.

■ **Caution.** *Carbon monoxide is an odorless and highly toxic gas. All preparations must be carried out in an efficiently operating fume hood.*

After completion the yellow solution is concentrated in a vacuum (10^{-2} torr) to one quarter of its original volume and subjected directly to column chromatography on Sephadex G-15. Yield: 856 mg (89%, based on Co); canary-yellow powder.

Anal. Calcd. for $C_{55}H_{55}CoNa_9O_{37}P_3S_9$; 1924.99: C, 34.20; H, 2.80; Co, 3.06; P, 4.80. Found: C, 34.07; H, 2.87; Co, 3.09; P, 4.76.

Properties

The yellow-colored substance shows hardly any sensitivity to air. For long-time storage, however, handling under inert gas atmosphere is recommended. The ^{31}P NMR (109.3 MHz, D_2O, 5°C) exhibits a singlet at δ 45.93 ppm. ^1H NMR (270 MHz, D_2O, 5°C): $\delta = -12.35$ [q, $^2J(PH) = 45$ Hz, 1H], 7.29 (br, 9H), 7.32 (m, 27H); The metal hydride stretch vibration is detectable in the IR spectra at 1953 (vst) cm^{-1}. Strong absorption are observable at 1904 (vst) and 1200 (vst) cm^{-1} which represent the *CO*- and *SO*-vibrations.

G. NONASODIUM-CHLORO-TRIS(3,3′,3″-PHOSPHINETRIYLBENZENESULFONATO)RHODIUM-NONAHYDRATE RhCl(TPPTS)$_3$·9H$_2$O

A: $RhCl_3 + 3P(C_6H_4SO_3Na)_3 \cdot 3H_2O + H_2O \rightarrow$

$$RhCl[P(C_6H_4SO_3Na)_3]_3 \cdot 9H_2O + O{=}P(C_6H_4SO_3Na)_3 + 2HCl$$

B: $RhCl[P(C_6H_5)_3]_3 + 3P(C_6H_4SO_3Na)_3 \cdot 3H_2O \rightarrow$

$$RhCl[P(C_6H_4SO_3Na)_3]_3 \cdot 9H_2O + 3P(C_6H_5)_3$$

C: $[(cod)RhCl]_2 + 6P(C_6H_4SO_3Na)_3 \cdot 3H_2O \rightarrow$

$$2RhCl[P(C_6H_4SO_3Na)_3]_3 \cdot 9H_2O + 2cod^*$$

* cod = cycloocta-1,5-diene.

Procedures

Method A. In a 50-mL, two-necked flask, a solution of 260 mg (1.0 mmol) of $RhCl_3 \cdot 3H_2O$ in 20 mL of water is stirred for about 3 h after addition of 5.87 g (10 mmol) of tppts dissolved in 10 mL of water. After completion, the resultant solution which contains tppts, the corresponding oxide tppots, and small amounts of the binuclear complex $[(\mu\text{-}Cl)Rh(tppts)_2]_2$ are separated by column chromatography on Sephadex G-15. The red fraction is collected and the solvent is removed in vacuo (10^{-2} torr). Yield: 1.46 g (73%).

Method B. In a Schlenk tube 100 mg (0.11 mmol) of chlorotris(triphenyl-phosphine)rhodium(I), $ClRh[P(C_6H_5)_3]_3$, is dissolved in a mixture of 20 mL of toluene and 10 mL of tetrahydrofuran. Then 20 mL of an aqueous solution of 1.87 g (3.3 mmol) of tppts is added to this solution to form a lower layer. After 12 h of vigorous stirring, the reaction mixture is transferred to a separation funnel. The aqueous phase is separated and washed twice, in each case with 5 mL of methylene chloride. The organic phases are discarded. The pure compound is obtained by column chromatography of the water phase on Sephadex G-15. Yield: 180 mg (82%).

Method C. 100 mg (0.4 mmol) of bis[μ-chloro{1,2,5,6-η^4-cyclooctadiene (1,5)}rhodium], $[RhCl(\eta^4\text{-}C_8H_{12})]_2^{23}$, is dissolved in 10 mL of methylene chloride and 0.68 g (1.2 mmol) of tppts in 10 mL of water is added. The two-phase system is stirred intensively for 30 min. Then the reaction mixture is transferred to a separation funnel. The aqueous phase is separated and the organic phase extracted twice, in each case with 5 mL of water. The combined aqueous phases are then washed twice with methylene chloride (2×5 mL). The water solution is transferred into a round-bottomed flask and the solvent is removed in a vacuum (10^{-2} torr). The raw product obtained is sufficiently pure for most reactions. Depending on the stoichiometry, small amounts of tppts or $(\eta^4\text{-}C_8H_{12})Rh_2(\mu\text{-}Cl)_2(tppts)_2$ are present, which can be removed by column chromatography on Sephadex G-15. Yield: 740 mg (93%); red glass-like solid.

Anal. Calcd. for $C_{54}H_{54}ClNa_9O_{36}P_3RhS_9$, 2005.7: C, 32.34; H, 2.70; P, 4.63; Rh, 5.13; Cl, 1.78. Found: C, 32.20; H, 2.71; P, 4.60; Rh, 5.15; Cl, 1.88.

Properties

The red solid is moderately air sensitive and should be handled and stored under inert gas atmosphere. The ^{31}P NMR (109.3 MHz, D_2O, 5°C) exhibits two groups of signals at δ 34.6 [dd, $^1J(RhP_A) = 144.1$ Hz, $^2J(P_AP_B) = 41$ Hz] and 52.8 ppm [dtr, $^1J(RhP_B) = 196$, $^2J(P_AP_B) = 41$]. The IR spectrum shows the characteristic SO-vibrations at 1206 (sh, vst), 1181 (vst), 1027 (vst), 741 (vst), 545 (vst), and 490 (vst) cm^{-1}. The Rh/tppts complexes have been widely used for hydroformylation,[1] hydrogenation,[24] and C–C coupling reactions.[25]

H. NONASODIUM-TRIS[3,3',3''-PHOSPHINETRIYLBENZENESULFONATO]PALLADIUM-NONAHYDRATE Pd(TPPTS)$_3 \cdot$9H$_2$O

A: $Pd[P(C_6H_5)_3]_4 + 3P(C_6H_4SO_3Na)_3 \cdot 3H_2O \rightarrow$

$$Pd[P(C_6H_4SO_3Na)_3]_3 \cdot 9H_2O + 4P(C_6H_5)_3$$

B: $K_2PdCl_4 + 3P(C_6H_4SO_3Na)_3 \cdot 3H_2O \xrightarrow{\text{NaBH}_4}$

$$Pd[P(C_6H_4SO_3Na)_3]_3 \cdot 9H_2O$$

Method A. In a 100-mL, two necked flask, 2.27 g (4 mmol) tppts dissolved in 20 mL of water is added with stirring to 0.46 g (0.4 mmol) of tetrakis (triphenylphosphine)palladium, $Pd[P(C_6H_5)_3]_4$, in 40 mL of toluene. The organic phase rapidly loses its color and stirring is continued for another 15 min to assure reaction completion. The reaction mixture is then transferred to a separation funnel. The phases are separated and the toluene phase is washed twice, in each case with 5 mL of water. The organic phases are discarded. The combined aqueous phases are filtered anaerobically on a glass frit (G4). The filtrate containing the target compound is transferred to a 50-mL flask and the solvent is removed from the filtrate in a vacuum (10^{-2} torr). The residue is purified by column chromatography on Sephadex G-25. Yield: 0.41 g (52%); brown powder.

Method B. In a 100-mL, two-necked flask equipped with magnetic stirrer and dropping funnel with gas inlet tube, 2.84 g (5 mmol) of tppts in 10 mL of water is added to a stirred solution of 0.32 g (1 mmol) of dipotassium tetrachloropalladate(II), $K_2[PdCl_4]$, dissolved in 10 mL of water. The reaction mixture turns brown. Then 170 mg (4.5 mmol) of sodium tetrahydridoborate, NaBH$_4$, in 10 mL of water is added dropwise over a period of 30 min. Stirring is continued for another 90 min. The water is removed in a vacuum. The solid residue is suspended with 5 mL of ethanol and filtered. The filtration residue is suspended again with 5 mL of ethanol and the slurry is filtered. The remaining residue is then purified by column chromatography on Sephadex G-25. Yield: 1.62 g (82%); brown powder.

Anal. Calcd. for $C_{54}H_{54}Na_9O_{36}P_3PdS_9$; 1973.77: C, 32.86; H, 2.76; O, 29.18; P, 4.71; Pd, 5.39; S, 14.62. Found: C, 32.35; H, 2.70, O, 29.95; P, 4.87; Pd, 5.30; S, 15.27.

Properties

The brownish-colored solid is air sensitive and should be handled and stored under inert gas atmospheres. Solutions are very air sensitive with precipitation of colloidal palladium. The ^{31}P NMR (109.3 MHz, D_2O, 5°C) exhibits a singlet at $\delta 22.6$ ppm. The IR displays the characteristic *SO*-vibrations at 1225 (sh, vst), 1200 (vst), 1039 (vst), and 622 (vst) cm^{-1}. The compound has been intensively utilized for carbonylation of benzylic chlorides,[26] aryl bromides,[27] and 5-hydroxymethylfurfural,[28] Heck-reactions,[29] allylic substitution reactions,[30] and oxidations.[16]

I. NONASODIUM-TRIS[3,3′,3″-PHOSPHINETRIYLBENZENESULFONATO]-NICKEL-NONAHYDRATE Ni(TPPTS)₃·9H₂O

$$NiCl_2 + 3P(C_6H_4SO_3Na)_3 \cdot 3H_2O \xrightarrow{NaBH_4} Ni[P(C_6H_4SO_3Na)_3]_3 \cdot 9H_2O$$

In a 50-mL, two-necked flask equipped with magnetic stirrer and dropping funnel with gas inlet tube, 71 mg (0.3 mmol) of $NiCl_2 \cdot 6H_2O$ and 0.85 g (1.5 mmol) of tppts is dissolved in a mixture of 10 mL of water and 10 mL of ethanol. The solution is cooled to -15°C in an isopropanol/dry-ice bath and 34 mg (0.9 mmol) of $NaBH_4$ (not cooled), dissolved in 5 mL of water and 5 mL of ethanol, is added dropwise over a period of 90 min. The reaction solution turns red and then reddish brown. After all the $NaBH_4$ has been added, the temperature of the solution is allowed to rise to room temperature over a period of 3 h. Afterward, the solvents are removed in a vacuum. The residue is purified by means of column chromatography on Sephadex G-15, the column being cooled to 0°C. Yield: 0.38 g (66%); reddish brown powder.

Anal. Calcd. for $C_{54}H_{54}Na_9NiO_{36}P_3S_9$: 1926.07: C, 33.67; H, 2.83; P, 4.82; Ni, 3.05. Found: C, 33.74; H, 2.87; P, 4.69; Ni, 3.49.

Properties

The compound is very sensitive toward air and temperature. Samples should be stored under argon atmosphere and kept frozen (-18°C). The ^{31}P NMR

(109.3 MHz, D_2O/C_2H_5OH 1:1, $- 30°C$) exhibits a singlet at $\delta 22.7$ ppm. The IR displays the characteristic vibrations of the sulfonato groups at 1222 (sh, vst), 1192 (vst), 1039 (vst), and 622 (vst) cm^{-1}. This compound underlined the versatility of the gel permeation chromatography, as the composition had to be corrected from Ni(tppts)$_4$ to the shown above, stoichiometry Ni(tppts)$_3$.[3,11] The compound was used for the hydrocyanation of 3-pentenenitrile.[31]

References

1. W. A. Herrmann and C. W. Kohlpaintner, *Angew. Chem.*, **105**, 1588 (1993); *Angew. Chem. Int. Ed. Engl.*, **32**, 1524 (1993).
2. (a) B. Cornils, J. Hibbel, W. Konkol, B. Lieder, J. Much, V. Schmid, and E. Wiebus, *Ger. Pat.* DE 3 234 701 (1982); *Chem. Abstr.*, **100**, 194022 (1984). (b) B. Cornils and E. Wiebus, *CHEMTECH*, **25**, 33 (1995). (c) B. Cornils and E. Wiebus, *Chem-Ing. Tech.*, **66**, 916 (1994).
3. E. G. Kuntz, *CHEMTECH*, **17**, 570 (1987).
4. R. Gärtner, B. Cornils, H. Springer, and P. Lappe, *U.S Pat.* US 4 483 802 (1984); *Chem. Abstr.*, **101**, 55331 (1984).
5. L. Lecomte, J. Triolet, D. Sinou, J. Bakos, and B. Heil, *J. Chromatogr.*, **408**, 416 (1987).
6. L. Lecomte and D. Sinou, *J. Chromatogr.*, **514**, 91 (1990).
7. B. Fontal, J. Orlewski, C. Santini, and J. M. Basset, *Inorg. Chem.*, **25**, 4320 (1986).
8. C. Larpent, R. Dabard, and H. Patin, *New. J. Chem.*, **12**, 907 (1988).
9. C. Larpent and H. Patin, *Appl. Organomet. Chem.*, **1**, 529 (1987).
10. W. A. Herrmann, J. A. Kulpe, W. Konkol, and H. Bahrmann, *J. Organomet. Chem.*, **389**, 85 (1990).
11. W. A. Herrmann, J. Kellner, and H. Riepl, *J. Organomet. Chem.*, **389**, 103 (1990).
12. (a) W. A. Herrmann, G. P. Albanese, R. B. Manetsberger, P. Lappe, and H. Bahrmann, *Angew. Chem.*, **107**, 893 (1995); *Angew. Chem. Int. Ed. Engl.*, **34**, 811 (1995). (b) W. A. Herrmann, R. B. Manetsberger, G. P. Albanese, P. Lappe, and H. Bahrmann, *Ger. Pat.* DE 4 321 512 (1993). (c) R. B. Manetsberger, *PhD. Thesis*, Techn. Univ. München (1994). (d) G. P. Albanese, *PhD. Thesis*, Techn. Univ. München, in preparation.
13. (a) W. A. Herrmann, W. R. Thiel, and J. G. Kuchler, *Chem. Ber.*, **123**, 1953 (1990). (b) W. A. Herrmann, W. R. Thiel, J. G. Kuchler, J. Behm, and E. Herdtweck, *Chem. Ber.*, **123**, 1963 (1990).
14. H.-J. Kneuper, M. Roeper, and R. Pacciello, EP 0 588 225 (1994); *Chem. Abstr.* **120**, 322.751 (1994).
15. W. A. Herrmann, J. Kulpe, W. Konkol, H. Bach, E. Wiebus, T. Müller, and H. Bahrmann, *U.S. Pat.* US 5 041 228 (1991).
16. W. A. Herrmann, J. Kulpe, J. Kellner, and H. Riepl, *U.S. Pat.* 5 057 618 (1991).
17. D. F. Shriver and M. A. Drezdzon, *The Manipulation of Air-Sensitive Compounds*, John Wiley and Sons, New York, 1986.
18. J. Kandzia, *Pharmacia, LKB aktuell*, **3**(1), 3 (1988) (published by LKB-Pharmacia).
19. T. Bartik, B. Bartik, B. E. Hanson, T. Glass, and W. Bebout, *Inorg. Chem.*, **31**, 2667 (1992).
20. J. Cornforth, R. H. Cornforth, and R. T. Gray, *J. Chem. Soc., Perkin Trans. I*, 2289 (1982)
21. (a) E. Monflier, A. Mortreux, and F. Petit, *Appl. Catal. A: General*, **102**, 53 (1993). (b) E. Monflier, and A. Mortreux, *J. Mol. Catal.*, **88**, 295 (1994).
22. I. Guo, B. E. Hanson, I. Toth, and M. E. Davis, *J. Organomet. Chem.*, **403**, 221 (1991).
23. G. Giordana and R. H. Crabtree, *Inorg. Synth.*, **19**, 218 (1979).

24. J. M. Grosselin, C. Mercier, G. Allmang, and F. Grass, *Organometallics*, **10**, 2126 (1991).
25. (a) G. Mignani, D. Morel, and Y. Colleuille, *Tetrahedron Lett.*, **26**, 6337 (1985). (b) C. Mercier, and P. Chabardes, *Pure and Appl. Chem.*, **66**, 1509 (1994).
26. C.W. Kohlpaintner and M. Beller, *Ger. Pat.*, DE 4 415 681 (1994).
27. F. Monteil and P. Kalck, *J. Organomet. Chem.*, **482**, 45 (1994).
28. G. Papadogianakis, L. Maat, and R. A. Sheldon, *J. Chem. Soc., Chem. Commun.*, 2659 (1994).
29. J. P. Genet, E. Blart, and M. Savignac, *Synlett*, 715 (1992).
30. (a) M. Safi and D. Sinou, *Tetrahedron Lett.*, **32**, 2025 (1991). (b) E. Blart, J. P. Genet, M. Safi, M. Savignac, and D. Sinou, *Tetrahedron*, **50**, 505 (1994).
31. E. G. Kuntz, *Ger. Pat.*, DE 2 700 904 (1977); *Chem. Abstr.* **87**, 117602 (1977).
32. (a) C. W. Kohlpaintner, *PhD. Thesis*, Techn. Univ. München (1992). (b) W. A. Herrmann, C. W. Kohlpaintner, R. B. Manetsberger, H. Bahrmann, and P. Lappe, *Eur. Pat.*, EP 571 819 (1992). (c) W. A. Herrmann, C. W. Kohlpaintner, R. B. Manetsberger, H. Bahrmann, and H. Kottmann, *J. Mol. Catal.*, **97**, 65 (1995).
33. (a) H.-J. Kleiner and D. Regnat, *Ger. Pat.*, DE 4 433 294 (1993). (b) H.-J. Kleiner and D. Regnat, *Ger. Pat.*, DE 4 338 826 (1993). (c) H. Bahrmann, K. Bergrath, H.-J. Kleiner, P. Lappe, C. Naumann, D. Peters, and D. Regnat, *J. Organomet. Chem.*, **520**, 97 (1996).
34. C. W. Kohlpaintner and M. Beller, *Ger. Pat.*, DE 4 415 682 (1994).

3. TRIS[TRIS(SODIUM *m*-SULFONATOPHENYL)-PHOSPHINO]PALLADIUM(0) ENNEAHYDRATE

$$PdCl_2 + 3P(C_6H_4\text{-}m\text{-}SO_3Na)_3 + CO + 10H_2O \rightarrow$$

$$[Pd\{P(C_6H_4\text{-}m\text{-}SO_3Na)_3\}_3 \cdot 9H_2O] + CO_2 + 2HCl$$

Submitted by G. PAPADOGIANAKIS,* L. MAAT,* and R. A. SHELDON*
Checked by C. J. BISHOFF[†] and B. D. DOGGETT[†]

Increasing interest has recently been focused on the coordination chemistry of trisulfonated triphenylphosphine, $P(C_6H_4\text{-}m\text{-}SO_3Na)_3$, (TPPTS)[1] owing to its utility as a water-soluble ligand for organometallic catalysis in water. For example, the Ruhrchemie/Rhône–Poulenc process for the hydroformylation of propene[2] and the Rhône–Poulenc process for the synthesis of vitamin E and A intermediates[3] employ a rhodium(I) TPPTS complex in an aqueous/organic two-phase system. Such biphasic catalysis combines the advantages of homogeneous and heterogeneous catalysis.[4]

More recently, a palladium TPPTS complex was shown[5,6] to catalyze carbonylations in aqueous media. The original method[1] of preparation of

* Laboratory of Organic Chemistry and Catalysis, Delft University of Technology, 2628 BL Delft, the Netherlands.
† Hoechst–Celanese Corporation, 1901 Clarkwood Road, Corpus Christi, TX 78469.

homoleptic palladium TPPTS complexes and subsequent modification thereof,[7] as well as other routes for the synthesis of the $[Pd\{P(C_6H_4\text{-}m\text{-}SO_3Na)_3\}_3 \cdot 9H_2O]$ complex,[6−8] lead to moderate yields (52–82%) and require either expensive reducing agents, such as sodium borohydride and palladium starting materials, or high excess of TPPTS to palladium precursor (P/Pd molar ratio of 12.5) and the use of organic solvents. Furthermore, all these procedures tend to yield products contaminated with potassium chloride, sodium hydroxide, boric acid and unreacted potassium tetrachloro palladate(II) or with the low-ligated zero-valent palladium TPPTS complex, $[Pd\{P(C_6H_4\text{-}m\text{-}SO_3Na)_3\}_2]$. Moreover, the separation of potassium chloride from transition metal TPPTS complexes, usually by gel permeation chromatography[9] using various Sephadex materials as adsorbents, is cumbersome.[7]

The route described below, which utilizes carbon monoxide to reduce the $[PdCl\{P(C_6H_4\text{-}m\text{-}SO_3Na)_3\}_3]^+$ intermediate[10] formed from palladium dichloride and TPPTS by stirring for 25 min in water at 25°C, results in the formation of $[Pd\{P(C_6H_4\text{-}m\text{-}SO_3Na)_3\}_3 \cdot 9H_2O]$ in 95–99% yield in 5 min at 25°C and a carbon monoxide pressure of only 2 bar. Further advantages of this route are that the byproducts (CO_2 and HCl) are gases which can easily be separated from product and that the use of a completely aqueous medium which has obvious environmental benefits.

Procedure

A 100-mL, single-necked, round-bottomed flask is equipped with a magnetic stirring bar and a dual-outlet adapter which is attached to high vacuum and a source of dry argon or N_2. The flask is evacuated, heated by a heat gun, and after cooling to room temperature filled with argon (provision is made for pressure relief through a silicone oil bubbler). This procedure is repeated. The flask is then charged, under argon, with a mixture of palladium dichloride* (35.5 mg, 0.2 mmol, as a fine powder), TPPTS $\cdot 3H_2O$ (746.9 mg, 1.2 mmol),† and 40 g of deoxygenated distilled demineralized water under argon, resulting in a brown slurry. [Note: The following procedure was used to prepare the water. Distilled demineralized water, 700 mL, was placed in a 1-L, single-necked, round-bottomed flask equipped with a dual-outlet connection

---

* $PdCl_2$ was purchased from Johnson Matthey GmbH (Karlsruhe, Germany).
† TPPTS is prepared according to the procedure of Hoechst AG Werk Ruhrchemie[11] with a purity of 99.3% (TPPTS-oxide: 0.7%). A similar synthesis is described in this volume and includes the synthesis of TPPTS. The checkers initially had problems with a sample of TPPTS which had been in solution for longer than 3 months and contained 21% TPPTS-oxide. When fresh TPPTS was used, the reaction worked very well and afforded 376.2 mg (95%) yield.

is deoxygenated in an ultrasound bath under high vacuum (trap is cooled by liquid nitrogen) for 1 h. During the deoxygenation the flask is disconnected from vacuum and the aqueous solvent is saturated with argon; this procedure is repeated several times. The time for deoxygenation of water using vacuum obtained by a water aspirator is more than 2 h.] The brown slurry is stirred under argon atmosphere at room temperature until the palladium dichloride is completely dissolved (25 min) to give a bright yellow solution of the cationic palladium chloro TPPTS $[PdCl\{P(C_6H_4\text{-}m\text{-}SO_3Na)_3\}_3]^+$ intermediate.[10] If at this point of the procedure oxygen is present in the reaction mixture, uncomplexed palladium dichloride catalyzes the oxidation of TPPTS to TPPTS oxide. Selected data for $[PdCl\{P(C_6H_4\text{-}m\text{-}SO_3Na)_3\}_3]^+$: $^{31}P\{^1H\}$ NMR spectrum [80.98 MHz, 25°C, D_2O] $\delta = +33.98$ ppm (t, 1P, P-*trans*-Cl), 30.76 (d, 2P, P-*cis*-Cl, $^2J_{P,P} = 14.6$ Hz); FT far-IR spectrum (PE pellet) ν_{max}/cm^{-1} 313 (m, $\nu PdCl$).

■ **Caution.** *Carbon monoxide is an odorless, colorless and very toxic gas. The reaction must be conducted in a well-ventilated fume hood. The glass autoclave must be covered by a protection cage during the work with carbon monoxide under pressure.*

Using a 60-mL hypodermic syringe equipped with a long needle which had been flushed with argon, the bright yellow solution (pH 3.52, measured at room temperature with a calibrated Aldrich Z11,344-1 microcombination pH electrode) is transferred to a sealed Hastelloy C autoclave, 300 mL of which (through a valve and the shortest connection pipe) had been previously evacuated and filled with argon. The Hastelloy C autoclave can be replaced by a glass autoclave which is safe under 2 bar carbon monoxide pressure. After five pressurizing–depressurizing cycles with carbon monoxide (purity 99.997%) under stirring to remove argon, the autoclave is pressured to 2 bar at 25°C. After 5 min of stirring, carbon monoxide is vented and the bright yellow mixture (pH 2.24) transferred with a hypodermic syringe to a 100-mL, single-necked, round-bottomed flask equipped with a magnetic stirring bar and a dual-outlet adapter attached to high vacuum and a source of dry argon. This flask was previously heated under vacuum and, after cooling to room temperature, is filled with argon. The procedure to remove air/oxygen is repeated. The solution was concentrated to ca. 2 mL under high vacuum, requiring ca. 3.5 h, (two additional traps filled with sodium hydroxide are connected to the oil pump to neutralize the hydrochloric acid formed) and chromatographed on Sephadex G-25 and eluted with deoxygenated distilled demineralized water to purify the product. Then 40 g of Sephadex G-25 (medium), purchased from Pharmacia Fine Chemicals AB, is degassed at room temperature for 3 h under high vacuum. During this period the flask was shaken several times, disconnected from vacuum, and filled with argon. This procedure is repeated three times. Then 400 mL of deoxygenated distilled demineralized water is added under argon to Sephadex and left to swell up for 18 h (provision was made for pressure relief through a silicon oil bubbler). Note: The Sephadex water mixture must not be stirred. The flask is only shaken several times. A glass column (diameter: 2.5 cm; length: 50 cm)* with a dual-outlet adapter is evacuated, and heated by a heat gun, and after cooling to room temperature, filled with argon. This procedure is repeated. The column is filled with the Sephadex gel under argon atmosphere. During the chromatography the bright yellow fraction containing pure $[Pd\{P(C_6H_4\text{-}m\text{-}SO_3Na)_3\}_3]$ is collected under argon in a 100-mL, single-necked, round-bottomed flask. This flask was previously twice degassed by heating under vacuum and after cooling to room temperature filled with argon. It is equipped with a magnetic stirring bar and a dual-outlet adapter attached to high vacuum and a source of dry argon. Note: The last part of the bright yellow ring on the Sephadex adsorbent is a mixture of $[Pd\{P(C_6H_4\text{-}m\text{-}SO_3Na)_3\}_3]$ with TPPTS (and small amounts of TPPTS oxide). The change from pure $[Pd\{P(C_6H_4\text{-}m\text{-}SO_3Na)_3\}_3]$ to the mixture with TPPTS can be seen from the inhomogenity (for a few seconds) of the first collected drop containing the mixture of $[Pd\{P(C_6H_4\text{-}m\text{-}SO_3Na)_3\}_3]$ with TPPTS in the solution of pure $[Pd\{P(C_6H_4\text{-}m\text{-}SO_3Na)_3\}_3]$. For a detailed discussion of the gel permeation technique used in this work see reference 9 or page 10 of this book. The bright yellow fraction is collected, the water is evaporated under high vacuum (ca. 3 h), and the obtained solid is dried at room temperature/0.01 torr/10 h to give 395 mg (99%) of $[Pd\{P(C_6H_4\text{-}m\text{-}SO_3Na)_3\}_3 \cdot 9H_2O]$.[†] $^{31}P\{^1H\}$ NMR spectrum [80.98 MHz, 25°C, D_2O, external reference 1% H_3PO_4] $\delta = +22.95$ ppm, sharp singlet ($\Delta v_{1/2} = 5.5$ Hz); IR (KBr pellet) v_{max}/cm^{-1} 3452 (br, vs, vO–H), 3068 (w, vC–H), 1632 (m, δO–H), 1463 (m, vC=C), 1401 (m, vP–C), 1205 (br, vs, vS–O), 1039 (vs, vS–O); Elemental analysis. Found: C, 32.30; H, 2.93; P, 4.46. Calcd. for $C_{54}H_{54}Na_9O_{36}P_3PdS_9$ (1973.81): C, 32.86; H, 2.76; P, 4.71%.

Properties

$[Pd\{P(C_6H_4\text{-}m\text{-}SO_3Na)_3\}_3 \cdot 9H_2O]$ is a brownish yellow solid which is relatively stable to air in the solid form. It is highly soluble in water and insoluble in most organic solvents. In aqueous solution it decomposes and in this

---

* Checkers' comment: A longer column, 2×100 cm, requiring 60 g of Sephadex, was found to perform better.
† The product was dried for a further one week at room temperature/0.01 torr prior to elemental analysis, affording 371 mg (94%) of dried product.

process changes color from bright yellow to red within a few minutes, from red to brown (1–2 h), and eventually to a precipitate of palladium black and a colorless solution (10–20 h). The $[Pd\{P(C_6H_4\text{-}m\text{-}SO_3Na)_3\}_3 \cdot 9H_2O]$ complex has been shown to catalyze the carbonylation of 5-hydroxymethylfurfural[5] and bromobenzene[6] to give 5-formylfuran-2-acetic acid and benzoic acid, respectively.

References

1. E. G. Kuntz, *Chemtech.*, **17**, 570 (1987).
2. B. Cornils and E. Wiebus, *Chemtech.*, **25**, 33 (1995).
3. C. Mercier and P. Chabardes, *Pure Appl. Chem.*, **66**, 1509 (1994).
4. J. Haggin, *Chem. Eng. News*, **72**(41), 28 (1994).
5. G. Papadogianakis, L. Maat, and R. A. Sheldon, *J. Chem. Soc., Chem. Commun.*, **1994**, 2659.
6. F. Monteil and Ph. Kalck, *J. Organomet. Chem.*, **482**, 45 (1994).
7. W. A. Herrmann, J. Kellner, and H. Riepl, *J. Organomet. Chem.*, **389**, 103 (1990).
8. W. A. Herrmann, J. A. Kulpe, J. Kellner, H. Riepl, H. Bahrmann, and W. Konkol, *Angew. Chem.*, **102**, 408 (1990).
9. W. A. Herrmann, J. A. Kulpe, W. Konkol, and H. Bahrmann, *J. Organomet. Chem.*, **389**, 85 (1990).
10. G. Papadogianakis, J. A. Peters, L. Maat, and R. A. Sheldon, *J. Chem. Soc., Chem. Commun.*, **1995**, 1105.
11. R. Gärtner, B. Cornils, H. Springer, and P. Lappe, *Ger. Offen. DE 32 35 030* (1982) to Ruhrchemie AG; *Chem. Abstr.*, **101**, 55331t (1984).

4. SULFONATED PHOSPHINES

Submitted by HAO DING,* BARBARA B. BUNN,* and BRIAN E. HANSON[†]
Checked by ROBERT W. ECKL,[†] CHRISTIAN W. KOHLPAINTNER,[†] and WOLFGANG A. HERRMANN[†]

Introduction

The commercial success of rhodium-trisulfonated triphenylphosphine (TPPTS) catalysts[1] has prompted considerable interest in TPPTS and other water-soluble ligands.[2] The potential for new applications for the synthesis of both bulk and fine chemicals in water has led to methods for the preparation of a wide variety of sulfonated phosphines including chiral phosphines[3] and

* Department of Chemistry, Virginia Polytechnic Institute and State University, Blacksburg, VA 24061-0212.
[†] Technische Universität München, Lichtenbergstrasse 4, D-85747 Garching, Germany.

surface-active phosphines.[4] We describe here a method for the sulfonation of the prototypical chelating phosphine, bis-diphenylphosphinoethane (DPPE), as well as phosphines tris-*p*-(3-phenylpropyl)phenylphosphine and (*S,S*)-2,4-bis{di[*p*-(3-phenylpropyl)phenyl]phosphino}pentane and their sulfonated analogs. The pendant group, *p*-(3-phenylpropyl)phenyl, is more easily sulfonated than aryl groups bound directly to phosphorus.

A. SYNTHESIS OF 1,2-(BIS[DI-*m*-SODIUMSULFONATO]-PHENYLPHOSPHINO)ETHANE

$$
\begin{array}{ccc}
\quad\;\; P\{C_6H_5\}_2 & & \quad\;\; P\{C_6H_4SO_3Na\}_2 \\
\;\,/ & & \;\,/ \\
CH_2 & \xrightarrow[\text{2. NaOH}]{\text{1. }H_2SO_4/SO_3} & CH_2 \\
| & & | \\
CH_2 & & CH_2 \\
\;\,\backslash & & \;\,\backslash \\
\quad\;\; P\{C_6H_5\}_2 & & \quad\;\; P\{C_6H_4SO_3Na\}_2 \\
\textbf{DPPE} & & \textbf{DPPETS}
\end{array}
$$

■ **Caution.** *Fuming sulfuric acid (oleum) is corrosive, a strong oxidant, and toxic. Protective clothing should be worn when handling oleum. The material should be used in an efficient fume hood.*

Direct sulfonation of phenylphosphines has proven to be a valuable route to a number of water-soluble phosphines.[1-3] Phosphorus bound directly to a phenyl ring is meta directing and deactivating for the sulfonation reaction. This requires severe reaction conditions, typically fuming sulfuric acid for several days. The reaction medium is strongly oxidizing and in the absence of additives to prevent oxidation yields significant quantities of sulfonated phenylphosphine oxides. Direct sulfonation of the chelating ligand bis-diphenylphosphinoethane, DPPE, yields its tetrasulfonated derivative, 1,2-(bis[di-*m*-sodiumsulfonato]phenylphosphino)ethane, DPPETS, in modest yield.[5]

All manipulations and syntheses are performed under prepurified argon or nitrogen using standard Schlenk techniques. All solvents are degassed by distillation under nitrogen. A dry three-necked, round-bottomed, 3-L flask equipped with a gas inlet, vented through a bubbler, and a magnetic stirrer is charged with 20 mL of concentrated H_2SO_4 and cooled to 0°C. DPPE* (10.0 g, 0.025 mol) is added slowly (1-g increments with stirring until completely dissolved), followed by the addition of 150 mL precooled fuming sulfuric acid (24%) over a 10-min period. The color of the reaction mixture

---

* Strem Chemical Co., Inc. Newburyport, MA 01950.

varies from tan to dark brown. The solution is allowed to warm to ambient temperature (about 8 h), following which stirring is no longer necessary. The gas inlet can be closed and the solution can sit quietly. Aliquots are taken every 24 h and worked up as follows. Approximately 1 mL of the reaction mixture is transferred to a small Schlenk vessel by pipette. The sample is chilled to 0°C and is neutralized by 30% NaOH with vigorous stirring. Methanol, about 25 mL, is added and the mixture is brought to reflux. The mixture is filtered while hot to remove the bulk of Na_2SO_4 formed. The solution is reduced in volume to about 1 mL under vacuum and analyzed by ^{31}P NMR spectroscopy to determine the extent of reaction.

When the NMR analysis shows that the tetrasulfonated product to be in near maximum concentration as determined by the intensity of the ^{31}P NMR signal at $\delta = -12.6$ ppm (typically, about four days), the reaction is quenched by cooling the mixture to 0°C. Precooled aqueous sodium hydroxide (20–30% w/w) is added with pressure-equalizing dropping funnel very slowly over 8–10 h. It is essential that the solution remain cold during this step. The final pH is adjusted to 8.5–9. Copious amounts of sodium sulfate are present in the flask at this point. The volume of the neutralized solution is reduced to about 20 mL by distillation. Approximately 800 mL of methanol is then added and the temperature is brought to reflux. After refluxing for 1 h, the solution is filtered hot. The remaining sodium sulfate is washed with hot $MeOH/H_2O$ (4:1 ratio) to recover all sulfonated phosphines from the sodium sulfate. Upon combination of the extracts, distillation to a volume of about 50 mL removes most of the methanol and leaves a water solution of all the phosphine products. The desired product, DPPETS, is obtained by fractional precipitation from a 10/1 methanol/water solution. This is achieved by the addition of 500 mL methanol to the aqueous solution of phosphines obtained above. The product crystallizes over a period of about 24 h and is collected by filtration. Further purification, to remove residual phosphine, is accomplished by dissolving the solid in 10 mL of water and adding 50 mL acetone to reprecipitate the solid. The yield of the final product is about 30%.

Properties

DPPETS is a slightly air-sensitive white solid that decomposes above 300°C. ^1H NMR $\delta(D_2O)$: 2.31(t, $J_{PH} = 4.8$ Hz, 4H); 7.49(d*, $J_{HH} = 7.8$ Hz, 8H); 7.81(t, $J_{HH} = 4.2$ Hz, 4H); 7.87(s br, 4H). ^{13}C NMR $\delta(D_2O)$: 25.37(s); 140.67(t, $J_{CP} = 5.8$ Hz); 137.70(t, $J_{CP} = 7.4$ Hz); 145.64(s); 132.15(t); 128.85(s). ^{31}P NMR $\delta(D_2O)$: -12.6(s)

Anal. Calcd. for the pentahydrate: $C_{26}H_{30}Na_4O_{17}P_2S_4$: C, 34.8%; H, 3.37%. Found: 34.7%; H, 3.43%.

B. TRISULFONATED TRIS-*p*-(3-PHENYLPROPYL)-PHENYLPHOSPHINE

The incorporation of 3-phenylpropylphenyl into a phosphine yields a phosphine that is relatively easy to sulfonate at the phenyl ring omega to phosphorus.[6] The greater reactivity leads to short sulfonation times, oleum is not required, and oxidation of the phosphine does not occur.

All manipulations and syntheses were performed under prepurified argon or nitrogen using standard Schlenk techniques. All solvents were degassed by distillation under nitrogen prior to use.

I. Synthesis of *p*-(3-phenypropyl)phenylchloride

1-Bromo-4-chlorobenzene,* (19.2, 0.1 mol) and 200 mL of diethyl ether are placed in a 500-mL, three-necked flask equipped with a reflux condensor, a pressure-equalizing dropping funnel, a gas inlet, and a magnetic stirrer. The flask is cooled to 0°C with an ice bath. Then 55.5 mL of 1.8 M butyl lithium in cyclohexane/diethylether is added slowly over a period of 1 h with stirring. The mixture then is allowed to come to room temperature and stirred for an additional hour. A solution of 3-phenylpropylbromide* (19.8 g, 0.1 mol) in 50 mL diethylether is added to the reaction flask dropwise and the mixture is heated at reflux temperature for four days. Lithium bromide is removed by filtration through a medium-porosity glass frit under nitrogen and the solvent is removed by distillation. The resulting viscous oil is vacuum distilled. The fraction collected at 150–155°C and 5–6 torr is *p*-(3-phenylpropyl)-phenylchloride. Yield: 15 g, 65%.

Properties

Colorless oil, ^{1}H NMR (CD${}_3$Cl): 1.98 {m, 2H}, 2.67 {m, 4H}, 7.12–7.46 {m, 9H}.

* Aldrich Chemical Co., Milwaukee, WI 53233.

II. Synthesis of the Tris-*p*-(3-phenylpropyl)phenylphosphine

Freshly chopped lithium metal (1.4 g, 0.2 mol) is added directly to a mixture of 100 mL diethylether and 50 mL THF in a 500-mL, three-necked flask equipped with a pressure-equalizing dropping funnel, an argon inlet, and an oil bubbler outlet. Then *p*-(3-phenylpropyl)phenylchloride* (23.0 g, 0.1 mol) in 100 mL diethylether is added dropwise at 5°C. The color of the reaction mixture should slowly change to deep red and be accompanied by the formation of a precipitate of lithium chloride. The mixture is stirred at room temperature for 10 h until almost all the lithium is consumed. Lithium chloride and any unreacted lithium metal are removed by filtration under argon through a medium-porosity frit equipped with a side arm. The deep red solution, collected in a 500-mL sidearm flask is cooled to 5°C and phosphorus trichloride (4.6 g, 0.033 mol) in 50 mL diethylether is added through a pressure-equalizing dropping funnel over a period of 1 h. The mixture is allowed to warm to room temperature and is stirred for an additional 10 h. Dry ice (10 g) is added slowly to the mixture to destroy any unreacted lithium reagent and the solution is filtered. The filtrate is pumped to dryness and redissolved in 100 mL of diethylether. The ether solution is washed three times with 50 mL degassed water and dried over $MgSO_4$. The diethylether is removed by vacuum distillation to leave a colorless oil. Yield: 17.5 g, 85%.

Properties

Colorless oil. ^{1}H NMR (CD_3Cl): 1.97 {q, J_{HH} = 7.5 Hz, 6H}, 2.66 {t, J_{HH} = 7.5 Hz, 12H}, 7.0–7.4 {m, 27H}. ^{31}P NMR (CD_3Cl): − 7.20 (s).

III. Sulfonation of Tris-*p*-(3-phenylpropyl)phenylphosphine

Tris-*p*-(3-phenylpropyl)phenylphosphine (5 g, 8.1 mmol) is placed into a 1000-mL flask, cooled to − 78°C, and 20 mL 96% H_2SO_4 is then added with stirring. With continued stirring, the mixture is allowed to come slowly to room temperature. After about 6 h, the brown reaction mixture is neutralized by slow addition of 20% NaOH. The pH is adjusted to 9 and the final volume is about 120 mL. Methanol (720 mL) is added and the mixture is heated to reflux for 30 min, filtered through a medium-porosity frit, and the precipitate is extracted with 200 mL of hot methanol. After two extractions in this manner, the two filtrates are combined and the volume is reduced to about 45 mL. Acetone (270 mL) is added to precipitate the white trisulfonated tris-*p*-(3-phenylpropyl)phenylphosphine which is collected and dried. Yield: 6.9 g, 92%.

---

* Aldrich Chemical Co., Milwaukee, WI 53233.

Properties

White solid; solubility in water 20 mg/mL. ^{1}H NMR (D_2O): 1.32 {br. s, 6H}, 2.11 {br. s, 12H}, 6.65 {br. s, 6H}, 6.73 {br. s, 6H}, 7.02 {br. s, 6H}, 7.50 {br. s, 6H}, 7.50 {br. s, 6H}, ^{31}P NMR (D_2O): − 8.41.

Anal. Calcd. for $C_{45}H_{48}Na_3O_{12}PS_3$: C, 55.33%; H, 4.92. Found: C, 55.19; H, 4.88.

C. (*S,S*)-2,4-BIS{DI[*p*-(3-[*p*-SODIUMSULFONATO]PHENYL-PROPYL)PHENYLPHOSPHINO}PENTANE (BDAPPTS)[6]

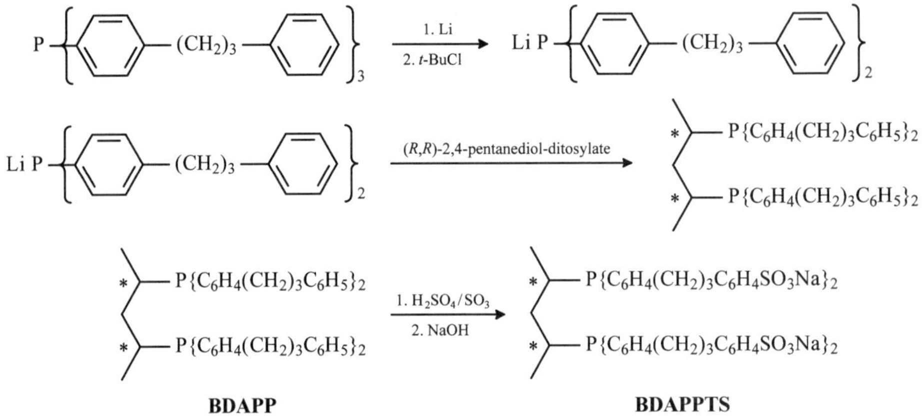

BDAPP   **BDAPPTS**

All manipulations and syntheses are performed under prepurified argon or nitrogen using standard Schlenk techniques. All solvents are degassed by distillation under nitrogen prior to use. The 2,4-pentanediol enantiomers and *p*-toluenesulfonyl chloride were obtained from Aldrich and used as received. The ditosylate of pentanediol was prepared from optically pure 2,4-pentanediol following a literature procedure as follows.[8]

I. Preparation of (−)-(2*R*,4*R*)-2,4-pentanediolditosylate

Para-toluenesulfonyl chloride (9.6 g, 50 mmol) is added slowly at 0°C to a mixture of (−)-(2*R*,4*R*)-2,4 pentanediol (2.1 g, 20 mmol) and dry pyridine (16 mL, 200 mmol). A suspension forms and the mixture is stirred for 16 h. Ether (50 mL) and 200 mL crushed ice is then added with vigorous stirring. The two phases separate and the organic layer is washed, in turn, with 50 mL portions of 5% aqueous HCl, 1% aqueous $NaHCO_3$, and distilled water and dried over $MgSO_4$. The solvent was removed under reduced pressure to yield 7.0 g of a white, crystalline solid, $[\alpha]^{20} = -6.5°$ (c 3.0, $CHCl_3$).[8]

II. Synthesis of (S,S)-2,4-Bis{di[p-(3-phenylpropyl)phenyl]phosphino}-pentane (BDAPP)

Finely chopped lithium metal (0.14 g, 0.02 mol) is suspended in 10 mL dry THF in a 250-mL sidearm flask and tris-p-(3-phenylpropyl)phenylphosphine (6.2 g, 0.01 mol) in 100 mL THF is added from a dropping funnel over a period of about 10 min with vigorous stirring. The resulting deep red solution is stirred at room temperature for an additional 2 h. Tertiary butylchloride (0.93 g, 0.01 mol) is added and the reaction mixture is brought to reflux for 15 min. The volume of the solution is reduced to about 10 mL and 80 mL dry degassed pentane is added. The reaction flask is then cooled to $-78°C$ to yield a dark red, viscous residue and a colorless liquid. The colorless liquid is decanted and the remaining residue is redissolved in 50 mL THF $\{^{31}P$ NMR (THF): $-25.97\}$. To this solution either (S,S) or (R,R) 2,4-pentanediol-ditosylate (2.0 g, 0.0049 mol) in 10 mL THF is added. The solvent is pumped off after 10 h to yield a pale yellow oil. The residue is redissolved in 40 mL diethyl ether and washed three times with 10 mL of water. The ether layer is dried over $MgSO_4$ and the volume reduced to about 10 mL; 30 mL methanol is added to yield a pale yellow oil. Yield: 3.2 g, 71%.

Properties

Yellow oil; ^1H NMR $\delta(CDCl_3)$: 0.87 {d of d, $J_{HH} = 6.7$ Hz, $J_{HP} = 15.4$ Hz, 6H}, 1.31 {pseudo q, $J_{HH} = J_{HH} = 6.9$ Hz, 2H}, 1.86 {m, 8H}, 2.37 {m, 2H}, 2.53 {m, 16H} 7.01–7.30 {m, 36H}; ^{31}P NMR $\delta(CDCl_3)$: -1.80 (s); δ(THF): -2.30.

III. Sulfonation of (S,S)-2,4-Bis{di[p-(3-phenylpropyl)phenyl]phosphino}-pentane

BDAPP (5 g, 5.5 mmol) is placed into a 1000-mL flask, cooled to $-78°C$; 20 mL 96% H_2SO_4 is then added with stirring. The mixture is allowed to come to room temperature while continuing stirring. The reaction mixture is neutralized by the slow addition of 20% NaOH after 6 h. The pH is adjusted to 9 and the final volume is about 120 mL. Methanol (720 mL) is added and the mixture is heated to reflux for 30 min. The reaction mixture is then filtered through a medium-porosity frit under nitrogen and the precipitate washed with 200 mL hot methanol. After two extractions, the filtrates are combined and the volume is reduced to about 45 mL. Acetone (270 mL) is added to precipitate white **BDAPPTS** which is collected and dried. Yield: 6.1 g, 84%.

Properties

White solid, $\alpha_D^{25} = -38.9°$ {(*S,S*)-BDAPPTS, c = 10 mg/mL, CH_3OH}; ^1H NMR: $\delta(CD_3OD)$: 0.82 {d of d, $J_{HH} = 6.5$, $J_{HP} = 15.5$, 6H}, 1.17 {m, 2H}, 1.81 {m, 8H}, 2.38 {m, 2H}, 2.52 {m, 16H}, 7.02 {m, 8H}, 7.14 {m, 8H}, 7.24 {m, 8H}, 7.64 {m, 8H}. ^{31}P NMR: $\delta(CD_3OD)$: $-1.11(s)$.

Anal. Calcd. for tetrahydrate: $C_{65}H_{74}Na_4O_{16}P_2S_4$: C, 56.05%; H, 5.37%. Found: C, 56.63%; H, 5.53%.

References

1. (a) E. Kuntz, *Chemtech*, **1987**, *17*, 570; (b) B. Cornils and E. Wiebus, *Chemtech*, **1995**, *25*(1), 33.
2. W. A. Herrmann and C. W. Kohlpaintner, *Angew. Chem. Int. Ed. Engl.*, **1993**, *32*, 1524.
3. Y. Amrani, L. Lecomte, D. Sinou, J. Bakos, I. Tóth, and B. Heil, *Organometallics*, **1989**, *8*, 542.
4. B. Fell and G. Pagadogianakis, *J. Mol. Catal.*, **1991**, *66*, 143.
5. T. Bartik, B. B. Bunn, B. Bartik, and B. E. Hanson, *Inorg. Chem.*, **1994**, *33*, 164.
6. H. Ding, B. E. Hanson, T. Bartik, and B. Bartik, *Organometallics*, **1994**, *13*, 3761.
7. H. Ding, B. E. Hanson, and J. Bakos, *Angew. Chem. Int. Ed. Engl.*, **1995**, *34*, 1645.
8. J. Bakos, I. Tóth, B. Heil, and L. Markó, *J. Organomet. Chem.*, **1985**, *279*, 23.

5. (*S,S*)-2,3-BIS[DI(*m*-SODIUMSULFONATOPHENYL)-PHOSPHINO]BUTANE (CHIRAPHOS_{TS}) AND (*S,S*)-2,4-BIS[DI(*m*-SODIUMSULFONATOPHENYL)-PHOSPHINO]PENTANE (BDPP_{TS})

Submitted by D. SINOU* and J. BAKOS†
Checked by D. J. DARENSBOURG‡ and F. BECKFORD‡

One of the most important advances in asymmetric synthesis during the last 20 years is the use of soluble organometallic chiral catalysts. However, one of

* Université Claude Bernard Lyon 1, CPE Lyon, 43, Boulevard du 11 Novembre 1918, 69622 Villeurbanne Cédex, France.
† Department of Organic Chemistry, University of Veszprem, 8201-Veszprem, Pf. 158, Hungary.
‡ Department of Chemistry, Texas A&M University, College Station, TX 77843-3255.

the central problems of such catalysis is the separation of the costly chiral catalyst from the reaction products. A very attractive approach is the use of metal complexes of chiral water-soluble ligands and a biphasic system, the organometallic catalyst being in the aqueous medium and the reactant(s) and the product(s) of the reaction in the organic phase.

A simple method for the hydrosolubilization of chiral diphosphines is the introduction of a functional group such as sulfonate. Such chiral-sulfonated diphosphines can be used to prepare homogeneous hydrogenation catalysts in situ for the asymmetric reduction of prochiral substrates.[1]

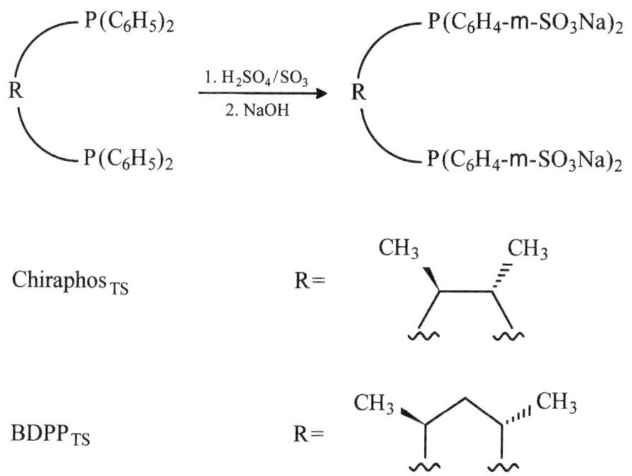

A. (*S,S*)-2,3-BIS[DI(*m*-SODIUMSULFONATOPHENYL)-PHOSPHINO]BUTANE {TETRA SODIUM[*S*-(*R**,*R**)]-3,3′,3″,3‴-[(1,2-DIMETHYL-1,2-ETHANEDIYL)-DIPHOSPHINYLIDYNE]TETRAKIS(BENZENESULFONATE)}

Procedure

■ **Caution.** *Sulfuric acid containing 30% SO_3, fuming sulfuric acid, is a severe caustic irritant product. Avoid any contact. All the reactions must be performed under an atmosphere of argon using standard Schlenk techniques and degassed solvents and reactants.*

The chiral diphosphine chiraphos* (1 g, 2.3 mmole) is dissolved under argon in 1 mL of sulfuric acid in a 50-mL Schlenk tube containing a Teflon-covered

* Available from Sigma. Aldrich Chimie, L'Isle d'Abean, France and Aldrich Chem. Co., Milwaukee, WI 53233.

magnetic stirring bar. The flask is cooled to 0°C and 10 mL of sulfuric acid containing 30% SO_3 is added very slowly under argon through a 20-mL dropping funnel with a pressure-equalization arm. The solution is stirred at room temperature for 2 days; after addition of 10 mL of sulfuric acid containing 30% SO_3, the solution is stirred again for 2 days. The mixture is then transferred very slowly under argon into a 500-mL Schlenk tube containing 100 g ice and a small amount of phenolphthalein, and the solution is carefully neutralized with a degassed solution of 50% sodium hydroxide, maintaining the temperature at 0°C. After decantation of the precipated sodium sulfate, the liquid phase is poured into 100 mL of degassed methanol contained in a 500-mL Schlenk tube; the residual solid is washed again with 100 mL of degassed methanol. After filtration of the precipitate and evaporation of the liquid phase under vacuum, the residue is dissolved under argon in the minimum amount of degassed water (5 mL) and the solution is poured into 20 mL methanol contained in a 50-mL Schlenk tube. Filtration of the solid and evaporation of the liquid under vacuum gives the tetrasulfonated diphosphine, which is dried in vacuo at room temperature. The yield is 1.6 g (70–80%). The sulfonated phosphine prepared by this methodology is pure enough for other purposes and especially its use in catalysis; however, further purification can be achieved by dissolving the compound under an argon atmosphere in the minimum amount of degassed water (5 mL), adding this solution to 20 mL of degassed methanol and evaporating, after filtration, the liquid phase.

Anal. (as the dioxide) Calcd. for $C_{28}H_{24}O_{14}P_2S_4Na_4 \cdot 8H_2O$: C, 33.27; H, 3.99; P, 6.13. Found: C, 33.0; H, 3.7; P, 5.9.

Properties

Tetrasulfonated chiraphos is a white solid containing some residual sodium sulfate and should be stored under argon. It can be used directly as obtained as a ligand in homogeneous catalysis. The compound is soluble only in water. The characteristic NMR data are: ^{31}P NMR (D_2O) δ ppm -9.6; ^{13}C NMR ($D_2O + H_2O_2$) phosphine oxide δ ppm 14.5 [d, CH_3, $^2J_{P-C} = 16.6$ Hz], 33.5 [d, CH, $^1J_{P-C} = 71$ Hz], 130.6 [d, C-2 ar, $^2J_{P-C} = 9.1$ Hz], 132.9 (s, C-4 ar), 132.9 [d, C-1 ar, $^1J_{P-C} = 97.6$ Hz], 133.2 [d, C-5 ar, $^3J_{P-C} = 12$ Hz], 135.9 [d, C-6 ar, $^2J_{P-C} = 9.9$ Hz], 146.0 [d, C-3 ar, $^3J_{P-C} = 10.3$ Hz].

The purity of the compound (e.g., the amount of mono-, di-, or trisulfonated diphosphine) can be determined by HPLC using the technique called "soap chromatography" with a single-wavelength (254 nm) detector.[2] Analytical separation was carried out on a 250×4.6 mm i.d. stainless-steel column packed with 5 μm Hypersil SAS (C_1) silica with a surface area of ca. 170 $m^2 g^{-1}$. The precolumn was filled with pellicular silica. Water-*n*-propanol (5:2) containing cetrimide (0.054 M) was used as the mobile phase, the flow rate was 0.6 mL/min and the pressure about 10^3 psi.

The tetrasulfonated chiraphos reacts in water with [Rh(COD)Cl]$_2$* in a ratio of 2:1 to give a rhodium complex [Rh(COD)(Chiraphos$_{TS}$)]Cl[1] [^{31}P(D$_2$O) δ ppm 57.4 (d, J_{Rh-P} = 148 Hz)] which catalyzes the enantioselective reduction of amino acid precursors in a two-phase system of water–ethyl acetate with ee in the range 81–88%.

B. (S,S)-2,4-BIS[DI(m-SODIUMSULFONATOPHENYL)-PHOSPHINO]PENTANE {TETRA SODIUM[S-(R*,R*)]-3,3',3'',3'''-[(1,3-DIMETHYL-1,2-PROPANEDIYL)DIPHOSPHINYLIDYNE]-TETRAKIS(BENZENESULFONATE)}

Procedure

The (S,S)-2,4-bis[di(m-sodiumsulfonatophenyl)phosphino]pentane may be prepared in 85% yield by the same methodology starting from (S,S)-2,4-bis(diphenylphosphino)pentane[†] using only 10 mL of sulfuric acid containing 30% SO$_3$ and with a reaction time of five days.

Anal. Calcd. for C$_{29}$H$_{26}$O$_{12}$P$_2$S$_4$Na$_4$.8H$_2$O: C, 35.08; H, 4.26; P, 6.24. Found: C, 35.1; H, 4.1; P, 6.1.

Properties

Tetrasulfonated BDPP is a white, air-stable solid and can be used directly as obtained as a ligand in homogeneous catalysis. The compound is soluble only in water. The characteristic NMR data are: ^{31}P NMR (D$_2$O) δ ppm 1.5, (D$_2$O + H$_2$O$_2$) phosphine oxide δ ppm + 45.4; ^{13}C NMR (D$_2$O) δ ppm 17.4 [d, CH$_3$, $^2J_{P-C}$ = 16.7 Hz], 27.4 (ps t, CH, $^1J_{P-C}$ + $^3J_{P-C}$ = 19.2 Hz), 38.3 (bs, CH$_2$), 128.8 [d, C-4 ar, $^4J_{P-C}$ = 2.6 Hz], 131.7 [d, C-5 ar, $^4J_{P-C}$ = 6.8 Hz], 132.4 [d, C-2 ar, $^2J_{P-C}$ = 21.7 Hz], 138.6 [d, C-6 ar, $^2J_{P-C}$ = 21.3 Hz], 138.7 [d, C-1 ar, $^1J_{P-C}$ = 14.2 Hz], 145.4 [d, C-3 ar, $^3J_{P-C}$ = 7.1 Hz].

The purity of the compound (e.g., the amount of mono-, di- or trisulfonated diphosphine) can also be determined by HPLC using the technique called "soap chromatography."[2]

---

* Available from Sigma-Aldrich Chimie, L'Isle d'Abeau, France.
†Available from Strem Chemicals, Inc., Bisheim, France.

The tetrasulfonated BDPP forms in water with [Rh(COD)Cl]$_2$* in a ratio of 2:1 a rhodium complex [Rh(COD)(BDPP$_{TS}$)]Cl[1] [[1](#)^{13}P (D$_2$O) δ ppm 29.3 (d, J_{Rh-P} = 144 Hz)] which catalyzes the enantioselective reduction of amino-acid precursors in a two-phase system with ee in the range 44–65%.

References

1. Y. Amrani, L. Lecomte, D. Sinou, J. Bakos, I. Toth, and B. Heil, *Organometallics*, **8**, 542 (1989).
2. L. Lecomte, J. Triolet, D. Sinou, J. Bakos, and B. Heil, *J. Chromatogr.*, **408**, 416 (1987).
   L. Lecomte and D. Sinou, *J. Chromatogr.*, **514**, 91 (1990).

6. 1,3,5-TRIAZA-7-PHOSPHATRICYCLO[3.3.1.1^{3,7}]DECANE AND DERIVATIVES

Submitted by DONALD J. DAIGLE[†]
Checked by TARA J. DECUIR[†], **JEFFREY B. ROBERTSON**[‡]
and DONALD J. DARENSBOURG[‡]

The use of aqueous solutions as media for reactions catalyzed by or-ganometallic and inorganic complexes is appealing to overcome product-separation problems, to minimize the use of expensive organic solvents, and to maximize catalyst recovery. Since triphenylphosphine and derivatives are most commonly used in organometallic chemistry and catalysis, the sulfona-tion of aryl groups giving entry to ionic water-soluble phosphine complexes has gained widespread use. The title complex, commonly referred to as 1,3,5-triaza-7-phosphaadamantane and abbreviated PTA, finds use as a neu-tral P-donor ligand, water soluble by virtue of H-bonding to the tertiary amine nitrogens. It has the advantages of air stability and a low steric demand. Furthermore, it typically binds more strongly to the metal centers than do the phenylsulfonated phosphine ligands. Indeed PTA has been characterized as an air-stable, water-soluble version of PMe$_3$.[1]

Published syntheses of PTA in yields of 40% have utilized various sources of nitrogen including hexamethylenetetraamine,[2] ammonia,[2] and ammonium acetate.[3] In those syntheses, phosphorus was introduced as the highly hygro-scopic *tris*(hydroxymethyl)phosphine (THP). In the synthesis described

* Available from Sigma-Aldrich Chimie, L'Isle d'Abeau, France.
† USDA, ARS, Southern Regional Research Center, New Orleans, LA 70124.
‡ Department of Chemistry, Texas A & M University, College Station, TX 77843.

below, the commercially available *tetrakis*(hydroxymethyl)phosphonium chloride, THPC, is used as phosphorus source.

A unique feature of phosphaadamantane is the reactivity of the nitrogen with alkylating agents such that ammonium salts are formed preferentially to phosphonium salts. The synthesis of the methylammonium salt of PTA as well as the phosphine oxide, O=PTA, are described below. The latter is achieved by the reaction of PTA with 30% H_2O_2. This harsh reagent is necessary because of the resistance of PTA, as well as its nitrogen analog, hexamethylenetetraamine, to mild oxidants.

The PTA ligand has recently been employed as a water-soluble ligand in a variety of studies, including catalytic biphasic hydrogenation reactions,[4,5] ligand-substitution reactions in metal clusters,[6] and enzyme-mediated oxo-transfer processes.[7]

A. 1,3,5-TRIAZA-7-PHOSPHATRICYCLO[3.3.1.1³·⁷]DECANE

$$P(CH_2OH)_4Cl + NaOH \longrightarrow P(CH_2OH)_3 + H_2O + CH_2O + NaCl$$

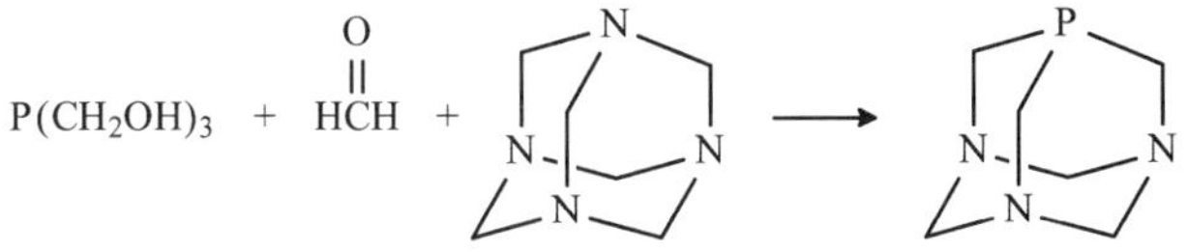

Procedure

To an open 800-mL beaker containing 124 g of ice and 238 g (1.25 mol) of THPC, *tetrakis*(hydroxymethyl)phosphonium chloride,* a solution of 63.9 g NaOH (50% w/w) is slowly added while manually stirring. After the resulting clear, colorless solution warms to room temperature, 450 g of formaldehyde (40%, 6.25 mol) is added. Then 140 g (1.00 mol) of hexamethylenetetraamine† is dissolved in the solution, which is subsequently allowed to stand overnight. The solution is then transferred to a large evaporating dish‡ and placed under a hood or draft of air for ease of evaporation. It is allowed to stand until the solution becomes approximately

---

* Pyroset TKCᴿ, Cyanamid Canada, Inc., Welland Plant, Garner Road, Niagara Falls, Ontario (905-374-5834). THPC utilized by the checkers was obtained from Aldrich Chemical Co., Milwaukee, WI 53233.
† Aldrich Chemical Co., Milwaukee, WI 53233.
‡ For more rapid evaporation the checkers divided the solution into several small portions, maintaining shallow covering of the bottom of the beaker.

90% solid. The solid is then filtered via a Büchner funnel, washed with several aliquots of cold ethanol, and allowed to dry in air. (This filtration process removes water, some NaCl, and any excess hexamethylenetetraamine.) Subsequently, the solid is dissolved in a minimum amount of chloroform (ca. 2 L), which allows for the separation of any residual NaCl. The solution is gravity filtered through filter paper, followed by removal of $CHCl_3$ by rotoevaporation. Recrystallization of PTA by dissolving the solid in hot ethanol followed by slow cooling to room temperature results in the formation of a white, crystalline solid which is analytically pure. This method yields 102 g (65%) with a minimum purity of 97% (NMR).[‡]

Anal. Calcd. for $C_6H_{12}N_3P$: C, 45.9%; H, 7.70%; N, 26.8%; P, 19.7; mol. wt., 157. Found: C, 45.6%; H, 7.70%; N, 26.5%; P, 19.8%; mol. wt., 158 (boiling point elevation, water).

Properties

The physical and chemical properties of PTA are similar to hexamethylenetetraamine; however, PTA is less soluble in water, methanol, and ethanol. PTA is a small, basic phosphine, similar in electronic and steric properties to PMe_3^1 and is air stable. Infrared spectrum (KBr, mμ): 3.52(m), 6.92(w), 7.02(w), 7.15(w), 7.4(w), 7.79(m), 8.12(s); 9.12(m), 9.67(w), 9.9(s), 10.0(s), 10.32(s), 10.57(m), 11.27(w), 12.37(m), 12.6(m), 13.51(m), 14.5(m). ^{1}H NMR in D_2O (resonances referenced against sodium-3-trimethylsilyl-1-propane sulfonate): δ 4.43, s (NCH_2N) and δ 3.9 (d, $J_{H-P} = 9$ Hz) (PCH_2N) in the ratio of 1:1. ^{31}P NMR: -98.3 ppm (in water and referenced to 85% H_3PO_4). PTA decomposes at $>260°C$.

B. 1,3,5-TRIAZA-7-PHOSPHATRICYCLO[3.3.1.1^{3,7}]DECANE 7-OXIDE

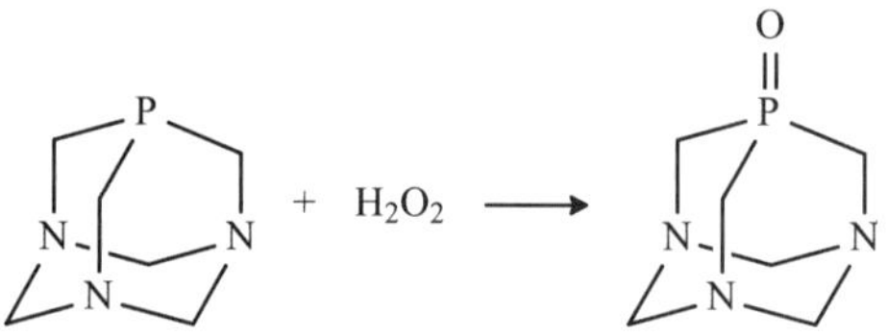

[‡] At one-tenth the scale, the checkers obtained a 58% yield of the recrystallized product.

Procedure

■ **Caution.** *30% H_2O_2 is corrosive and should be handled with care.*

In a 100-mL beaker, containing a magnetic stir bar, 0.5 g, 3.2×10^{-3} mole, of PTA is dissolved in 20 mL of methanol. Over a period of 10 min a solution consisting of 0.5 g (0.5 mL) 30% H_2O_2 in 20 mL of ethanol is slowly added to the solution containing PTA by means of a Pasteur pipette. The resulting solution, which contained a small amount of white precipitate, is stirred for 20 min. The solvent is then removed by evaporation under a hood or in a draft of air at room temperature to complete dryness. The product is recrystallized from hot ethanol, producing 0.44 g (80% yield) of O=PTA.

Anal. Calcd. for $C_6H_{12}N_3OP$: C, 41.62%; H, 6.99%; N, 24.3%; P, 17.9%; mol. wt., 173. Found: C, 41.8%; H, 7.02%; N, 24.2%; P, 18.0%; mol. wt., 169 (boiling point elevation, water).

Properties

The solubility of the phosphine oxide, O=PTA, is slightly less in polar solvents than the parent PTA. The infrared spectrum (KBr, mμ) exhibited bands at 3.15(w), 7.0(m), 7.25(w), 7.35(w), 7.6(w), 7.85(s), 8.15(m), 8.62(s), 9.16(m), 9.65(w), 10.0(s), 10.3(s), 10.6(w), 11.08(s), 12.4(s), 12.82(w), 13.5(w), 13.7 μ(w). ^{1}H NMR resonances in D_2O referenced against sodium-3-trimethylsilyl-1-propane sulfonate: δ 4.28 (s) and δ 4.0 (doublet, $J_{H-P} = 11$ Hz) in ratio of 1:1. ^{31}P NMR: -2.49 ppm (vs. 85% H_3PO_4, measured on a Varian 200 MHz spectrometer in water). The oxide O=PTA decomposes at $>260°C$.

C. 1-METHYL-3,5-DIAZA-1-AZONIA-7-PHOSPHATRICYCLO-[3.3.1.1³,⁷]DECANE IODIDE

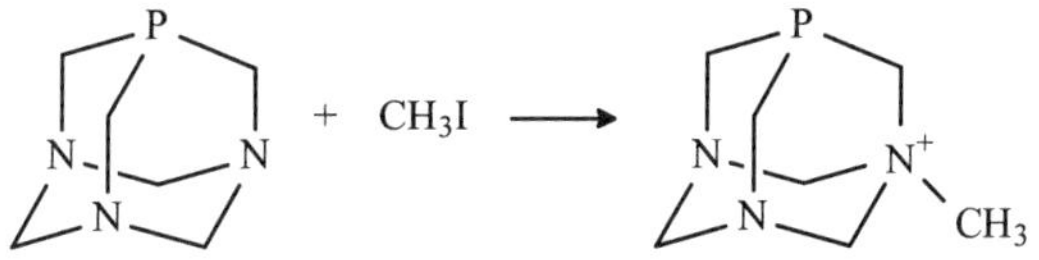

Procedure

■ **Caution.** *Methyl iodide is a known carcinogen. Protective gloves should be worn and the material handled with care.*

In a 100-mL, round-bottomed flask fitted with a reflux condenser, PTA (0.5 g, 3.2×10^{-3} mol) and methyl iodide (0.46 g, 3.2×10^{-3} mol) are dissolved in 60 mL of acetone. The mixture is then refluxed for 1 h, during which a white precipitate forms. The solution is filtered in air through a Büchner funnel and subsequently washed with acetone, resulting in a crude yield of 0.91 g of the methylammonium salt.* The product is recrystallized by heating a 50-mL solution of methanol/ethylacetate (50:50) containing the $[PTAMe^+]I^-$ until all solid material dissolves. Upon cooling in an ice bath, white, flaky crystals form. The product was then filtered in air through a Büchner funnel and washed with ethyl acetate.

Anal. Calcd. for $C_7H_{15}IN_3P$: C, 28.1%; H, 5.06%; N, 14.1%; P, 10.4%; I, 42.4%. Found: C, 28.1%; H, 5.06%; N, 14.3%; P, 10.4%; I, 42.4%. The purified methylammonium salt melted at 204°C.

Properties

Similar to the parent phosphine PTA, 1-methyl-3,5-diaza-1-azonia-7-phospha-tricyclo$[3.3.1.1^{3,7}]$decane iodide, is soluble in polar solvents. Owing to the nature of the ammonium salt functionality, its insolubility in organic non-polar solvents such as esters is much greater than for PTA. The infrared spectrum (KBr, mμ) exhibited bands at 3.5(m), 3.55(w), 6.97(s), 7.23(m), 7.3(w), 7.5(w), 7.65(s), 7.85(m), 7.95(m), 8.05(m), 8.17(m), 8.33(w), 8.8(m), 9.04(s), 9.25(s), 9.35(m), 9.38(s), 9.57(w), 9.67(w), 10.1(w), 10.26(s), 10.3(w), 10.51(w), 10.65(w), 10.96(s), 11.2(m), 11.7(w), 11.85(w), 12.23(m), 12.35(s), 12.9(w), 13.1(s), 13.65(m), 13.75(m), and 14.65 μ(w). The ^1H NMR spectrum in dimethyl sulfoxide-d_6 had a complex set of peaks from δ 5.1 to δ 3.78 and a singlet at δ 2.71 in a ratio of 4:1.

References

1. D. J. Daigle and M. Y. Darensbourg, *Inorg. Chem.*, **14**, 1217 (1975); J. R. Delerno, L. M. Trefonas, M. Y. Darensbourg, and R. J. Majeste, *Inorg. Chem.*, **15**, 816 (1976).
2. D. J. Daigle, A. B. Pepperman, Jr., and S. L. Vail, *J. Heterocyclic Chem.*, **11**, 407 (1974).
3. E. Fluck and J. E. Forster, *Chem. Zeitung*, **99**, 246 (1975).

* Checkers' comment: On the same scale, we obtained 0.872 g of the crude product. After recrystallization, the yield was 66%. The NMR spectra were recorded in d_6-DMSO on 200 MHz instruments. The ^{31}P NMR exhibited a single peak at -86.12 ppm; ^1H NMR showed a complex pattern between 3.7 and 5.0 ppm, and a singlet at 2.61 ppm, in a ratio of 4.4:1, respectively.

4. D. J. Darensbourg, F. Joo, M. Kannisto, A. Katho, J. H. Reibenspies, and D. J. Daigle, *Inorg. Chem.*, **33**, 200 (1994).
5. D. J. Darensbourg, N. W. Stafford, F. Joo, and J. H. Reibenspies, *J. Organometal. Chem.*, **488**, 99 (1996).
6. D. M. Saysell, C. D. Borman, C.-H. Kwak, and A. G. Sykes, *Inorg. Chem.*, **35**, 173 (1996).
7. B. E. Schultz, R. Hille, and R. H. Holm, *J. Am. Chem. Soc.*, **117**, 827 (1995).

Chapter Two

BIOMIMETIC AND SPECIAL PROPERTY LIGANDS

7. 2,2′:6′,2″-TERPYRIDINE

Submitted by DONALD L. JAMESON* and LISA E. GUISE*
Checked by CAROL A. BESSEL[†] and KENNETH TAKEUCHI[‡]

The 2,2′:6′,2″-terpyridine molecule is a member of the polypyridine class of ligands and is widely used in transition metal chemistry as a meridionally coordinating, tridentate chelate.[1] Focus on terpyridine stems, at least in part, from the interesting photochemical and redox properties of the related ligands 2,2′-bipyridine and 1,10-phenanthroline.[2] The redox and photochemical properties of supramolecular terpyridine complexes of Ru(II) and Os(II) have been the subject of a recent review.[3] Several groups are currently engaged in exploring polypyridine ligands as structural elements in the assembly of helical metal complexes.[4,5] Terpyridine is the simplest of the homologous series of polypyridine ligands which are capable of forming helical complexes.[5]

The strategies for the synthesis of terpyridine can be divided into two groups. A "nonrational" strategy, where three pyridine rings are coupled usually gives rise to low yields and complex mixtures of oligopyridines and regioisomers, but has the advantage of generally using cheap starting materials in a one-step reaction sequence. In this way, terpyridine was first isolated in very low yields by Morgan and Burstall as one of several products from the

* Department of Chemistry, Gettysburg College, Gettysburg, PA 17325.
[†] Department of Chemistry, Villanova University, Villanova, PA 19085.
[‡] Department of Chemistry, SUNY at Buffalo, Buffalo, NY 14214.

oxidative dehydrogenation of pyridine effected by iron(III) chloride.[6] Raney nickel will also effect this coupling reaction to give mainly 2,2′-bipyridine, in addition to small amounts of terpyridine.[7] The Ullman reaction (an oxidative coupling) of appropriately halogenated pyridine intermediates gives terpyridine (as well as other oligopyridines) in relatively low yields.[8] More "rational" strategies involve the construction of the central pyridine by the aza ring closure of 1,5-diketone intermediates. The four-step "Kronke-type" synthesis developed by Constable is an example of this strategy and affords terpyridine in an overall yield of 30–35%.[9] Potts and co-workers have developed a three-step synthesis of terpyridine by the aza ring closure of a 1,5-enedione.[10] A thioether substituent, which ends up in the 4′ position of the ring closure product, is the source of both an advantage and a drawback to this strategy. The thioether provides an entry into terpyridine ligands bearing a variety of substituents in the 4′ position and provides increased solubility, but its removal to provide the parent ligand involves a very inefficient reduction reaction. A second drawback of this strategy is the formation of the 1,5-enedione occurs with the release of methanethiol, whose smell pervades both the product (even after purification) and the reaction wastes.

The present approach to terpyridine is a two-step procedure based on a variation of the Potts strategy.[11] In this sequence, the 1,5-enedione formation is accompanied by the loss of dimethyl amine and the ring closure gives terpyridine directly, thus avoiding both drawbacks of the Potts method. The synthesis of the intermediate enaminone (**1**) has been previously reported.[12]

A. β-(DIMETHYLAMINO)VINYL 2-PYRIDYL KETONE

1

Procedure

■ **Caution.** *All procedures should be performed in a well-ventilated hood.*

A solution of 153.0 g (1.26 mole) of 2-acetylpyridine (Aldrich) and 180.0 g (1.41 mole) of *N,N*-dimethylformamide dimethylacetal (Aldrich, 94% pure) in 250 mL of toluene are heated to reflux in a 1-L, round-bottomed flask. Methanol is gradually removed by fractional distillation through a 20-cm

Vigreux column. When no more methanol distills over (ca. 24 h), the dark brown reaction mixture is reduced in volume by the distillation of 100 mL of toluene. While the flask is still hot, 300 mL of cyclohexane are added with vigorous stirring and a yellow, crystalline solid comes out of solution. At this point, the copious amount of solid makes stirring difficult. The mixture is allowed to stand overnight and the yellow solid is collected by suction filtration. The solid is washed with 2:1 cyclohexane/toluene (ca. 300 mL) and, then cyclohexane and is allowed to air dry. The solid is then dried in a vacuum dessicator (25°C, 24 h) to remove the last traces of *N,N*-dimethylformamide dimethylacetal (a small amount of sublimation of the product will occur). Failure to completely remove the last traces of *N,N*-dimethylformamide dimethylacetal results in the gradual darkening of the product over several months time. Yield: 190.3 g (86% based on 2-acetylpyridine),* mp 127–128°C.

Anal. Calcd. for $C_{10}H_{12}N_2O$: C, 68.15; H, 6.87; N, 15.90. Found: C, 68.16; H, 6.97; N, 15.81.

Properties

The *β*-(dimethylamino)vinyl 2-pyridyl ketone is a yellow, crystalline solid which is stable for long periods of time at room temperature. The compound is soluble in polar organic solvents and chlorinated solvents, slightly soluble in toluene, and nearly insoluble in aliphatic hydrocarbons. The ^{1}H NMR (80 MHz, CDCl$_3$) exhibits signals at δ 3.07 (br s, 6 H), 6.45 (d, 1 H, J = 12.8 Hz), 7.34 (d of d of d, 1 H, J = 7.4, 4.8, 1.4 Hz), 7.78 (d of t, 1 H, J = 7.7, 1.8 Hz), 7.90 (d, 1 H, J = 12.8 Hz), 8.15 (d of d of d, 1 H, J = 7.8, 1.3, 1 Hz), 8.62 (d of d of d, 1 H, J = 4.8, 1.8, 0.9 Hz).

B. 2,2′:6′,2″-TERPYRIDINE

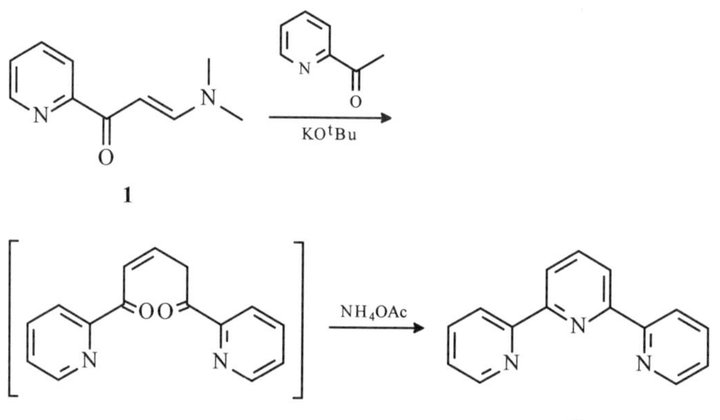

1

2

* The checkers performed this reaction on one-eighth the scale and reported a yield of 80%.

A flame-dried, 2-L, three-necked flask containing a large (2.5-in. egg) magnetic stirring bar is purged with nitrogen. One liter of anhydrous THF (Aldrich Sure-Seal) is transferred by cannula to the flask and 50.0 g (0.419 mole) of potassium *t*-butoxide (Aldrich, 95%) is added followed by 24.2 g (0.200 mole) of 2-acetylpyridine. The solution is stirred at room temperature for 2 h, whereupon a white solid forms. The *β*-(dimethylamino)vinyl 2-pyridyl ketone (35.2 g, 0.200 mole) is added in a single portion and the solution, which gradually turns a very deep red, is stirred at room temperature for 40–60 h. The reaction mixture is cooled in ice and a solution of 150 g (1.95 mole) of ammonium acetate in 500 mL of acetic acid is added with stirring, whereupon the color changes from deep red to brown. Methanol (100 mL) is added,* the flask is fitted with a distillation head, the mixture is heated to reflux, and THF is gradually removed by distillation over an 8- to 10-h period. The distillation is continued until the head temperature reaches 115°C. The dark, almost black, mixture is poured into 1 L of water and the acetic acid is carefully neutralized by the portionwise addition of solid sodium carbonate (vigorous effervescence!). Neutralization is judged to be complete when the effervescence ceases. At this point, a black, tarry material is present in the reaction mixture. Toluene (300 mL) and 50 g of filter aid are added and the mixture is stirred and heated to 80°C for 2 h. The flask is cooled to room temperature and the mixture is filtered through a pad of filter aid. The black solid† in the funnel is broken up into a powder with the aid of a spatula and the pad is washed with 3 × 100 mL of toluene. The organic layer of the filtrate is separated and the aqueous layer is extracted with 2 × 150 mL of toluene. The combined organic layers are stirred with 60 g of activity I neutral alumina and filtered, and the filtrate is evaporated to a dark brown oil on a rotary evaporator. The oil is dissolved in 100 mL of methylene chloride and 100 g of activity III alumina is added. The slurry is evaporated to dryness on a rotary evaporator and the resulting brown powder is added to the top of a column (45 × 4 cm) of 400 g of activity III neutral alumina. The column is then eluted with 20:1 cyclohexane/ethyl acetate. A light yellow fraction is collected, followed by a fraction which is slightly contaminated by an orange band. The yellow band is evaporated to dryness and recrystallized from 200 mL of 10:1 hexanes/cyclohexane‡ to give 22.45 g of 2,2':6',2"-terpyridine (mp 87–88°C) as an off-white solid.

---

* The addition of methanol is essential to prevent clogging of the condenser with solid ammonium acetate during the distillation.
† The dark, tarry byproduct of this reaction can conveniently be removed from glassware with 2M sulfuric acid.
‡ The cyclohexane in the recrystallization solvent reduces the tendency for the product to come out of solution as an oil.

Evaporation of the filtrate and trituration of the oily solid with 10:1 hexanes/cyclohexane affords an additional 2.80 g (mp 85–86°C) for an overall yield of 25.25 g (54%). Yields for this step are typically in the range 45–55%. The filtrate of this second crop and the orange band from the chromatography still contain more of this valuable ligand. Further chromatography does not appreciably improve this material, as the major impurities coelute with the terpyridine. This additional amount of terpyridine can be isolated from the mixture as $Fe(terpy)_2(PF_6)_2$ and the ligand separated from the complex by the method of Constable.[9] A total of 3.40 g of the iron(II) complex (which corresponds to 1.95 g of terpyridine) can be isolated in this manner. Including this product increases the overall yield to 58%.*

Anal. Calcd. for $C_{15}H_{11}N_3$: C, 77.23; H, 4.76; N, 18.01. Found: C, 77.17; H, 4.91; N, 17.80.

Properties

2,2′:6′,2″-terpyridine is a white, crystalline solid which is soluble in virtually all organic solvents save alkanes from which it can be recrystallized. The ^1H NMR (80 MHz, $CDCl_3$) exhibits signals at δ 7.29 (d of d of d, 2 H, J = 7.4, 4.8, 1.3 Hz), 7.83 (t of m, 2 H), 7.93 (t, 1 H, J = 7.4 Hz), 8.39–8.65 (m, 4 H), 8.68 (d of d of d, 2 H, J = 4.7, 2.0, 0.9 Hz). The ^{13}C NMR (20 MHz, $CDCl_3$) exhibits signals at δ 121.0, 121.1, 123.7, 136.7, 137.8, 149.1, 155.4, 156.3. The crystal structure of the free ligand has been determined.[13]

References

1. E. C. Constable, *Adv. Inorg. Chem. Radiochem.*, **30**, 69 (1986).
2. A. Juris, V. Balzani, F. Barigelletti, S. Campagna, P. Belser, and A. von Zelewsky, *Coord. Chem. Rev.*, **84**, 85 (1988).
3. J.-P. Sauvage, J.-P. Collin, J.-C. Chambron, S. Guillerez, C. Coudret, V. Balzani, F. Barigelletti, L. De Cola, and L. Flamigni, *Chem. Rev.*, **94**, 993 (1994).
4. E. C. Constable, *Tetrahedron*, **48**, 10013 (1992).
5. K. T. Potts, K. A. Gheysen Raiford, and M. Keshavarz-K, *J. Am. Chem. Soc.*, **115**, 2793 (1993).
6. G. Morgan and F. H. Burstall, *J. Chem. Soc.*, 1649 (1937).
7. G. M. Badger and W. H. F. Sasse, *J. Chem. Soc.*, 616 (1956).
8. F. H. Burstall, *J. Chem. Soc.*, 1662 (1938).
9. E. C. Constable, M. D. Ward, and S. Corr, *Inorg. Chim. Acta*, **141**, 201 (1988).
10. K. T. Potts, M. J. Cipullo, P. Ralli, and G. Theodoridis, *J. Org. Chem.*, **47**, 3027 (1982). K. T. Potts, M. J. Cipullo, P. Ralli, G. Theodoridis, and P. Winslow, *Org. Synth.*, **64**, 189 (1985).
11. D. L. Jameson and L. E. Guise, *Tetrahedron Lett.*, **32**, 1999 (1991).
12. Y. Lin and S. A. Lang, *J. Heterocycl. Chem.*, **14**, 345 (1977).
13. C. A. Bessel, R. F. See, D. L. Jameson, M. R. Churchill, and K. J. Takeuchi, *J. Chem. Soc., Dalton Trans.*, 3223 (1992).

* The checkers performed this reaction on 1/5 the scale and reported a yield of 40%.

8. POLY(1-PYRAZOLYL)ALKANE LIGANDS

Submitted by DONALD L. JAMESON* and RONALD K. CASTELLANO*
Checked by DANIEL L. REGER[†] and JAMES E. COLLINS[†]
(PARTS A, B, E, F)
WILLIAM B. TOLMAN[‡] and CHRISTOPHER J. TOKAR[‡] (PARTS C AND D)

The polypyrazolylborates, a class of uninegative bidentate and tridentate ligands developed by Trofimenko in the early 1970s, remain the most important class of pyrazole-containing ligand.[1] Trofimenko also devised several routes for the preparation of polypyrazolylalkanes, neutral analogs of the polypyrazolylborates.[2] Although their coordination chemistry was largely unexplored for several years, they are now utilized, particularly for the stabilization of a variety of organometallic complexes.

Some examples of metal complexes which are supported by bidentate bis(pyrazolyl)alkanes (L') are Pd(II)(L')allyl$^+$,[2] Mo(II)(L')(CO)$_2$X(allyl),[3] Sn(IV)(L')(R)Cl$_3$,[4] Nb(III)(L')(alkyne)Cl$_3$,[5] Pd(II)(L')MeX, and Pd(II)(L')Me$_2$.[6] Among the complexes supported by the tridentate tris(pyrazolyl)methanes (L'') are Pt(IV)(L'')Me$_3$,[7] Pd(IV)(L'')R$_3^+$,[7,8] Ru(III)(L'')(O)(OAc)2(L'')Ru(III),[9] W(IV)(L'')(CR)(CO)$_2^+$,[10] Nb(III)(L'')(alkyne)Cl$_2^+$,[5] and a variety of substituted molybdenum carbonyls.[2]

The bis(pyrazolyl)alkanes are synthesized by two routes. Where two pyrazole rings are bridged by a methylene, the ligands are prepared by displacement of the halides of CH$_2$X$_2$ by pyrazolate anions (Method A). When X = Br or I, the organohalide is used in a stoichiometric amount[2,3] and a cosolvent is used. When X = Cl, methylene chloride is both the electrophile and the solvent.[11,12] Recent preparations of these ligands exploit the aid of a phase-transfer catalyst.[3,11,12] When the pyrazole rings are bridged by larger alkyl groups (most commonly isopropylidene), the ligands are prepared by the transamination of ketals (Method B).[2] Examples of both methods are given below for parent ligands **A** (following closely the method of Trofimenko[2]) and **D** (following the method of Elguero et al.[11]).

The recent report of a chiral bis(1-pyrazolyl)methane provides the first example of L' not based on pyrazole or 3,5-dimethylpyrazole.[13,14] The substituted bis(pyrazolyl)methane was prepared by Method A, which gave

* Department of Chemistry, Gettysburg College, Gettysburg, PA 17325.
[†] Department of Chemistry and Biochemistry, University of South Carolina, Columbia, SC 29208.
[‡] Department of Chemistry, University of Minnesota, Minneapolis, MN 55455.

rise to all three possible regioisomers. A different approach to bis(pyrazolyl)methanes bearing unsymmetrically substituted pyrazoles is presented below for ligands **B** and **C**. Method B is conducted under equilibrating conditions and, with the use of an unsymmetrically substituted pyrazole, gives rise to the single isomer in which steric interactions between the pyrazole substituent and the gem-dimethyl groups are minimized. The synthesis of (4S,7R)-7,8,8-trimethyl-4,5,6,7-tetrahydro-4,7-methano-2-indazole from (1R)-(+)-camphor given below is a slight modification of a literature procedure[16] and is an example of how pyrazoles can be prepared from ketones.

Tris(1-pyrazolyl)methane was first prepared by Hückel and Bretschneider by the reaction of sodium pyrazolate with chloroform.[17] Trofimenko prepared the tris(3,5-dimethyl-1-pyrazolyl)methane by a similar method.[2] More recent preparations have utilized both liquid–liquid[18] and solid–liquid[12] phase-transfer catalysis. The preparation of tris(1-pyrazolyl)methane (**E**) given below is a modification of the solid–liquid phase-transfer procedure.[12]

The preparation of ligand **F** represents an approach to tris(1-pyrazolyl)methanes bearing unsymmetrically substituted pyrazoles. The conditions of the nucleophilic displacement reaction (kinetic control) give rise to a complex mixture of regioisomers. Treatment of this mixture to equilibrating conditions (refluxing toluene, catalytic acid) results in the conversion to a single regioisomer in which steric interactions between the pyrazole substituent and the methine proton are minimized.

A. 2,2-BIS(1-PYRAZOLYL)PROPANE

Procedure

A 100-mL, round-bottomed flask is charged with 17.02 g (0.250 mole) of pyrazole,* 10,42 g (0.100 mole) of 2,2-dimethoxypropane,* (distilled from sodium hydride and stored under N_2), 20–30 mg of p-toluenesulfonic acid monohydrate,* and a 1-in. egg-stirring bar. The flask is fitted with a 20-cm Vigreux column topped by a short-path distillation apparatus. The mixture is heated to reflux and methanol (about 8 mL) is slowly removed by distillation.

* Aldrich Chemical Co., Milwaukee, WI 53233.

The reaction is cooled to room temperature and approximately 50 mL of toluene and 35 mL of 5% sodium carbonate solution are added. The organic layer is separated, extracted with 3×50 mL of 5% sodium carbonate solution, dried over $MgSO_4$, and filtered. The toluene is removed on a rotary evaporator and the residue dissolved in 20 mL of methylene chloride. The concentrated solution is absorbed at the top of a column of alumina (2×20 cm; ca. 100 g of activity III) and eluted with 150 mL of CH_2Cl_2 to remove any unreacted pyrazole. The solvent is removed on a rotary evaporator and the solid is recrystallized from 100 mL hexanes to give 10.28 g of the product as white crystals melting at 84–86°C. A second crop of product (0.60 g) can be obtained by concentration of the filtrate, giving an overall yield of 10.88 g (62%) based on 2,2-dimethoxypropane.

Anal. Calcd. for $C_9H_{12}N_4$: C, 61.35; H, 6.87; N, 31.78. Found: C, 61.06; H, 6.78; N, 31.78.

Properties

The 2,2-bis(1-pyrazolyl)propane is a white, crystalline solid that is soluble in most organic solvents except hydrocarbons, in which it is slightly soluble. The ligand is stable in basic and neutral solutions, but in acidic protic solvents it is susceptible to solvolysis, with decomposition to pyrazole and acetone. The ^1H NMR (80 MHz, $CDCl_3$) exhibits signals at δ 2.27 (s, 6H), 6.38 (t, 2H, J = 2.2 Hz), 7.38 (d, 2H, J = 2.5 Hz), 7.53 (d, 2H, J = 1.6 Hz). The ^{13}C NMR (20 MHz) exhibits signals at δ 28.0, 77.1 (partially obscured by the $CDCl_3$), 106.2, 127.0, 139.5.

B. 2,2-BIS(3-PHENYL-1-PYRAZOLYL)PROPANE

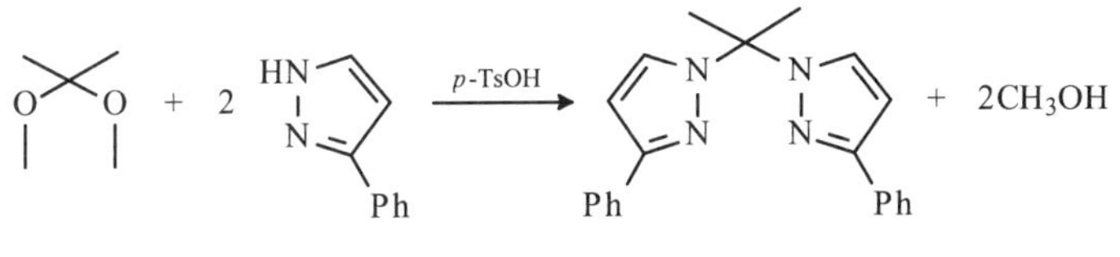

Procedure

A 250-mL, round-bottomed flask is charged with 20–30 mg of *p*-toluenesulfonic acid hydrate,* 100 mL of toluene, and a 1-in. egg-stirring bar. The acid

* Aldrich Chemical Co., Milwaukee, WI 53233.

is dehydrated by distilling off about 10 mL of solvent. The reaction flask is cooled under N_2 and 60 mL of cyclohexane is added followed by 5.00 g (34.7 mmole) of 3(5)-phenylpyrazole.[19] The mixture is gently heated until the solid dissolves and 1.72 g (16.5 mmole) of 2,2-dimethoxypropane is added. The flask is fitted with a Soxhlet apparatus in which the thimble is filled with 20 g of 4-Å molecular sieves (activated at 400°C for 5 h). The reaction is heated to reflux for approximately 24 h. The progress of the reaction is monitored by TLC (alumina/toluene). Four spots may be observed at various times during the reaction: unreacted phenylpyrazole has an R_f of about 0, the methoxypyrazolylpropane derivative has an R_f of between 0.4 and 0.5, the desired bispyrazolylpropane has an R_f of between 0.6 and 0.7, and an elimination product, 2-(3-phenyl-1-pyrazolyl)propene, has an R_f of between 0.7 and 0.8. Attempts to convert all of the methoxypyrazolylpropane to **B** using longer reaction times resulted in a gradual increase in the amount of the pyrazolylpropene at the expense of product yield. The reaction mixture is cooled to room temperature and quenched with an equal volume of saturated sodium carbonate solution and the two-phase mixture is stirred for 15 min. The organic layer is separated and the aqueous layer is extracted with an additional 50 mL of toluene. The combined organic layers are dried over $MgSO_4$, filtered through a cotton plug, and rotary evaporated to 10–15 mL. The concentrated solution is absorbed on a column of activity IV neutral alumina (100 g) and eluted with toluene (150–200 mL). The toluene fractions containing the product (TLC; alumina/toluene) are rotary evaporated and recrystallized from hot cyclohexane to afford 4.44 g (82%) of white crystals (mp 111.5–112.5°C).

Anal. Calcd. for $C_{21}H_{20}N_4$: C, 76.80; H, 6.14; N, 17.06. Found: C, 76.61; H, 6.04; N, 17.04.

Properties

2,2-Bis(3-phenyl-1-pyrazolyl)propane is a white, crystalline solid which is soluble in most organic solvents. The compound is moderately soluble in hot aliphatic hydrocarbons, but sparingly soluble in cold, making these ideal solvents for recrystallization. The ligand is stable in basic and neutral solutions, but in acidic protic solvents, it is susceptible to solvolytic decomposition to acetone and 3(5)-phenylpyrazole. The ^1H NMR (80 MHz, $CDCl_3$) exhibits signals at δ 2.34 (s, 6H), 6.53 (d, 2H, J = 2.5 Hz), 7.29–7.37 (m, 6H), 7.42 (d, 2H, J = 2.5 Hz), 7.76–7.88 (m, 4H). The ^{13}C NMR (20 MHz, $CDCl_3$) exhibits signals at δ 28.0, 77.8, 103.5, 125.8, 127.7, 128.5, 133.6, 151.4.

C. 2,2-BIS[(4*S*,7*R*)-7,8,8-TRIMETHYL-4,5,6,7-TETRAHYDRO-4,7- METHANO-2-INDAZOLYL]PROPANE

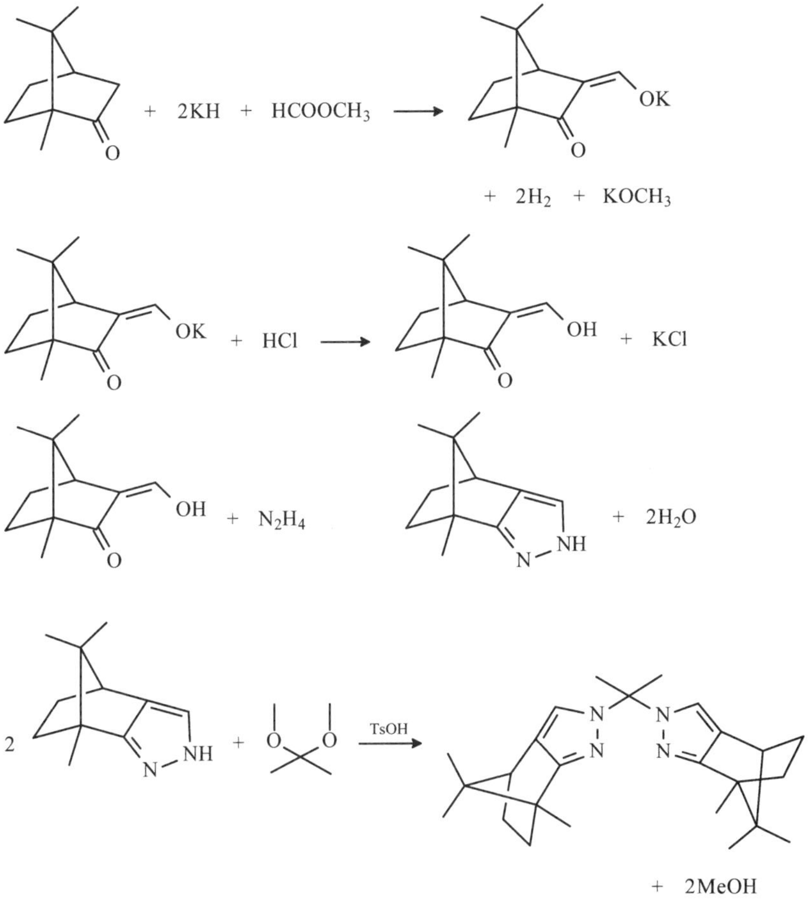

+ 2KH + HCOOCH₃ ⟶

+ 2H₂ + KOCH₃

+ HCl ⟶ + KCl

+ N₂H₄ ⟶ + 2H₂O

2 + TsOH ⟶

+ 2MeOH

Procedure

(1R)-2-hydroxymethylenecamphor. A 3-L, three-necked flask is fitted with a nitrogen t-tube, a septum stopper, and the wide neck with a rubber stopper. The flask is flushed with N₂ and charged with 1.5 L of anhydrous 1,2-dimethoxyethane* (Aldrich/Sure-Seal, transferred by cannula), 150 g of 35% potassium hydride in mineral oil suspension* (52 g of KH; 1.3 mole), and

* Aldrich Chemical Co., Milwaukee, WI 53233.

a 3.5-in. egg-stirring bar.* The flask is cooled in ice and 80.0 g (0.526 mole) of camphor[†] is added in five equal portions with stirring over a 30-min period. The addition is accompanied by the vigorous evolution of hydrogen. The mixture is allowed to warm to room temperature and is stirred for an additional 6 h. The flask is again cooled in ice and 100 g (103 mL, 2.9 mole) of anhydrous methyl formate (Aldrich/Sure-Seal) is added by syringe in five equal portions over a 24-h period. The reaction is allowed to gradually reach room temperature between the additions of methyl formate. The reaction is very mildly exothermic and the solid which is deposited makes stirring somewhat difficult. Following the addition of the last portion of methyl formate, the mixture is stirred at room temperature for an additional 48 h. The mixture is cooled in ice and quenched by the careful addition of 25 mL of methanol. The mixture is transferred, in portions, to a 2-L, round-bottomed flask and the solvent is removed on a rotary evaporator. Water (700 mL) is added to dissolve the solid residue and mixture is extracted with 3×250 mL of petroleum ether (to remove the mineral oil and any neutral organic byproducts). The aqueous layer, which contains the sodium salt of the product, is gently aerated for 10 h to remove any traces of organic solvent, treated with 5 g of activated charcoal, and filtered. The clear solution is cooled in ice and slowly treated with 300 mL of 6 M HCl (**caution**). A voluminous, light yellow precipitate comes out of solution and the mixture is placed in the refrigerator (ca. 5°C) overnight. The product is isolated by filtration, washed with 3×300 mL of water, and air dried to afford 75.5 g (80%) of an off-white solid (mp 64–68°C) which can be used directly in the next step.[‡]

■ **Caution.** *Hydrazine hydrate is highly toxic. The procedure should be carried out in a well-ventilated hood and gloves should be worn.*

(4S,7R)-7,8,8-Trimethyl-4,5,6,7-tetrahydro-4,7-methano-2-indazole. A solution of 40.0 g (0.222 mole) of the hydroxymethylene compound in 250 mL of absolute ethanol in a 1-L, round-bottomed flask is treated with 20 mL (21 g, 0.42 mole) of hydrazine hydrate.[§] The flask is fitted with a condenser and the solution is heated to reflux for 48 h, after which the volume is reduced to

---

* The checkers used a 5-L, three-necked round-bottomed flask and a mechanical stirrer.
[†] Aldrich Chemical Co., Milwaukee, WI 53233.
[‡] The checkers noted (by NMR) a second minor component in the crude solid. They suggest recrystallization of the crude solid from warm hexanes. The submitters noted both components and identified them (by NMR) as the *anti* and *syn* isomers of the hydroxy group. Both isomers are converted to the pyrazole upon treatment with hydrazine hydrate.
[§] The checkers found that freshly opened or freshly distilled hydrazine hydrate gave the maximum yield of high-quality product. Failure to use pure hydrazine hydrate results in contamination of the product with a gummy, orange impurity.

150 mL using a rotary evaporator. Water (400 mL) is added and the volume is reduced to 300 mL by rotary evaporator, resulting in the precipitation of a light yellow solid. The solid is suction filtered and air dried. The crude product is purified by dissolving the solid in 150 mL of ethyl acetate, absorbing the solution on a column (5 × 10 cm) of activity III neutral alumina, and eluting with ethyl acetate (500 mL). The eluent is evaporated and the resulting solid is recrystallized from hot cyclohexane. The crystalline solid is filtered and washed with cyclohexane to afford 35.1 g (90%) of the product melting at 152–154°C.*

Anal. Calcd. for $C_{11}H_{16}N_2$: C, 74.96; H, 9.16; N, 15.89. Found: C, 75.03; H, 8.93; N, 15.89.

2,2-Bis[(4S,7R)-7,8,8-Trimethyl-4,5,6,7-tetrahydro-4,7-methano-2-indazolyl]propane. A 250-mL, round-bottomed flask is charged with 20–30 mg of *p*-toluenesulfonic acid hydrate,[†] 100 mL of toluene, and a 1-in. egg-stirring bar. The acid is dehydrated by distilling off about 10 mL of solvent. The reaction flask is cooled under N_2 and 60 mL of cyclohexane is added, followed by 7.17 g (40.7 mmole) of (4S,7R)-7,8,8-trimethyl-4,5,6,7-tetrahydro-4,7-methano-2-indazole. The mixture is gently heated until the solid dissolves and 2.02 g (19.4 mmole) of 2,2-dimethoxypropane is added by syringe. The flask is fitted with a Soxhlet apparatus in which the thimble is filled with 20 g of 4-Å molecular sieves (activated at 400°C for 5 h).[‡] The reaction mixture is heated to reflux for approximately 24 h. Progress of the reaction is monitored by TLC (alumina/toluene). After 24 h the TLC reveals unreacted pyrazole ($R_f \approx 0$), the product ($R_f \approx 0.4$), and a very small amount of pyrazolylpropene ($R_f \approx 0.6$). The reaction mixture is cooled to room temperature and quenched with an equal volume of saturated sodium carbonate solution and the resulting two-phase mixture is stirred for 15 min. The organic layer is separated and the aqueous layer is extracted with an additional 50 mL of toluene. The combined organic layers are dried over $MgSO_4$, filtered and rotary evaporated to 10–15 mL. The concentrated solution was absorbed on a column of activity IV neutral alumina (100 g) and eluted with toluene (150–200 mL). Fractions containing the desired product were combined, evaporated, and recrystallized from hot hexanes to afford 5.96 g (78%) of white crystals (mp 172–174°C).

---

* The checkers reported a 27% yield of product melting at 145–147°C.
† Aldrich Chemical Co., Milwaukee, WI 53233.
‡ The checkers used crushed 4-Å molecular sieves which were dried under vacuum at 250°C.

Anal. Calcd. for $C_{17}H_{24}N_4$: C, 71.79; H, 8.51; N, 19.70. Found: C, 71.40; H, 8.60; N, 19.58.

Properties

(1R)-2-hydroxymethylenecamphor is an off-white, crystalline solid which is soluble in all organic solvents save aliphatic hydrocarbons. The material which precipitates from aqueous solution in the preparation above is sufficiently pure to proceed to the next step. When stored for long periods at room temperature, the compound will slowly decompose and take on a gummy consistency. This gummy solid may still be converted to the pyrazole in slightly lower, but still satisfactory, yields. The ^{1}H NMR (80 MHz, CDCl$_3$) exhibits signals at δ 0.84 (s, 3H), 0.93 (s, 3H), 0.97 (s, 3H), 1.0–2.2 (m, 4H), 2.43 (d, 1H, J = 3.6 Hz), 6.77 (s, 1H), 7.8 (br s, 1H). The (4S,7R)-7,8,8-trimethyl-4,5,6,7-tetrahydro-4,7-methano-2-indazole is a white, crystalline solid which is soluble in most organic solvents. The compound is moderately soluble in hot aliphatic hydrocarbons, but sparingly soluble in cold, making these ideal solvents for recrystallization. The ^{1}H NMR (80 MHz, CDCl$_3$) exhibits signals at δ 0.66 (s, 3H), 0.96 (s, 3H), 1.33 (s, 3H), 1.0–2.3 (m, 4H), 2.77 (d, 1H, J = 3.6 Hz), 7.08 (s, 1H), 10.5 (br s, 1H). The 2,2-bis[(4S,7R)-7,8,8-trimethyl-4,5,6,7-tetrahydro-4,7-methano-2-indazolyl]propane is a white, crystalline solid which is soluble in most organic solvents. The compound is moderately soluble in hot aliphatic hydrocarbons, but sparingly soluble in cold, making these ideal solvents for recrystallization. The ligand is stable in basic and neutral solutions, but in acidic protic solvents, it decomposes to acetone and (4S,7R)-7,8,8-trimethyl-4,5,6,7-tetrahydro-4,7-methano-2-indazole. The ^{1}H NMR (80 MHz, CDCl$_3$) exhibits signals at δ 0.59 (s, 6 H), 0.91 (s, 6H), 1.26 (s, 6H), 1.0–2.1 (m, 8H), 2.18 (s, 6H), 2.66 (m, 2H), 6.79 (s, 2H). The ^{13}C NMR (20 MHz) has signals at d 10.8, 19.2, 20.6, 27.8, 28.5, 33.9, 47.3, 50.4, 60.2, 76.7, 119.4, 127.1, 166.1. The [a]$_D$ = $-$ 15°C (c 0.100 g/mL, CHCl$_3$).

D. BIS(1-PYRAZOLYL)METHANE

+   2NaCl

Procedure

A solution of 15.0 g (0.221 mole) of pyrazole and 2.0 g (5.9 mmole) of tetrabutylammonium hydrogen sulfate in 250 mL of methylene chloride is treated with 100 mL of 50% aqueous sodium hydroxide (ca. 2 mole). The solution is heated to reflux with rapid stirring for 14 h. The mixture is cooled in ice and sufficient water is added to dissolve the solid in the aqueous layer. The mixture is transferred to a separatory funnel and the organic layer is removed. (Note: whether the organic layer is the top or bottom depends on how much water is added.) The aqueous layer is then washed with 3×60 mL of methylene chloride. The combined organic layers are stirred with 20 g of solid sodium bicarbonate, filtered, and evaporated to a volume of 50 mL. This light yellow solution is absorbed on a 5×10 cm column of neutral, activity III alumina.* The column is eluted with methylene chloride (250 mL) and the eluent is evaporated to about 50 mL. Cyclohexane (300 mL) is added and the mixture is evaporated to 150 mL. The resulting white, crystalline solid is filtered, washed with cyclohexane, and dried to yield 14.79 g (90%) of the ligand (mp 108–109°C). A second crop (1.25 g, mp 108–109°C) may be obtained by evaporating the filtrate to a volume of 25 mL. The total yield is 16.04 g (98%).

Anal. Calcd. for $C_7H_8N_4$: C, 56.75; H, 5.45; N, 37.80. Found: C, 56.96; H, 5.31; N, 37.59.

Properties

Bis(1-pyrazolyl)methane is a white, crystalline compound which is soluble in most organic solvents and sparingly soluble in alkanes. The ligand is stable in basic and neutral solutions, but in acidic protic solvents, it decomposes to formaldehyde and pyrazole. The ^{1}H NMR (80 MHz, $CDCl_3$) exhibits signals at δ 6.23 (d of d, 2H, J = 2.4, 1.9 Hz), 6.25 (s, 2H), 7.51 (d of d, 2H, J = 1.8, 0.4 Hz), 7.61 (d of d, 2H, J = 2.4, 0.5 Hz). The ^{13}C NMR (20 MHz, $CDCl_3$) exhibits signals at δ 65.0, 106.8, 129.3, 140.4.

E. TRIS(1-PYRAZOLYL)METHANE

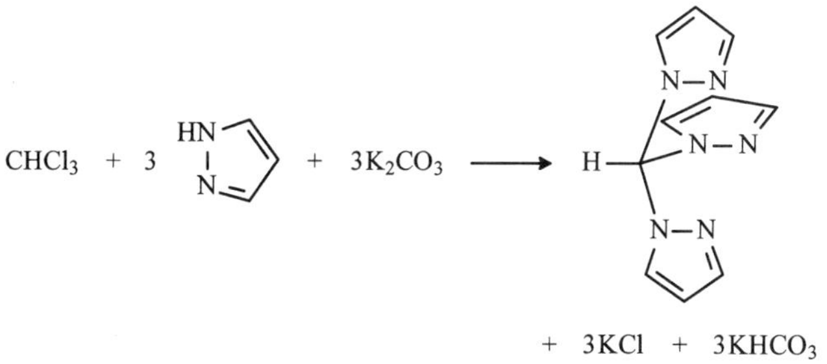

$$CHCl_3 \; + \; 3 \; \text{HN} \; + \; 3K_2CO_3 \longrightarrow \text{H}$$

$$+ \; 3KCl \; + \; 3KHCO_3$$

* The checkers found that alumina was not sufficient to remove the $NBu_4^nHSO_4$ and used silica gel as the chromatography adsorbent.

Procedure

A 500-mL, round-bottomed flask, fitted with a gas inlet tube, is charged with 100 g (0.724 mole) of anhydrous potassium carbonate and heated under vacuum using the gentle flame of a Bunsen burner. Liberation of water from the solid potassium carbonate is visible. The flask is cooled under vacuum, the vacuum is released under N_2, and 21.0 g (0.309 mole) of pyrazole, 4.00 g (0.0118 mole) of tetrabutylammonium hydrogen sulfate, and a 2-in. egg-stirring bar are rapidly added to the flask. The flask is capped with a septum and 250 mL of anhydrous chloroform (Aldrich, Sure Seal)* is added via cannula. The septum is replaced by a condenser fitted with a nitrogen line t-tube and the mixture is heated to reflux with rapid stirring for 48 h. The dark brown solution is filtered and the K_2CO_3/KCl cake is washed with methylene chloride until the filtrate is essentially colorless. The filtrate is evaporated on a rotary evaporator and the dark brown oil is treated with 300 mL of 0.5% sodium carbonate solution. The mixture is reduced in volume by another 25 mL to ensure that all the organic solvent is driven off. Celite (5 g) and sand (15 g) are added to help disperse the gummy residue and the flask is heated to 80°C with stirring for 1 h. The solids are allowed to settle and the still-warm supernatant is decanted. The solids are extracted with two more 200-mL portions of 0.5% sodium carbonate and the combined aqueous phases are warmed, treated with activated charcoal, and filtered. The pale yellow filtrate is evaporated to about 100 mL on a rotary evaporator. The suspension is allowed to stand at room temperature overnight and the white crystalline solid is collected by filtration, washed with water, and air dried. A first crop yields 9.01 g (mp 103–105°C) and evaporation of the filtrate affords a second crop of 1.42 g (mp 102–104°C). The total yield is 10.43 g (47%). The melting point of the product can be raised to 105–106°C by recrystallization from cyclohexane.

Anal. Calcd. for $C_{10}H_{10}N_6$: C, 56.07; H, 4.71; N, 39.22. Found: C, 55.86; H, 4.60; N, 39.01.

Properties

Tris(1-pyrazolyl)methane is a white, crystalline compound which is soluble in most organic solvents, but sparingly soluble in hydrocarbons. The ligand is stable in basic and neutral solutions. In acidic protic solvents, decomposition to formic acid and pyrazole occurs readily. The ^1H NMR (80 MHz, $CDCl_3$) exhibits signals at δ 6.35 (d of d, 3H, J = 2.5, 1.8 Hz), 7.57 (d of d, 3H, J = 2.4,

* Aldrich Chemical Co., Milwaukee, WI 53233.

0.4 Hz), 7.66 (d of d, 3H, J = 1.7, 0.6 Hz), 8.43 (s, 1H). The ^{13}C NMR (20 MHz, CDCl$_3$) exhibits signals at δ 83.4, 107.2, 129.4, 141.6.

F. TRIS(3-PHENYL-1-PYRAZOLYL)METHANE

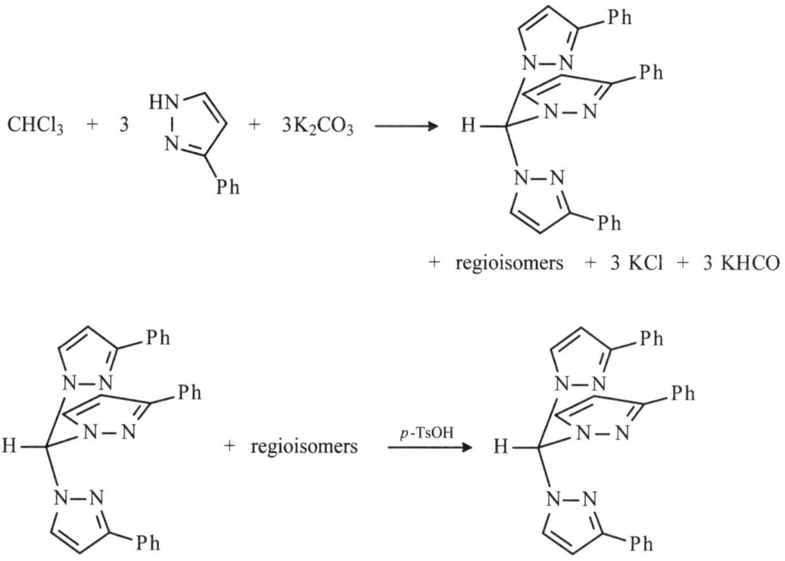

$$\text{CHCl}_3 \ + \ 3 \ \text{[pyrazole]} \ + \ 3\text{K}_2\text{CO}_3 \ \longrightarrow \ \text{[product]}$$

+ regioisomers + 3 KCl + 3 KHCO

+ regioisomers $\xrightarrow{p\text{-TsOH}}$

Procedure

A 500-mL, round-bottomed flask, fitted with a gas inlet tube, is charged with 80 g (0.58 mole) of anhydrous potassium carbonate and heated under vacuum using the gentle flame of a Bunsen burner. Liberation of water from the solid potassium carbonate is visible. The flask is cooled under vacuum, the vacuum is released under N$_2$ and 29.0 g (0.201 mole) of (3)5-phenylpyrazole;[19] 3.00 g (0.0089 mole) of tetrabutylammonium hydrogen sulfate and a 2-in. egg-stirring bar are rapidly added to the flask. The flask is capped with a septum and 250 mL of anhydrous chloroform (Aldrich, Sure Seal)* is added via cannula. The septum is replaced by a condenser fitted with an N$_2$ line t-tube and the mixture is heated to reflux with rapid stirring for 72 h. The dark brown solution is filtered and the K$_2$CO$_3$/KCl cake is washed with methylene chloride until the filtrate is essentially colorless. Thin-layer chromatography (alumina/toluene) indicates the presence of several product spots. The filtrate

---

* Aldrich Chemical Co., Milwaukee, WI 53233.

is evaporated on a rotary evaporator, resulting in a brown oil. Anhydrous *p*-toluenesulfonic acid (50 mg; dried 4 h under vacuum at 100°C) in toluene (250 mL) is added to the brown oil and the mixture is heated to reflux under an N_2 atmosphere for 24 h. Thin-layer chromatography (alumina/toluene) now indicates a single product spot ($R_f = 0.5$) in addition to unreacted phenylpyrazole ($R_f = 0.1$). After cooling, 5 mL of 5% sodium carbonate is added and the mixture is stirred for 30 min. The product precipitates out of the brown mixture as fine needles. Cyclohexane (100 mL) is added, the flask is chilled in ice, and the product is filtered and washed with 2×50 mL of 1:1 toluene/cyclohexane followed by 2×50 mL of hexanes. The off-white solid is dissolved in a minimum amount of methylene chloride and absorbed on a column (5×20 cm) of activity III neutral alumina. The column is eluted with toluene (500 mL) and the fractions containing the product are combined and evaporated on a rotary evaporator. The solid residue is recrystallized by dissolving in hot toluene (100 mL) and adding cyclohexane (100 mL). The precipitate is isolated by suction filtration and rinsed with cyclohexane to give 13.27 g (45%) of fine, white needles melting at 177–178°C. A second crop of product is isolated by evaporating the crude, brown filtrate from the first filtration to 100 mL, absorbing the solution on a 5×15 cm column of activity III neutral alumina, and eluting with toluene. The fractions containing product are combined with the filtrate from the first crop and the solvent is evaporated on a rotary evaporator. Recrystallization from hot 1:1 toluene/cyclohexane gives an additional 3.71 g of ligand (mp 177–178°C). Combined yield: 16.98 g (57%).*

Anal. Calcd. for $C_{28}H_{22}N_6$: C, 76.00; H, 5.02; N, 18.98. Found: C, 76.08; H, 5.01; N, 18.98.

Properties

Tris(3-phenyl-1-pyrazolyl)methane is a white, crystalline compound which is soluble in most organic solvents, but is sparingly soluble in hydrocarbons. The ligand is stable in basic and neutral solutions, but in acidic protic solvents, it readily decomposes to formic acid and the free pyrazole. The ^{1}H NMR (80 MHz, CDCl$_3$) exhibits signals at δ 6.65 (d, 3H, J = 2.5 Hz), 7.31–7.45 (m, 9H), 7.67 (d, 3H, J = 2.5 Hz), 7.75–7.87 (m, 6H), 8.49 (s, 1H). The ^{13}C NMR (20 MHz, CDCl$_3$) exhibits signals at δ 83.5, 104.6, 126.1, 128.5, 128.6, 130.8, 132.7, 153.6.

---

* The checkers obtained an overall yield of 26%.

References

1. S. Trofimenko, *Chem. Rev.*, **93**, 943 (1993). S. Trofimenko, *Prog. Inorg. Chem.*, **34**, 115 (1986). K. Niedenzu and S. Trofimenko, *Top. Curr. Chem.*, **131**, 1 (1986).
2. S. Trofimenko, *J. Am. Chem. Soc.*, **92**, 5118 (1970).
3. V. S. Joshi, A. Sarkar, and P. R. Rajamohanan, *J. Organomet. Chem.*, **409**, 341 (1991).
4. G. G. Lobbia, F. Bonati, A. Cingolani, D. Leonesi, and A. Lorenzotti, *J. Organomet. Chem.*, **359**, 21 (1989).
5. J. Fernández-Baeza, F. A. Jalón, A. Otero, and Mª. E. Rodrigo-Blanco, *J. Chem. Soc., Dalton Trans.*, 1015 (1995).
6. P. K. Byers and A. J. Canty, *Organometallics*, **9**, 210 (1990).
7. P. K. Byers, A. J. Canty, B. W. Skelton, and A. H. White, *Organometallics*, **9**, 826 (1990).
8. D. G. Brown, P. K. Byers, and A. J. Canty, *Organometallics*, **9**, 1231 (1990).
9. A. Llobet, M. E. Curry, H. T. Evans, and T. J. Meyer, *Inorg. Chem.*, **28**, 3131 (1989).
10. P. K. Byers and F. G. A. Stone, *J. Chem. Soc., Dalton Trans.*, 3499 (1990).
11. S. Juliá, P. Sala, J. del Mazo, M. Sancho, C. Ochoa, J. Elguero, J.-P. Fayet, and M.-C. Vertut, *J. Heterocyclic Chem.*, **19**, 1141 (1982).
12. S. Juliá, J. Mª del Mazo, L. Avila, and J. Elguero, *Org. Prep. Proc. Int.* , **16**, 299 (1984).
13. M. Bovens, A. Togni, and L. M. Venanzi, *J. Organomet. Chem.*, **451**, C28 (1993).
14. In this context it is worth noting that bis(pyrazolyl)methanes are isosteric with the bis(oxazolyl)methanes, which have been successfully exploited in asymmetric catalysis.[15]
15. A. Pfaltz, *Acc. Chem. Res.*, **26**, 339 (1993).
16. D. D. LeCloux, C. J. Tokar, M. Osawa, R. P. Houser, M. C. Keyes, and W. B. Tolman, *Organometallics*, **13**, 2855 (1994).
17. W. Hückel and H. Bretschneider, *Berichte*, **70**, 2024 (1936).
18. F. De Angelis, A. Gambacorta, and R. Nicoletti, *Synthesis*, 798 (1976).
19. S. Trofimenko, J. C. Calabrese, and J. S. Thompson, *Inorg. Chem.*, **26**, 1507 (1987).

9. TRIS[*N*-(3-*tert*-BUTYL)PYRAZOLYL]METHANE

Submitted by DANIEL L. REGER,* JAMES E. COLLINS,*
DONALD L. JAMESON,† and RONALD K. CASTELLANO†
Checked by ALLAN J. CANTY‡ and HONG JIN‡

An important advance in the development of the chemistry using poly(pyrazolyl)borate ligands was the introduction of "second-generation" ligands bearing very bulky substituents on the pyrazolyl rings. A major advantage of these ligands is that they prevent the formation of L_2M species.[1] Analogous advantages exist with the tris(pyrazolyl)methane family of ligands.

* Department of Chemistry and Biochemistry, University of South Carolina, Columbia, SC 29208.
† Department of Chemistry, Gettysburg College, Gettysburg, PA 17325.
‡ Chemistry Department, University of Tasmania, Tasmania 7001, Australia.

The tris(*N*-(3-*tert*-butyl)pyrazolyl)methane ligand, with an extremely bulky 3-substituent, favors the formation of tetrahedral complexes[2] in much the same way as observed in the chemistry of the hydrotris(3-*tert*-butyl-pyrazolyl)borate ligand.[1]

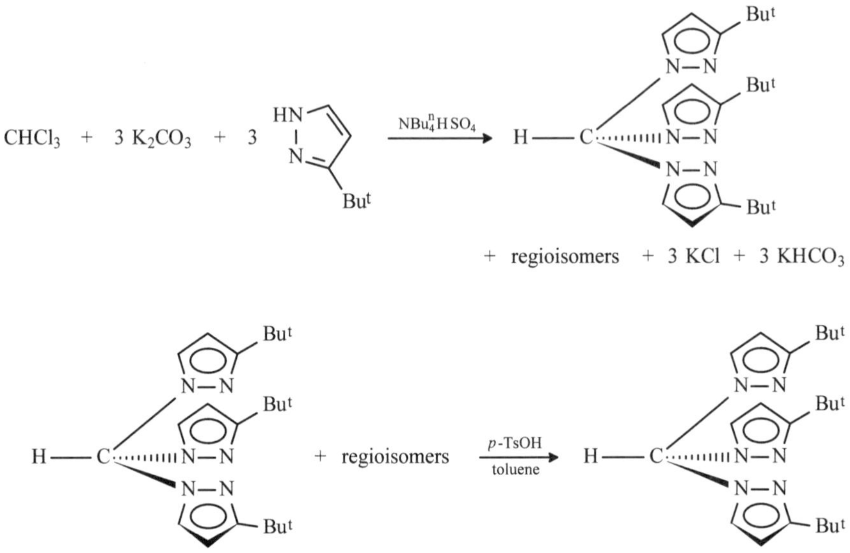

CHCl3 + 3 K2CO3 + 3
HN
N
But
NBu4nHSO4
H
C
N—N
But
+ regioisomers + 3 KCl + 3 KHCO3
+ regioisomers
p-TsOH
toluene

Procedure

[HC(3-Bu^{t}pz)$_3$]. A 500-mL Schlenk flask is charged with K_2CO_3 (55.6 g; 0.403 mol) and the solid is dried under vacuum by the gentle heat of a Bunsen burner (or by a heating mantle at ca. 500°C for ca. 3 h). The reduced pressure is maintained until the flask cools to room temperature. A solid mixture of 3-Bu^{t}pzH[3] (10.0 g; 0.0805 mol) and $NBu^n_4HSO_4$ (1.37 g; 4.03 mmol) as well as a 1-in. egg-shaped stirring bar is then added to the flask against a counter stream of nitrogen. Chloroform (400 mL, freshly distilled from P_4O_{10} and kept under nitrogen) is added by cannula transfer or syringe. The flask is fitted with a reflux condenser and warmed to gentle reflux under nitrogen for 48 h. After this period, the yellow solution is filtered and the remaining solid is washed with CH_2Cl_2 (or $CHCl_3$, 3 × 25 mL). All organic extracts are combined and the halocarbons are removed under reduced pressure on a rotovap. The remaining yellow oil is dissolved with toluene (100 mL) and a catalytic amount of predried *p*-toluenesulfonic acid (0.05 g; 0.3 mmol) is added. The resulting solution is warmed to reflux for 24 h. After cooling to room temperature, the golden toluene solution is neutralized with 5%

Na$_2$CO$_3$(aq) (100 mL) and subsequently washed with distilled water (3 × 100 mL). The homogeneous organic layer is then dried overnight using MgSO$_4$. The mixture is filtered and the solid washed with toluene (2 × 20 mL). Toluene is then removed under reduced pressure to about 25 mL prior to loading onto a column of silica gel (230–400 mesh; 5 cm × 10 cm). The toluene extract (ca. 300 mL) is collected until the yellow color of the filtrate is barely visible. A darker band remains on the column. Removal of the toluene under reduced pressure (first by rotovap, then by vacuum pump) gives a yellow oil that provides a clean ^{1}H NMR spectrum of the desired product. Dissolution of this oil in pentane and removal of the pentane under vacuum yields 5.95 g (57.9%) of a pale yellow powder (mp 58–61°C).*

Anal. Calcd. for C$_{22}$H$_{34}$N$_6$: C, 69.07; H, 8.96; N, 21.97. Found: C, 68.88; H, 8.93; N, 22.13.

Properties

HC(3-Bu^{t}pz)$_3$ is soluble in all organic solvents and can be crystallized from cold pentane. ^{1}H NMR (CDCl$_3$): δ8.17 (s; 1; *H*C(3-Bu^{t}pz)$_3$); 7.15 (d; 3; 5-*H* in pz); 6.12 (d; 3; 4-*H* in pz); 1.26 (s; 27; C(C*H*$_3$)$_3$). ^{13}C NMR (CDCl$_3$): δ163.7 (3-*C* in pz); 129.1 (5-*C* in pz); 103.2 (4-*C* in pz); 83.3 (*H*C(3-Bu^{t}pz)$_3$); 32.2 (*C*(CH$_3$)$_3$); 30.4 (C(*C*H$_3$)$_3$).

References

1. (a) S. Trofimenko, *Chem Rev.*, **93**, 943 (1993). (b) N. Kitajima and W. B. Tolman, *Prog. Inorg. Chem.*, **43**, 419 (1995).
2. (a) D. L. Reger, J. E. Collins, A. L. Rheingold, and L. M. Liable-Sands, *Organometallics*, **15**, 2029 (1996). (b) D. L. Reger, J. E. Collins, A. L. Rheingold, L. M. Liable-Sands, and G. P. A. Yap, *Organometallics*, **16**, 349 (1997).
3. S. Trofimenko, J. C. Calabrese, and J. S. Thompson, *Inorg. Chem.*, **26**, 1507 (1987).

* The checkers had problems with solidification of an oil at this stage and suggest the following alternative to the pentane step. Petroleum ether (60–80°C, 2 mL) was added and the resulting yellow solution was kept at −20°C overnight to give a pale yellow, crystalline solid. After decantation and drying in a vacuum, a wet, pale yellow solid containing a minor amount of oil was obtained. The solid was then transferred onto a No. 3 filter paper in a Buchner filtration funnel, and reduced vacuum was applied to remove traces of oil. A white, crystalline solid (5.3 g, 52%) was obtained after drying in a vacuum (mp 57–60°C).

10. TRIS[2-(1,4-DIISOPROPYLIMIDAZOLYL]PHOSPHINE

Submitted by THOMAS N. SORRELL* and WILLIAM E. ALLEN*
Checked by GERARD PARKIN† and CLARE KIMBLIN†

Tripodal ligands have found many applications in transition metal chemistry, especially to bind metal ions in ways that mimic coordination sites in metalloproteins. Tris(pyrazolyl)borates are the most commonly used chelates having a tripodal arrangement of donor atoms,[1] but 1,4,9-triazacyclononane and its derivatives have also been utilized extensively, especially in the last several years.[2-4] There are fewer examples in which tris(imidazolyl)carbinol[5,6] or tris(imidazolyl)phosphine[7-10] ligands have been employed for these purposes, although it has been suggested that their charge (neutral) and binding group (imidazole ring) might be well suited for biomimetic applications.[10,11]

One obstacle that has prevented a wider use of tris(imidazolyl)phosphine ligands has been the relative paucity of appropriately substituted imidazole precursors. This is especially true for imidazoles that bear bulky groups like isopropyl and *tert*-butyl fragments, which could provide hindered coordination environments around a metal ion. We recently outlined a general method for making 1,4-disubstituted imidazoles regiospecifically,[12] and at least two other routes have since been reported.[13,14] Therefore, it is now possible to create a diverse range of tripodal ligands that can be applied to specific objectives within the larger field of coordination chemistry.

The procedure described below can be applied to several derivatives of the parent tris(imidazolyl)phosphine ligand. The details are provided for preparing tris(1,4-diisopropylimidazolyl)phosphine, but the same route can be used to synthesize tris(1-isopropyl-4-*tert*-butylimidazolyl)phosphine, tris(1-isopropyl-4-phenylimidazolyl)phosphine, and tris(1-isopropyl-4-methylimidazolyl)phosphine, starting with commercially available 1-bromopinacolone, 2-bromoacetophenone, and 1-chloroacetone, respectively.

A. 1,4-DIISOPROPYLIMIDAZOLE

* Department of Chemistry, The University of North Carolina at Chapel Hill, Chapel Hill, NC 27599-3290.
† Department of Chemistry, Columbia University, New York, NY 10027.

Procedure

Under a dinitrogen atmosphere, a 300-mL, two-necked, round-bottomed flask containing a magnetic stir bar and fitted with a pressure-equalizing dropping funnel and an inert gas inlet is charged with a solution of 2-aminopropane (10.6 g, 0.180 mol) in dry diethyl ether (70 mL). The flask and its contents are cooled in a dry ice—isopropanol bath to −78°C; then a solution of 1-bromo-3-methyl-2-butanone* (10.0 g, 0.0606 mol)† in diethyl ether (20 mL) is added dropwise from the dropping funnel to the stirred solution of amine over 15 min

■ **Caution.** *1-bromo-3-methyl-2-butanone is a severe lachrymator so all manipulations in this section should be performed in a good fume hood.*

After stirring for 1 h at −78°C, the reaction mixture is allowed to warm to room temperature. Stirring is continued for several hours, until precipitation of the HBr salts appears complete. The contents of the flask are then poured into a separatory funnel. Twenty milliliters of 15% aqueous NaOH is added, and the mixture is shaken until the white solid dissolves. The ether layer is subsequently washed with two 50-mL portions of water and two 50-mL portions of saturated aqueous sodium chloride solution. The organic phase is dried over $MgSO_4$.

While the ether solution of the aminoketone is drying, formamide (35 mL) is added to a 300-mL, three-necked flask, which is fitted with an air-cooled condenser, a pressure-equalizing dropping funnel that is topped with an inert atmosphere inlet tube, and a thermometer, which extends into the liquid. The formamide is heated with stirring to 180°C under a slow stream of dinitrogen, which exhausts through the top of the condenser.

■ **Caution.** *Formamide is a teratogen. This reaction should be conducted in a good fume hood.*

As the formamide is being heated, the ether solution of 1-isopropylamino-3-methyl-2-butanone is filtered and concentrated at room temperature with use of a rotary evaporator to yield a light yellow oil. This oil is transferred to the addition funnel, from which it is added dropwise to the hot formamide over

---

* M. Gaudry and A. Marquet, *Org. Synth. Coll. Vol. 6*, **1988**, 193. This material slowly isomerizes at room temperature to 3-bromo-3,3-dimethyl-2-butanone so it should be used shortly after it is prepared.
† The checkers carried out this reaction on a larger scale, using 200 g (1.21 mol) of 1-bromo-3-methyl-2-butanone. They obtained the 1,4-diisopropylimidazole in 23% yield (42.5 g). The crude product was purified by distillation rather than by flash chromatography.

1 h. The flow of dinitrogen is adjusted so that a slow stream of water vapor can be observed as a white mist exiting from the top of the condenser. If the dinitrogen stream is too rapid, the formamide evaporates from the flask and the volume begins to decrease noticeably. The mixture is allowed to stir for an additional 3 h at 180°C, then it is cooled to room temperature.

The dark brown solution is treated with an equal volume of water and 20 mL of 15% aqueous NaOH. The mixture is extracted twice with two 200-mL portions of toluene. These are combined and washed with two 100-mL portions of water and 50 mL of saturated aqueous sodium chloride solution. The organic phase is dried over sodium sulfate, filtered, and concentrated with use of a rotary evaporator to yield a yellow-brown oil. The oil is purified by flash chromatography[15] on a 2 × 15 cm column of silica using 9:1 (v:v) ethyl acetate–methanol as the eluent ($R_f = 0.39$) and a flow rate of about 10 mL/min. The combined fractions are concentrated to an oil with use of a rotary evaporator, and the oil is subsequently distilled with use of a Kugelrohr apparatus.

Anal. Calcd. for $C_9H_{16}N_2$: C, 71.0; H, 10.6; N, 18.4. Found: C, 70.9; H, 10.5; N, 18.4. The yield of 1,4-diisopropylimidazole is 4.34 g (47%).

Properties

1,4-Diisopropylimidazole is a colorless, hygroscopic oil, bp 101–104°C (13 torr). It is best stored in an inert atmosphere in the refrigerator. Its proton NMR spectrum in $CDCl_3$ shows resonances at $\delta 1.26$ [6H, d, $J = 6.8$ Hz, 4-CH(CH$_3$)$_2$], 1.46 [6H, d, $J = 6.7$ Hz, 1-CH(CH$_3$)$_2$], 2.88 [1H, sept, $J = 6.8$ Hz, 4-CH(CH$_3$)$_2$], 4.27 [1H, sept, $J = 6.7$ Hz, 1-CH(CH$_3$)$_2$], 6.65 (1H, t, $J = 1.1$ Hz, imidazolyl C$_5$H), and 7.43 (1H, d, $J = 1.1$ Hz, imidazolyl C$_2$H).

B. TRIS[2-(1,4-DIISOPROPYLIMIDAZOLYL)]PHOSPHINE

Procedure

A 250-mL, three-necked, round-bottomed flask containing a stir bar is dried at 120°C and then allowed to cool under dinitrogen after attaching a gas inlet tube. The flask is subsequently fitted with a septum and an addition funnel containing 1,4-diisopropylimidazole (6.0 g, 39 mmol)* in diethyl ether (50 mL). A syringe is used to transfer a solution of 1.6 M n-butyllithium in hexanes (25 mL, 40 mmol) to the flask, which is then lowered into a dry ice–isopropanol bath. Dropwise addition of the imidazole solution to the cold, stirred solution of butyllithium should be completed over 20 min, producing a yellow reaction mixture. After 1 h, the solution is allowed to warm to room temperature over 45 min, during which time the color darkens to a golden orange. The flask is cooled to $-78°C$ again, and another addition funnel containing freshly distilled PCl_3 (1.4 g, 10 mmol) in 10 mL of diethyl ether is quickly exchanged for the first. Dropwise addition of the PCl_3 solution over 1 h is accompanied by precipitation of a white solid. After the addition is complete, the mixture is allowed to stir for 2 h at $-78°C$ and then is warmed to room temperature over 1 h. Concentrated NH_4OH (60 mL) is added and the biphasic mixture is stirred for 1 h, still under dinitrogen. The solution is then transferred to a separatory funnel in the air, and the layers are separated. The organic layer is washed sequentially with two 50-mL aliquots of water and 50 mL of saturated aqueous sodium chloride solution. It is then dried over $MgSO_4$, filtered, and evaporated with the use of a rotary evaporator to give an oily, off-white solid. This residue is triturated with 20 mL of hexanes; then the solid is filtered and washed with two 20-mL portions of hexanes and dried under vacuum. The yield of tris[2-(1,4-diisopropylimidazolyl)]phosphine is 2.3 g (47%).

Properties

Tris[2-(1,4-diisopropylimidazolyl)]phosphine is a white solid which is slightly air sensitive, so it should be stored in an inert atmosphere or under vacuum. Its ^1H NMR spectrum in $CDCl_3$ shows resonances at δ 1.14 [18H, d, $J = 6.6$ Hz, 4-CH(CH$_3$)$_2$], 1.20 [18 H, d, $J = 6.9$ Hz, 1-CH(CH$_3$)$_2$], 2.87 [3H, sept, $J = 6.6$ Hz, 4-CH(CH$_3$)$_2$], 4.62–4.84 (3H, m, 1-CH(CH$_3$)$_2$], and 6.82 (3H, s, imidazolyl C$_5H$). Its ^{31}P NMR spectrum in $CDCl_3$ has a single peak at $\delta - 61.1$ ppm relative to H_3PO_4, which is referenced to external P(OMe)$_3$ at 141 ppm. A small percentage of the product is tris[2-(1,4-diisopropylimidazolyl)]phosphine oxide. It can be distinguished by a doublet in the ^1H NMR spectrum at 6.93 ppm and by a resonance in the ^{31}P NMR spectrum at -3.98 ppm.

---

* The checkers carried out this reaction on a larger scale, using 31 g (0.20 mol) of 1,4-diisopropylimidazole. They obtained tris[2-(1,4-diisopropylimidazolyl)phosphine] in 20% yield (5.1 g).

Acknowledgment

Financial support for this research was provided by the National Science Foundation (CHE #9100280).

References

1. S. Trofimenko, *Chem. Rev.*, **93**, 943 (1993).
2. A. Niemann, U. Bossek, K. Wieghardt, C. Butzlaff, A. X. Trautwein, and B. Nuber, *Angew. Chem. Int. Ed. Engl.*, **31**, 311 (1992).
3. B. Mauerer, J. Crane, J. Schuler, K. Wieghardt, and B. Nuber, *Angew. Chem. Int. Ed. Engl.*, **32**, 289 (1993).
4. S. Mahapatra, J. A. Halfen, E. C. Wilkinson, L. Que, and W. B. Tolman, *J. Am. Chem. Soc.*, **116**, 9785 (1994).
5. R. Breslow, J. T. Hunt, R. Smiley, and T. Tarnowski, *J. Am. Chem. Soc.*, **105**, 5337 (1983).
6. T. N. Sorrell and A. S. Borovik, *J. Am. Chem. Soc.*, **109**, 4255 (1987).
7. R. S. Brown, M. Zamkanei, and J. L. Cocho, *J. Am. Chem. Soc.*, **106**, 5222 (1984).
8. T. N. Sorrell, W. E. Allen, and P. S. White, *Inorg. Chem.*, **34**, 952 (1995).
9. W. E. Lynch, D. M. Kurtz, S. Wang, and R. A. Scott, *J. Am. Chem. Soc.*, **116**, 11030 (1994).
10. C. Kimblin, W. E. Allen, and G. Parkin, *J. Chem. Soc., Chem. Commun.*, 1813 (1995).
11. C. Kimblin, V. J. Murphy, and G. Parkin, *Chem. Commun.*, 235 (1996).
12. T. N. Sorrell and W. E. Allen, *J. Org. Chem.*, **59**, 1589 (1994).
13. D. A. Horne, K. Yakushijin, and G. Büchi, *Heterocycles*, **39**, 139 (1994).
14. G. Shapiro and B. Gomez-Lor, *Heterocycles*, **41**, 215 (1995).
15. W. C. Still, M. Kahn, and A. Mitra, *J. Org. Chem.*, **43**, 2923 (1978).

11. TRIS[(2-PYRIDYL)METHYL]AMINE (TPA) AND (+)-BIS[(2-PYRIDYL)METHYL]-1-(2-PYRIDYL)-ETHYLAMINE (α-METPA)

Submitted by JAMES W. CANARY,* YIHAN WANG,* and RICHARD ROY JR.*
Checked by LAWRENCE QUE, JR.[†] and HIROYUKI MIYAKE[†]

The ligand tris[(2-pyridyl)methyl]amine (TPA, **1**) was first reported by Anderegg and Wenk in 1967.[1] Modifications of the original synthesis have been reported by several laboratories.[2,3] The compound is made by alkylation of picolyl amine with picolyl chloride, both of which are commercially available. The reaction is conducted in water with slow addition of 10 M NaOH. The product may be isolated as the perchlorate salt[1,2] but extractive

* Department of Chemistry, New York University, New York, NY 10003.
† Department of Chemistry, University of Minnesota, Minneapolis, MN 55455.

isolation followed by distillation[3] is preferable for both convenience and safety reasons.[4]

TPA forms coordination complexes with a variety of metal ions. Potentiometric studies of TPA complexes with Mn(II), Fe(II), Co(II), Ni(II), Cu(II), Zn(II), Cd(II), Pb(II), Hg(II), and Ag(I) have been reported.[5] Spectroscopic and/or crystallographic studies have also been reported for TPA complexes with alkali metal cations,[6,7] V(IV),[8] Co(III),[2] Fe(III),[9,10] Mn(III),[11] Cu(I),[12] Cu(II),[13,14] and Zn(II).[15] Copper[13,16-18] and iron[19,20] complexes of TPA have been used to model biological oxygen binding and activation.

The ligand bis[(2-pyridyl)methyl]-1-(2-pyridyl)ethylamine (α-MeTPA, **2**, R = Me) was first reported by Hojland et al.[3] The method of preparation from 2-pyridylethylamine is similar to that for TPA. It was obtained also from TPA by reaction with a strong base followed by alkylation with MeI.[6] It was recently prepared in enantiomerically pure form[21,22] and it was shown that the ligand forms propeller-like complexes with Zn[II] and Cu[II] ions that have overall pseudo-C_3 symmetry.[21]

A. TRIS[(2-pyridyl)methyl]amine (TPA)

1 (tpa)

■ **Caution.** *As picolyl chloride·HCl is a severe irritant, protective clothing and gloves should be worn.*

Procedure

To a 200-mL, round-bottomed flask equipped with magnetic stirring, 2-(aminomethyl)pyridine (Acros, 8 mL, 80 mmol) is added to a solution of picolyl chloride·HCl* (25.6 g, 160 mmol) in distilled water (40 mL). Sodium hydroxide (31.0 mL of a 10 M solution, 320 mmol) is added dropwise (approximately 5 drops/min) over a period of 2 h using an addition funnel. The flask is then placed in a 70°C oil bath for 30 min. The solution is removed from the oil bath and is cooled to room temperature. The resulting dark red solution is transferred to a separatory funnel and extracted with three 150-mL portions of $CHCl_3$. The combined extracts are dried over anhydrous Na_2SO_4. After filtration, the $CHCl_3$ is removed on a rotary evaporator. The

* Aldrich Chemical Co., Milwaukee, WI 53233.

residue is then purified by short path vacuum distillation at 200°C and 0.01 mm Hg. The condenser is not cooled with water but is wrapped with fiberglass wool to retain heat and prevent solidification of the product during the distillation. The yield is 11.3 g (51%).

Properties

The product prepared by this procedure possesses a light yellow color. It is soluble in water, ethanol, THF, and benzene. Its bp is 170–173°C at 0.01 mm Hg. After distillation, the solid gives a mp of 73–77°C. ^{1}H NMR spectra (200 MHz) reveals δ3.9 (s, 6H), 7.1–8.5 (m, 12H).

B. *S*-(−)-α-MeTPA

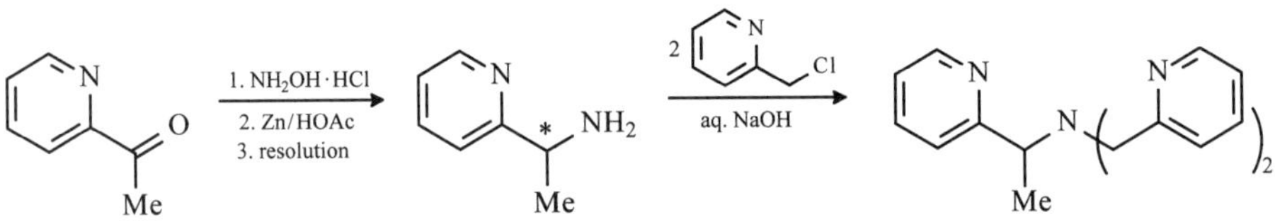

Procedure

Methyl 2-pyridyl ketoxime. A solution of 2-acetylpyridine (**3**, Acros, 27.2 g, 0.225 mole) and hydroxylamine hydrochloride (18.0 g, 0.259 mole) in 100 mL of pyridine is stirred magnetically and heated to reflux in a 250-mL, round-bottomed flask fitted with a water-cooled reflux condenser for 6 h. After removing the solvent on a rotary evaporator, 400 mL of H_2O is added, resulting in precipitation of the oxime in the form of a white precipitate which is isolated by vacuum filtration. The filtrate is extracted three times with 100 mL of ether. The white solid is dissolved in ether and combined with the extracts, and then dried with Na_2SO_4. The solvent is evaporated using a rotary evaporator. The crude product (30.0 g, 97%) is used directly in the next step without further purification.

R,S-(±)-1-(2-pyridyl)ethyl amine. In a 2-L, three-necked reaction flask equipped with overhead mechanical stirring, the oxime (30.0 g, 0.220 mole) and acetic acid (270 mL) are dissolved in 400 mL of 95% EtOH. Zinc dust (270 g) is added in 10 g increments over 2 h. The reaction mixture is stirred with an overhead mechanical stirrer at ambient temperature for 12 h. The undissolved zinc is removed by filtration and washed with 150 mL of EtOH. The filtrate is concentrated in vacuo. Several 150-mL portions of H_2O are added and evaporated to remove AcOH. The mixture is neutralized and then made strongly basic (pH 12 or higher) by addition of saturated aqueous KOH solution (200 mL or more, depending on the quantity of residual acetic acid) until the $Zn(OH)_2$ precipitate redissolves and a brown/yellow oil appears on the surface of the mixture. The mixture is transferred to a separatory funnel and extracted with three 150-mL portions of ether, and the combined ether fractions are dried over Na_2SO_4. After removal of the solvent, the crude product is distilled using a Kügelrohr apparatus, yielding **4** (23.6 g, 88%, bp 43°C/0.1 mm Hg).

S-(−)-1-(2-Pyridyl)ethylamine.[22] To a 1-L Erlenmeyer flask 34.5 g (0.230 mole) of L-(+)-tartaric acid in 220 mL of boiling 95% EtOH is added 23.6 g (0.193 mole) of racemic 1-(2-pyridyl)ethylamine in 300 mL of hot 95% EtOH. As the solution cools to ambient temperature and is allowed to sit for several hours, a white precipitate forms. Four recrystallizations of this precipitate from 100 mL of 95% EtOH gives 12.9 g of the (+)-acid tartrate salt of *S*-(−) amine (>99% ee). The salt is neutralized by dissolving in 250 mL of 2 N NaOH and extracting with four 150-mL portions of ether. The organic layer is dried over solid NaOH and then concentrated by rotary evaporation.

   The % ee of the tartrate salt of the amine may be determined by HPLC using a CROWNPAK CR column (Chiral Technologies, Inc.*) The eluent is pH 1.0 $HClO_4$, and the flow rate is 0.6 mL/min.

S-(−)-Bis((2-pyridyl)methyl)-1-(2-pyridyl)ethylamine (α-MeTPA, 2). In a 50-mL, round-bottomed flask equipped with magnetic stirring, picolyl chloride hydrochloride (4.83 g, 29.4 mmol) is dissolved in 7.4 mL of water. Amine **4** (1.80 g, 14.7 mmol) is added. The mixture is warmed to 70°C and 10 M NaOH solution (5.7 mL) is added slowly. Stirring is continued at this temperature for another 4 h. After cooling to ambient temperature, the reaction mixture is transferred to a separatory funnel and extracted with three 50-mL portions of $CHCl_3$. The combined organic extracts are dried over Na_2SO_4. The solvent is removed in vacuo and crude product is chromatographed on basic alumina (CH_2Cl_2/EtOAc = 3:2, R_f = 0.27). The yield of pure **2** is 2.40 g (53.6%).

Anal. Calcd for $C_{19}H_{20}N_4$: C, 74.97; H, 6.62; N, 18.41. Found: C, 74.69; H, 6.86; N, 18.25.

* Chiral Technologies, Inc. Exton, PA 19341.

Properties

The ligand α-MeTPA is similar to TPA in appearance and character, except that it melts at lower temperatures and thus is a waxy solid at room temperature. ^{1}H-NMR (300 MHz, CDCl$_3$): 8.57–8.47 (m, 3H), 7.67–7.47 (m, 6H), 7.15–7.07 (m, 3H), 4.05 (q, J = 6.9 Hz, 1H), 3.97 (d, J_{gem} = 14.9 Hz, 2H), 3.76 (d, J_{gem} = 14.9 Hz, 2H), 1.56 (d, J = 6.9 Hz, 3H). $[\alpha]_D^{25}$ = -85.5 (MeCN, c = 1.0).

The authors have obtained the complex S-($-$)-[Zn(α-MeTPA)Cl]ClO$_4$ as a white crystalline material by mixing equimolar, methanolic solutions of ligand, Zn(ClO$_4$)$_2$,[4] and NaCl and chilling the resultant solution in the refrigerator. ^{1}H NMR (300 MHz, CD$_6$COCD$_6$): δ 9.19–9.10 (m, 3H), 8.33–8.20 (m, 3H), 7.94 (d, J = 8.0 Hz, 1H), 7.84–7.76 (m, 5H), 4.83 (d, J_{gem} = 16.7 Hz, 1H), 4.53 (d, J_{gem} = 16.4 Hz, 1H), 4.38 (q, J = 6.8 Hz, 1H), 4.21 (d, J_{gem} = 16.7 Hz, 1H), 4.17 (d, J_{gem} = 16.4 Hz, 1H), 1.94 (d, J = 6.8 Hz, 3H). The ^{1}H NMR spectra of the free ligand and the complex differ by the significant downfield shifts in the complex and by the appearance of an additional AB coupling system due to slow tertiary nitrogen atom inversion on the ^{1}H NMR time scale. $[\alpha]_D^{25}$ = -139.0 (MeCN, c = 1.0).

References

1. G. Anderegg and F. Wenk, *Helv. Chim. Acta*, **50**, 2330 (1967).
2. J. B. Mandel, C. Maricondi, and B. E. Douglas, *Inorg. Chem.*, **27**, 2990 (1988).
3. F. Hojland, H. Toftlund, and S. Yde-Andersen, *Acta Chem. Scand. A*, **37**, 251 (1983).
4. **Caution**: Perchlorate salts of metal complexes with organic ligands are potentially explosive. They should be handled in small quantities (e.g., <100 mg) and with caution. *Chem. Eng. News*, **61**, 4 (1983).
5. G. Anderegg, E. Hubmann, N. G. Podder, and F. Wenk, *Helv. Chim. Acta*, **60**, 123 (1977).
6. S. K. Brownstein, P. Y. Plouffe, C. Bensimon, and J. Tse, *Inorg. Chem.*, **33**, 354 (1994).
7. C.-L. Chuang, M. Frid, and J. W. Canary, *Tetrahedron Lett.*, **36**, 2909 (1995).
8. H. Toftlund, S. Larsen, and K. S. Murray, *Inorg. Chem.*, **30**, 3964 (1991).
9. R. A. Leising, B. A. Brennan, L. Que, Jr., B. G. Fox, and E. Münck, *J. Am. Chem. Soc.*, **113**, 3988 (1991).
10. A. Hazell, K. B. Jensen, C. J. McKenzie, and H. Toftlund, *Inorg. Chem.*, **33**, 3127 (1994).
11. P. A. Goodson, A. R. Oki, J. Glerup, and D. J. Hodgson, *J. Am. Chem. Soc.*, **112**, 6248 (1990).
12. R. R. Jacobson, Z. Tyeklar, and K. D. Karlin, *Inorg. Chim. Acta*, **181**, 111 (1991).
13. K. D. Karlin, J. C. Hayes, S. Juen, J. P. Hutchinson, and J. Zubieta, *Inorg. Chem.*, **21**, 4106 (1982).
14. J. Zubieta, K. D. Karlin, and J. C. Hayes, in *Copper Coordination Chemistry: Biochemical and Inorganic Perspectives*, K. D. Karlin and J. Zubieta, Eds. Adenine Press, New York, 1983, pp. 97–108.
15. N. N. Murthy and K. D. Karlin, *J. Chem. Soc., Chem. Commun.*, 1236 (1993).
16. K. D. Karlin, N. Wein, B. Jung, S. Kaderli, P. Niklaus, and A. D. Zuberbühler, *J. Am. Chem. Soc.*, **115**, 9506 (1993).

17. A. Nanthakumar, S. Fox, N. N. Murthy, K. D. Karlin, N. Ravi, B. H. Huynh, R. D. Orosz, E. P. Day, K. S. Hagen, and N. J. Blackburn, *J. Am. Chem. Soc.*, **115**, 8513 (1993).
18. N. Wei, D.-H. Lee, N. N. Murthy, Z. Tyeklár, K. D. Karlin, S. Kaderli, B. Jung, and A. D. Zuberbühler, *Inorg. Chem.*, **33**, 4625 (1994).
19. T. Kojima, R. A. Leising, S. Yan, and L. Que, Jr., *J. Am. Chem. Soc.*, **115**, 11328 (1993).
20. Y. Zang, T. E. Elgren, Y. Dong, and L. Que, Jr., *J. Am. Chem. Soc.*, **115**, 811 (1993).
21. J. W. Canary, C. S. Allen, J. M. Castagnetto, and Y. Wang, *J. Am. Chem. Soc.*, **117**, 8484 (1995).
22. H. E. Smith, L. J. Schaad, R. B. Banks, C. J. Wiant, and C. F. Jordan, *J. Am. Chem. Soc.*, **95**, 811 (1973).

12. C_2-SYMMETRIC 1,4-DIISOPROPYL-7-*R*-1,4,7-TRIAZACYCLONONANES

Submitted by JASON A. HALFEN* and WILLIAM B. TOLMAN*
Checked by KARL WEIGHARDT[†]

1,4,7-triazacyclononane (TACN) based ligands have been widely utilized in the synthesis of high- and low-valent organometallics as well as for the preparation of models of metalloprotein active sites.[1] Because of the kinetic and thermodynamic stability which these macrocycles impart upon their complexes, they have become highly valued as supporting ligands in much the same fashion as the cyclopentadienyl anion (Cp^-)[2] and hydrotris(pyrazolyl)borates $(HBpz_3^-)$.[3] Accordingly, several groups have reported syntheses of the parent macrocycle, 1,4,7-triazacyclononane.[4]

The attachment of coordinating and noncoordinating groups to the TACN framework is most often achieved through alkylation of the 2° nitrogen atoms of the parent macrocycle, resulting in C_3-symmetric *N,N',N''*-trisubstituted ligands. Simple alkyl groups (e.g., methyl,[5] isopropyl,[6] or benzyl[7]) as well as a variety of functionalized substituents such as phenolate,[8] thiolate,[9] alkoxide,[10] pyridyl,[11] or imidazolyl[12] have been added to the TACN ring in this way, generating a battery of ligands possessing divergent steric and electronic properties and coordination environments.

The selective alkylation of only one or two TACN nitrogen atoms has only recently received attention, as this procedure requires the three equivalent reaction sites to be differentiated from each other. One approach to this type of functionalization relies upon the formation[13] and subsequent hydrolysis of

* Department of Chemistry, University of Minnesota, Minneapolis, MN 55455.
[†] Max-Planck-Institut für Strahlenchemie, Stiftstr. 34-36, D-45470, Muelheim an der Ruhr, Germany.

a tricyclic orthoamide, yielding a single *N*-formyl protected nitrogen atom. The unprotected nitrogen atoms may then be derivatized, and following hydrolysis of the single *N*-formyl group, *N,N'*-dialkylated triazacyclononanes may be isolated.[14] Alternatively, *p*-toluenesulfonyl (tosyl) groups may be used to differentiate the three nitrogen atoms, as selectively tosylated TACN rings are accessible via the partial hydrolysis of 1,4,7-tris(*p*-toluenesulfonyl)-1,4,7-triazacyclononane[15] or by the statistical tosylation of the parent macrocycle.[16]

Our own interest in the investigation of copper complexes supported by sterically hindered triazacyclononanes[17] has led us to develop an efficient, high-yield synthesis of the dialkylated macrocycle 1,4-diisopropyl-1,4,7-triazacyclononane, a hindered analog of the previously reported 1,4-dimethyl-substituted variant.[18] The route to this dialkylated macrocycle has the advantages of proceeding in higher overall yield through fewer synthetic steps than the alternate methodology discussed above. The single 2° nitrogen atom in 1,4-diisopropyl-1,4,7-triazacyclononane may be further derivatized by reaction with a variety of electrophiles, yielding ligands which have proven to be highly valuable for the preparation of models of metalloprotein active sites.[19] We present below the synthesis of 1,4-diisopropyl-1,4,7-triazacyclononane as well as an example of its use in the preparation of C_2-symmetric 1,4-diisopropyl-7-*R*-trisubstituted ligands. The particular derivative described herein, 1,4-diisopropyl-7-(2-pyridylmethyl)-1,4,7-triazacyclononane, represents a unique marriage of the properties of the triazamacrocycle with those of another ligand used widely in bioinorganic modeling studies, tris(2-pyridylmethyl)amine (TPA).[20]

A. 1-(*p*-TOLUENESULFONYL)-1,4,7-TRIAZACYCLONONANE[15]

■ **Caution.** *HBr (g) is vigorously evolved during the initial heating of the reaction mixture. Care should be taken to vent the reaction apparatus through* *a vessel filled with water to trap the HBr. A positive pressure of N$_2$ prevents water from backfilling into the reaction apparatus. The gas evolution ceases after several hours at 90°C.*

Procedure

A 1000-mL, three-necked, round-bottomed flask equipped with a reflux condenser, a gas inlet, and a magnetic stirring bar is charged with 1,4,7-tris(p-toluenesulfonyl)-1,4,7-triazacyclononane (47.10 g, 79.6 mmol)[4] and phenol (56.0 g, 595 mmol). After establishing a nitrogen atmosphere in the apparatus, a solution of HBr in glacial acetic acid (33%, 600 mL)[21] is added, and the mixture gently heated with stirring to 90°C (see caution above). A colorless precipitate of 1-(p-toluenesulfonyl)-1,4,7-triazacyclononane·2HBr appears within 2–4 h. After heating for 36 h, the mixture is cooled to room temperature and filtered. The solid is washed with Et$_2$O (200 mL) and then dissolved in 1 M NaOH (500 mL). The pink aqueous mixture is then extracted with CHCl$_3$ (3 × 300 mL) and the organic extracts dried (MgSO$_4$) and evaporated under reduced pressure. Drying the resultant oil under vacuum for 2 h affords the pure product, 18.45 g (82%).[15]

Properties

1-(p-Toluenesulfonyl)-1,4,7-triazacyclononane is a colorless crystalline solid that is soluble in most polar organic solvents. The methodology described above is based on that reported by Sessler and co-workers[15] and reproducibly proceeds in yields ranging from 65 to 85%, depending on the purity of the original 1,4,7-tris(p-toluenesulfonyl)-1,4,7-triazacyclononane. The ^{1}H NMR spectral data for the compound are as follows (200 MHz, CDCl$_3$): δ 7.67 (d, $J = 8.2$ Hz, 2H), 7.29 (d, $J = 8.2$ Hz, 2H), 3.19–3.14 (m, 4H), 3.08–3.04 (m, 4H), 2.67 (s, 4H), 2.41 (s, 3H), 1.71 (br. s, 2H) ppm.

B. 1,4-DIISOPROPYL-7-(p-TOLUENESULFONYL)-1,4,7-TRIAZACYCLONONANE

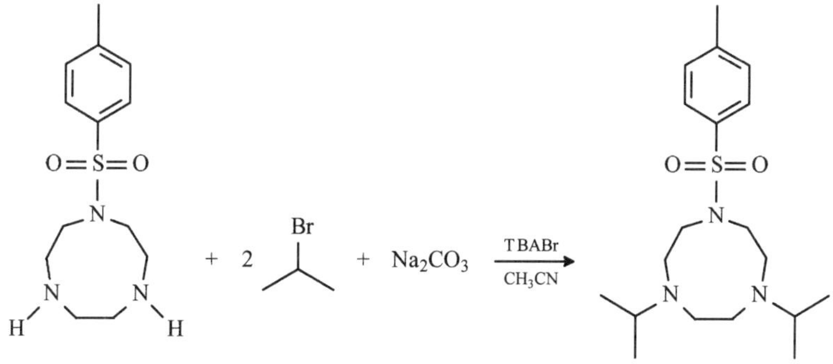

Procedure

A 250-mL round-bottomed flask equipped with a reflux condenser and a magnetic stirring bar is charged with 1-(p-toluenesulfonyl)-1,4,7-triazacyclononane (11.61 g, 0.041 mol), 2-bromopropane (15.5 ml, 0.164 mol), and CH_3CN (50 mL, distilled from CaH_2 under N_2). Na_2CO_3 (17.3 g) is added as a solid to the above colorless solution, followed by tetrabutylammonium bromide (0.1 g), and the resultant mixture is heated at reflux under N_2 for 18 h. After cooling to room temperature, the yellow mixture is filtered and the filter cake is washed with $CHCl_3$ (50 mL). The combined filtrates are evaporated under reduced pressure to yield a yellow oil with a small amount of a white solid. This residue is redissolved in $CHCl_3$ (30 mL) and then washed with 1M NaOH (30 mL). The organic layer is removed and the aqueous phase is extracted with $CHCl_3$ (3 × 10 mL). The combined $CHCl_3$ layers are dried ($MgSO_4$) and the solvent is removed under reduced pressure to yield the product, 13.56 g (90%).

Anal. Calcd. for $C_{19}H_{33}N_3O_2S$: C, 62.09; H, 9.05; N, 11.43. Found: C, 61.95; H, 9.05; N, 11.43.

Properties

1,4-Diisopropyl-7-(p-toluenesulfonyl)-1,4,7-triazacyclononane is a tan solid (mp 58–59°C) which is soluble in most organic solvents. The product has been stored for over 1 year unprotected from air without detectable decomposition. The NMR spectral data for the compound are as follows: ¹H NMR (300 MHz, CDCl₃): δ 7.66 (d, J = 8.4 Hz, 2H), 7.26 (d, J = 8.4 Hz, 2H), 3.28–3.25 (m, 4H), 2.85–2.82 (m, 4H), 2.75 (heptet, J = 6.6 Hz, 2H), 2.43 (s, 4H), 2.39 (s, 3H), 0.91 (d, J = 6.6 Hz, 12H) ppm; $^{13}C\{^1H\}$ NMR (75 MHz, CDCl₃) δ 142.54, 136.29, 129.33, 126.94, 53.53, 53.51, 52.11, 50.17, 21.28, 18.08 ppm.

C. 1,4-DIISOPROPYL-1,4,7-TRIAZACYCLONONANE

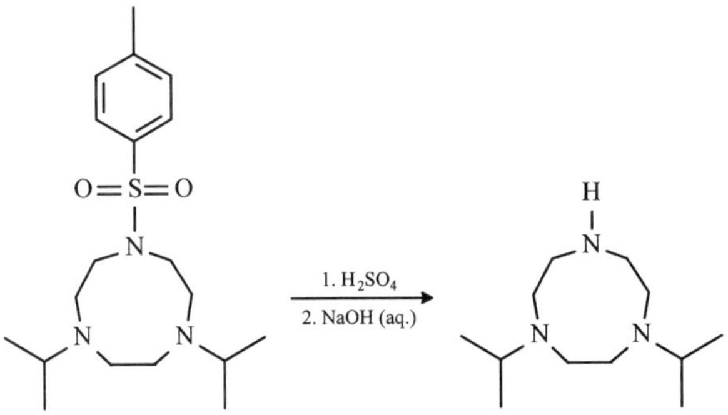

■ **Caution.** *Contact with concentrated H$_2$SO$_4$ can cause severe burns. Precautions should be exercised when conducting this reaction.*

■ **Caution.** *The neutralization conducted at the end of the reaction is very exothermic. The aqueous base should be added very slowly to minimize spattering.*

Procedure

A 125-mL, pear-shaped flask equipped with a nitrogen inlet and a magnetic stirring bar is charged with 1,4-diisopropyl-7-(p-toluenesulfonyl)-1,4,7-triaza-cyclononane (13.50 g, 0.0367 mol) and 18 M H$_2$SO$_4$ (50 mL). This mixture is heated at 120°C under N$_2$ for 24 h.[22] After cooling to room temperature, the dark mixture is poured into crushed ice (100 g) and aqueous NaOH (3 M) is *cautiously* added while cooling the mixture in an ice bath until the pH of the mixture exceeds 11. The mixture is extracted with CHCl$_3$ until the organic extracts are colorless (ca. 500 mL total). The CHCl$_3$ solution is dried (MgSO$_4$) and the solvent is removed under reduced pressure to yield an amber oil which is purified by vacuum distillation (80–82°C, 0.01 torr) to yield the pure product, 6.54 g (84%).

Anal. Calcd. for C$_{12}$H$_{27}$N$_3$: C, 67.55; H, 12.75; N, 19.69. Found: C, 66.79; H, 12.65; N, 19.69.

Properties

1,4-diisopropyl-1,4,7-triazacyclononane is a colorless oil which is soluble in most organic solvents. It is readily protonated in water but may be extracted from basic aqueous media. This amine becomes discolored when stored in air, but is indefinitely stable when stored under an inert atmosphere. The NMR spectral data for the compound are as follows: ^{1}H NMR (300 MHz, CDCl$_3$): δ 3.08 (br. s, 1H), 2.82 (heptet, J = 6.6 Hz, 2H), 2.64–2.58 (m, 4H), 2.56–2.50 (m, 4H), 2.44 (s, 4H), 0.96 (d, J = 6.6 Hz, 12H) ppm; ^{13}C{^{1}H} NMR (75 MHz, CDCl$_3$) δ 52.72, 48.90, 47.86, 47.03, 18.49 ppm.

D. 1,4-DIISOPROPYL-7-(2-PYRIDYLMETHYL)-1,4,7-TRIAZACYCLONONANE

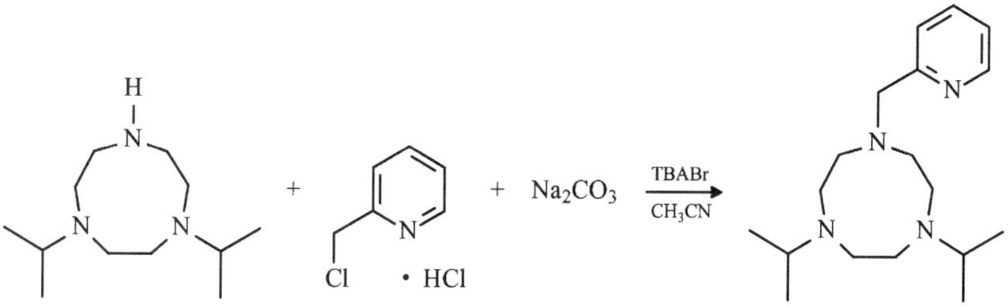

Procedure

A 50-mL, pear-shaped flask equipped with a reflux condenser and a magnetic stirring bar is charged with 1,4-diisopropyl-1,4,7-triazacyclononane (0.718 g, 3.36 mmol) and distilled CH_3CN (10 mL). 2-Picolyl chloride hydrochloride* (0.553 g, 3.37 mmol) is added as a solid, followed by Na_2CO_3 (1.45 g) and tetrabutylammonium bromide (30 mg). The resultant yellow mixture is heated at reflux under N_2 for 8 h. After cooling to room temperature, the mixture is poured into water (40 mL), brought to pH >11 with aqueous NaOH, and extracted with $CHCl_3$ (3 × 30 mL). The $CHCl_3$ phase is dried ($MgSO_4$) and the solvent is removed under reduced pressure to yield the crude product as an amber to orange oil. Kugelrohr distillation (118–122°C, 0.01 torr) affords the pure product, 0.927 g (90%).

Anal. Calcd. for $C_{18}H_{32}N_4$: C, 71.04; H, 10.59; N, 18.40. Found: C, 71.04; H, 10.49; N, 18.43.

Properties

1,4-Diisopropyl-7-(2-pyridylmethyl)-1,4,7-triazacyclononane is a light yellow oil which is soluble in most organic solvents. Although the ligand is readily protonated in water, it may be extracted from strongly basic aqueous media. As the ligand becomes discolored upon exposure to air, it is best stored under an inert atmosphere. The NMR spectral data for the compound are as follows: ^1H NMR (300 MHz, $CDCl_3$) δ 8.50–8.47 (m, 1H), 7.63–7.54 (m, 2H), 7.12–7.07 (m, 1H), 3.82 (s, 2H), 2.94–2.91 (m, 4H), 2.86 (heptet, $J = 6.6$ Hz, 2H), 2.63–2.60 (m, 4H), 2.56 (s, 4H), 0.93 (d, $J = 6.6$ Hz, 12H) ppm; $^{13}C\{^1H\}$ NMR (75 MHz, $CDCl_3$) δ 161.05, 148.78, 136.20, 123.05, 121.59, 63.89, 55.62, 54.78, 52.82, 52.26, 18.32 ppm.

References

1. P. Chaudhuri and K. Wieghardt, *Prog. Inorg. Chem.*, **35**, 329 (1987).
2. T. J. Marks, *Prog. Inorg. Chem.*, **24**, 51 (1978). *Ibid*, **25**, 223 (1979).
3. S. Trofimenko, *Chem. Rev.*, **93**, 943 (1993). N. Kitajima and W. B. Tolman, *Prog. Inorg. Chem.*, **43**, 419 (1995).
4. T. J. Atkins, J. E. Richman, and W. F. Oettle, *Org. Synth.*, **58**, 86 (1978). D. W. White, B. A. Karcher, P. A. Jacobsen, and J. G. Verkade, *J. Am. Chem. Soc.*, **47**, 412 (1982). G. H. Searle and R. J. Geue, *Aust. J. Chem.*, **37**, 959 (1984).
5. K. Wieghardt, P. Chaudhuri, B. Nuber, and J. Weiss, *Inorg. Chem.*, **21**, 3086 (1982).
6. G. Haselhorst, S. Stoetzel, A. Strassburger, W. Walz, K. Wieghardt, and B. Nuber, *J. Chem. Soc., Dalton Trans.*, 83 (1993). J. Sessler, J. W. Sibert, and V. Lynch, *Inorg. Chim. Acta*, **216**, 89 (1994).

* Aldrich Chem. Co., Milwaukee, WI 53233.

7. T. Beisel, B. S. P. L. Della Vedova, K. Weighardt, and R. Boese, *Inorg. Chem.*, **29**, 1736 (1990).
8. U. Auerbach, U. Eckert, K. Wieghardt, B. Nuber, and J. Weiss, *Inorg. Chem.*, **29**, 938 (1990). D. A. Moore, P. E. Fanwick, and M. J. Welch, *Inorg. Chem.*, **28**, 1504 (1989).
9. U. Bossek, D. Hanke, K. Wieghardt, and B. Nuber, *Polyhedron*, **12**, 1 (1993). D. A. Moore, P. E. Fanwick, and M. J. Welch, *Inorg. Chem.*, **29**, 672 (1990).
10. B. A. Sayer, J. P. Michael, and R. D. Hancock, *Inorg. Chim. Acta*, **77**, L63 (1983).
11. L. Christiansen, D. N. Hendrickson, H. Toftlund, S. R. Wilson, and C. L. Xie, *Inorg. Chem.*, **25**, 2813 (1986). K. Wieghardt, E. Schöffmann, B. Nuber, and J. Weiss, *Inorg. Chem.*, **25**, 4877 (1986).
12. M. DiVaira, F. Mani, and P. Stoppioni, *J. Chem. Soc., Chem. Commun.*, 126 (1989).
13. T. J. Atkins, *J. Am. Chem. Soc.*, **102**, 6365 (1980).
14. A. J. Blake, I. A. Fallis, R. O. Gould, S. Parsons, S. A. Ross, and M. Schröder, *J. Chem. Soc., Chem. Commun.*, 2467 (1994). G. R. Weisman, D. J. Vachon, V. B. Johnson, and D. A. Gronbeck, *J. Chem. Soc., Chem. Commun.*, 886 (1987).
15. J. L. Sessler, J. W. Sibert, and V. Lynch, *Inorg. Chem.*, **29**, 4143 (1990).
16. K. Wieghardt, I. Tolksdorf, and W. Herrmann, *Inorg. Chem.*, **24**, 1230 (1985).
17. J. A. Halfen, S. Mahapatra, M. M. Olmstead, and W. B. Tolman, *J. Am. Chem. Soc.*, **116**, 2173 (1994). S. Mahapatra, J. A. Halfen, and W. B. Tolman, *J. Chem. Soc., Chem. Commun.*, 1625 (1994). J. A. Halfen and W. B. Tolman, *J. Am. Chem. Soc.*, **116**, 5475 (1994). S. Mahapatra, J. A. Halfen, E. C. Wilkinson, L. Que, Jr., and W. B. Tolman, *J. Am. Chem. Soc.*, **116**, 9785 (1994). S. Mahapatra, J. A. Halfen, E. C. Wilkinson, G. Pan, C. J. Cramer, L. Que, Jr., and W. B. Tolman, *J. Am. Chem. Soc.*, **117**, 8865 (1995). J. A. Halfen, S. Mahapatra, E. C. Wilkinson, A. J. Gengenbach, V. G. Young, Jr., L. Que, Jr., and W. B. Tolman, *J. Am. Chem. Soc.*, **118**, 763 (1996). J. A. Halfen, S. Mahapatra, E. C. Wilkinson, S. Kaderli, V. G. Young, Jr., A. D. Zuberbühler, L. Que, Jr., and W. B. Tolman, *Science*, **271**, 1397 (1996).
18. C. Flassbeck and K. Wieghardt, *Z. Anorg. Allg. Chem.*, **608**, 68 (1992).
19. R. P. Houser, J. A. Halfen, J. V. G. Young, N. J. Blackburn, and W. B. Tolman, *J. Am. Chem. Soc.*, **117**, 10745 (1995). J. A. Halfen, V. G. Young, Jr., W. B. Tolman, *Angew. Chem., Int. Ed. Engl.*, **35**, 1687 (1996). J. A. Halfen, J. V. G. Young, and W. B. Tolman, *J. Am. Chem. Soc.*, **118**, 10920 (1996). S. Mahapatra, J. A. Halfen, E. C. Wilkinson, G. Pan, X. Wang, V. G. Young, Jr., C. J. Cramer, L. Que, Jr., and W. B. Tolman, *J. Am. Chem. Soc.*, **118**, 11555 (1996) S. Mahapatra, J. A. Halfen, and W. B. Tolman, *J. Am. Chem. Soc.*, **118**, 11575 (1996). S. Mahapatra, V. G. Young, Jr., and W. B. Tolman, *Angew. Chem., Int. Ed. Engl.*, **36**, 130 (1997). J. A. Halfen, B. A. Jazdzewski, S. Mahapatra, L. M. Berreau, E. C. Wilkinson, L. Que, Jr., and W. B. Tolman, *J. Am. Chem. Soc.*, **119**, 8217 (1997).
20. For example, see Y. Dong, H. Fujii, M. P. Hendrich, R. A. Leising, G. Pan, C. R. Randall, E. C. Wilkinson, Y. Zang, L. Que, Jr., B. G. Fox, K. Kauffmann, and E. Münck, *J. Am. Chem. Soc.*, **117**, 2778 (1995). Z. Tyeklàr, R. R. Jacobsen, N. Wei, N. N. Murthy, J. Zubieta, and K. D. Karlin, *J. Am. Chem. Soc.*, **115**, 2677 (1993).
21. Solutions of HBr/CH_3CO_2H are available commercially (e.g., Acros Organics, 711 Forbes Avenue, Pittsburgh, PA 15219; 800-227-6701) in concentrations ranging from 30 to 33%.
22. While extended reaction times (up to three days) have not been found to be detrimental, it is *critical* that the temperature of the reaction be maintained at 120°C. Above 130°C, *N*-dealkylation of the macrocyclic ring accompanies hydrolysis of the tosylamide group, yielding a nearly intractable mixture of mono- and dialkylated ligands.

13. N-(2-HYDROXYETHYL)-3,5-DIMETHYLPYRAZOLE, A DINUCLEAR COPPER COMPLEX, AND N-(2-p-TOLUENESULFONYLETHYL)-3,5-DIMETHYLPYRAZOLE

Submitted by ELISABETH BOUWMAN* and WILLEM L. DRIESSEN*
Checked by JEEHEE KANG† and MARCETTA Y. DARENSBOURG†

Biomimetic inorganic systems are designed and synthesized to obtain insight into the relation between the structure and activity of the active sites in metalloproteins. The metal ions in the enzymes are often coordinated by histidine residues of the protein backbone. In low-molecular-weight model compounds, the ligands used should therefore contain one or more groups closely resembling the coordinating side chains of the peptide molecules. Functional groups which have been used to mimic the histidine side chain are benzimidazole, imidazole, pyridine, and pyrazole. To incorporate pyrazole groups into larger ligand systems, N-hydroxymethyl- and N-hydroxyethyl-pyrazoles have been used as starting materials.

N-Hydroxymethylpyrazoles are readily formed through the condensation of hydrazine with either pentanedione[1] or 1,1,3,3-tetraethoxypropane,[2] followed by reaction with paraformaldehyde.[3,4] N-hydroxyethylpyrazoles are prepared in a one-step synthesis through the condensation of hydroxyethyl-hydrazine with pentanedione[5] or 1,1,3,3-tetraethoxypropane.[6] Hydroxyethylpyrazoles do not react with thiol or amine groups, but they may be activated by conversion of the hydroxy group to a tosylate function.[5,7] Incorporation of pyrazole groups into chelating ligands can also be achieved through the reaction of the sodium pyrazolate with alkyl chlorides.[8]

The present paper describes the synthesis of N-(2-hydroxyethyl)-3,5-dimethylpyrazole and the synthesis of the tosylated compound N-(2-p-toluenesulfonylethyl)-3,5-dimethylpyrazole which can be used for the synthesis of larger chelating ligands. The tosylation is carried out in a water/acetone mixture, unlike most classical tosylations, which are performed in pyridine.[7] A high yield of pure tosylated product is obtained from this reaction. A water/acetone mixture as the solvent for the synthesis of other tosylates may very well be also successful. Since the compound N-(2-hydroxyethyl)-3,5-dimethylpyrazole may itself act as a didentate N,O ligand in coordination compounds with transition metal ions,[9] an example using Cu(II) is provided below.

* Leiden Institute of Chemistry, Gorlaeus Laboratories, Leiden University, P.O. Box 9502, 2300 RA Leiden, The Netherlands.
† Department of Chemistry, Texas A & M University, College Station, TX 77843.

Starting Materials

Solvents and reagents (2,4-pentanedione, 2-hydroxyethylhydrazine, copper bromide, *p*-toluenesulfonylchloride) are available from the common commercial sources.

A. *N*-(2-HYDROXYETHYL)-3,5-DIMETHYLPYRAZOLE

Procedure

In a one-necked, 500-mL, round-bottomed flask with a magnetic stirrer bar, 50 g (0.5 mole) 2,4-pentanedione is dissolved in 150 mL of absolute ethanol. The flask is fitted with a pressure-equalizing dropping funnel containing 38 g (0.5 mole) of 2-hydroxyethylhydrazine dissolved in 50 mL of absolute ethanol. The flask is immersed in an ice-salt bath,* and the 2-hydroxyethyl-hydrazine is added slowly while stirring, thereby keeping the temperature of the reaction mixture below 15°C. After all of the reagent is added, the cooling bath is removed and the reaction mixture is stirred for 1 h at room temperature. The solvent is then removed using a rotary evaporator, while the temperature of the water bath is kept below 50°C. To the remaining yellow-orange oil, 200 mL of diethylether and 20 mL of acetone are added. The flask is stoppered and placed in the freezer (-20°C) where a white (or slightly yellow) crystalline product forms. The crystals are filtered and washed with a small amount of diethylether.† Three crops of crystals can be obtained by evaporating the filtrate followed by addition of ether/acetone in diminishing quantities. The overall yield is approximately 58 g (78%). The compound is recrystallized from diethylether, which dissolves the *N*-(2-hydroxyethyl)-3,5-dimethylpyrazole but not the orange oil impurity. Mp 76.5°C.

Anal. Calcd. for $C_7H_{12}N_2O$: C, 59.98%; H, 8.63%; N, 19.98%. Found: C, 59.99%; H, 8.55%; N, 19.89%.

Properties

The product *N*-(2-hydroxyethyl)-3,5-dimethylpyrazole is easily obtained as large, colorless crystals. Varying the ratio of ether and acetone used in

---

* Checkers' comment: An ice (0°C) bath is sufficient.
† Checkers' comment: Precooling diethylether minimizes product loss.

crystallization may produce coloration of the product. Although the NMR of this yellow product finds it to be fairly pure, the small impurity negatively affects the tosylation reaction. If the yellow compound is kept for a longer period of time, it slowly turns brown and eventually turns into a brown oil. ^{1}H NMR (CDCl$_3$, internal standard TMS): 2.16 (s, 3H, pzCH_3), 2.22 (s, 3H, pzCH_3), 3.90 (d·t, 2H, CH_2OH), 4.01 (t, 2H, pzCH_2), 4.61 (t, 1H, OH), 5.77 (s, 1H, pzCH) ppm. ^{13}C NMR: 10.8 and 13.1 (pzCH_3), 49.8 (pzCH_2), 61.2 (CH_2OH), 104.6 (pzCH), 139.4 and 147.4 (PzCCH$_3$) ppm. IR (KBr pellet): 3252 (s,br), 2981 (w), 2953 (m), 2919 (m), 2872 (m), 1550 (vs), 1458 (vs,br), 1418 (s), 1388 (m), 1309 (s), 1269 (m), 1203 (w), 1127 (m), 1060 (vs), 1025 (s), 987 (m), 819 (m), 785 (s), 699 (s), 661 (s), 636 (w), 478 (m) cm^{-1}.

B. BIS-[BROMO-(*N*-OXYETHYL-3,5-DIMETHYLPYRAZOLE)-COPPER(II)]

$$2CuBr_2 + 2C_7H_{12}N_2O + 2(C_2H_5)_3N \rightarrow$$
$$[Cu(C_7H_{11}N_2O)Br]_2 + 2(C_2H_5)_3NHBr$$

Procedure

In a 100-mL beaker, 0.44 g (2 mmole) copper bromide is dissolved in 10 mL of absolute ethanol with heating. In another 100-mL beaker, 0.2 g (2 mmole) triethylamine is added to a solution of 0.28 g (2 mmole) *N*-(2-hydroxyethyl)-3,5-dimethylpyrazole in 10 mL absolute ethanol. The two solutions are heated until boiling before adding the copper solution to the ligand solution. The reaction mixture is filtered through a filter paper into a 100-mL Erlenmeyer flask to remove solid impurities. From the dark brown filtrate a dark green complex with formula [Cu(*N*-oed)Br]$_2$ is obtained almost immediately. A higher yield is obtained when the flask is left open for a few days. (*N*-oed = deprotonated N,O ligand). The crystals are filtered and washed with a small amount of ethanol.

Anal. Calcd. for $C_{14}H_{22}Br_2Cu_2N_4O_2$: C, 29.7%; H, 3.9%; Cu, 22.5%. Found: Cu 22.1%.

Properties

The compound [Cu(*N*-oed)Br]$_2$ readily crystallizes from the reaction mixture, resulting in dark green, X-ray quality single crystals.[9] The two copper ions in the dinuclear unit are 3.0 Å apart, bridged by the two oxo atoms from the ligands. Each copper ion is in a square planar O$_2$NBr environment, with

Cu–O distances of 1.92 Å, a Cu–N distance of 1.94 Å, and Cu–Br distance of 2.35 Å. The copper ions are antiferromagnetically coupled and therefore the compound is epr-silent. *N*-Hydroxyethyl-3,5-dimethylpyrazole may also yield complexes with the neutral (nondeprotonated) molecule, in which the hydroxo group is coordinated but not bridging.[9] These complexes have to be prepared from acidic reaction mixtures. IR of [Cu(N-oed)Br]$_2$ as a KBr pellet: 2927 (m), 2868 (m), 1636 (w, br), 1548 (s), 1478 (s), 1437 (m), 1390 (s), 1348 (w), 1299 (s), 1225 (s), 1139 (m), 1104 (m), 1062 (vs), 973 (w), 896 (s), 777 (m), 622 (m), 378 (w), 354 (m) cm^{-1}. Electronic absorption maxima (diffuse reflectance): 14.5×10^3, 17.4×10^3, 27.0×10^3 cm^{-1}.

C. *N*-(2-*p*-TOLUENESULFONYLETHYL)-3,5-DIMETHYLPYRAZOLE

Procedure

A 1-L, round-bottomed flask is charged with 42 g (0.3 mole) *N*-(2-hydroxyethyl)-3,5-dimethylpyrazole and 57 g (0.3 mole) *p*-toluenesulfonyl chloride suspended in 300 mL of acetone and 300 mL of distilled water. This suspension is stirred with a magnetic stirrer bar and cooled with an ice/salt bath.* The flask is fitted with a pressure-equalizing dropping funnel containing 12 g (0.3 mole) of sodium hydroxide dissolved in 150 mL distilled water. The sodium hydroxide solution is added dropwise over 2.5 h. The ice bath is subsequently removed and the reaction mixture is stirred for 1 h at room temperature. To evaporate the acetone, air is passed over the solution while the reaction mixture is vigorously stirred, resulting in the precipitation of a white solid. The product is filtered, washed with distilled water, and then stored in a vacuum desiccator under vacuum and in the presence of P$_2$O$_5$ to absorb water. The yield of the dry, white solid is approximately 79 g (90%). The compound can be recrystallized from diethylether. Mp 64°C.

---

* Checkers' comment: An ice (0°C) bath is sufficiently cold.

Anal. Calcd. for $C_{14}H_{18}N_2O_3S$: C, 57.12%; H, 6.16%; N, 9.52%. Found: C, 57.02%; H, 6.17%; N, 9.45%.

Properties

When pure recrystallized *N*-(2-hydroxyethyl)-3,5-dimethylpyrazole is used in the tosylation reaction, a white, granular solid is obtained. This product is quite pure and stable. However, when the starting material is slightly impure, a yellow to brown solid might result, which, according to the NMR, is still relatively pure, but not stable. The compound gradually turns darker brown, and eventually turns into a brown oil. *N*-(2-*p*-Toluenesulfonylethyl)-3,5-dimethylpyrazole is a starting material for a wide range of chelating ligands. The compound may be reacted with primary or secondary amines[10] or with thiols[5] in water with sodium hydroxide, but it has been found that a selective reaction with a thiol in presence of an amine, or a reaction with a primary amine to obtain a monoadduct, is best carried out in a tetrahydro-furan–water mixture using sodium hydroxide.[11,12]

^{1}H NMR ($CDCl_3$, internal standard TMS): 2.10 (s, 3H, pzCH_3), 2.21 (s, 3H, pzCH_3), 2.42 (s, 3H, phCH_3), 4.19 (t, 2H, pzCH_2), 4.31 (t, 2H, CH_2OTs), 5.71 (s, 1H, pzCH), 7.27 (d, 2H, phenyl), 7.61 (d, 2H, phenyl) ppm. ^{13}C NMR: 10.9 and 13.3 (PzCH_3), 21.5 (PhCH_3), 47.0 (pzCH_2), 68.5 (CH_2OTs), 105.1 (pzCH), 127.7 and 129.7 (phCH), 132.1 (phCCH$_3$), 139.9 and 148.0 (pzCCH$_3$), 144.7 (phCSO$_2$) ppm. IR (KBr pellet): 2945 (w), 2918 (w), 1597 (m), 1552 (s), 1492 (w), 1443 (m), 1423 (m), 1364 (m), 1352 (vs), 1309 (m), 1298 (w), 1266 (w), 1188 (vs), 1170 (vs), 1096 (m), 1026 (m), 1007 (s), 919 (s,br), 808 (s), 789 (m), 766 (s), 661 (vs), 578 (s), 553 (vs), 482 (m), 442 (m) cm^{-1}.

References

1. R. H. Wiley and P. E. Hexner, *Org. Synth. Coll. Vol.*, **4**, 351 (1963).
2. R. G. Jones, *J. Am. Chem. Soc.*, **71**, 3994 (1949).
3. I. Dvoretzky and G. H. Richter, *J. Org. Chem.*, **15**, 1285 (1950).
4. W. L. Driessen, *Recl. Trav. Chim. Pays-Bas*, **101**, 441 (1982).
5. W. G. Haanstra, W. L. Driessen, J. Reedijk, U. Turpeinen, and R. Hämäläinen, *J. Chem. Soc., Dalton Trans.*, 2309 (1989).
6. I. L. Finar and K. Utting, *J. Chem. Soc.*, 5272 (1960).
7. P. J. Machin, D. N. Hurst, R. M. Bradshaw, L. C. Blaber, D. T. Burden, and R. A. Melarange, *J. Med. Chem.*, **27**, 503 (1984).
8. T. N. Sorrell and M. R. Malachowski, *Inorg. Chem.*, **22**, 1883 (1983).
9. W. L. Driessen, S. Gorter, W. G. Haanstra, L. J. J. Laarhoven, J. Reedijk, K. Goubitz, and F. R. Seljée, *Recl. Trav. Chim. Pays-Bas*, **112**, 309 (1993).
10. W. G. Haanstra, W. L. Driessen, R. A. G. de Graaff, G. C. Sebregts, J. Suriano, J. Reedijk, U. Turpeinen, R. Hämäläinen, and J. S. Wood, *Inorg. Chim. Acta*, **189**, 243 (1991).

11. P. M. van Berkel, W. L. Driessen, F. J. Parlevliet, J. Reedijk, and D. C. Sherrington, *Eur. Pol. J.*, **33**, 129 (1996).
12. W. L. Driessen, R. M. de Vos, A. Etz, and J. Reedijk, *Inorg. Chim. Acta*, **235**, 127 (1995).

14. *N*-(2-MERCAPTOETHYL)-3,5-DIMETHYLPYRAZOLE

Submitted by ELISABETH BOUWMAN* and JAN REEDIJK*
Checked by JEEHEE KANG and MARCETTA Y. DARENSBOURG†

In general, chelating ligands containing pyrazole and thioether groups can readily be prepared from *N*-[2-(*p*-toluenesulfonyl)ethyl]-3,5-dimethyl-pyrazole and thiols.[1] The synthesis of a potentially pentadentate N_2S_2O chelating ligand starting from the dithiol 1,3-dimercapto-2-propanol, how-ever, proved to be tedious,[2] because the dithiol is neither commercially available nor can it be prepared with sufficient purity from 1,3-dichloro-2-propanol and sodium trithiocarbonate.[3] An alternative route to the ligand was therefore developed[2,4] by reaction of *N*-(2-mercaptoethyl)-3,5-dimethyl-pyrazole with 1,3-dichloro-2-propanol.[2] The synthesis of *N*-(2-mercapto-ethyl)-3,5-dimethylpyrazole proceeds without problems, with a reasonable yield.

Starting Materials

The preparation of *N*-(2-*p*-toluenesulfonylethyl)-3,5-dimethylpyrazole is de-scribed elsewhere in this volume.[5] 1,1,3,3-Tetramethyl-2-thiourea and other reagents are available from the usual commercial sources.

* Leiden Institute of Chemistry, Gorlaeus Laboratories, Leiden University, P.O. Box 9502, 2300 RA Leiden, The Netherlands.
† Department of Chemistry, Texas A & M University, College Station, Texas 77843.

■ **Caution.** *Concentrated hydrochloric acid should be handled with care.*

Procedure

Into a 500-mL, round-bottomed flask fitted with a reflux condenser, a mixture of 41.2 g (0.14 mole) N-(2-p-toluenesulfonylethyl)-3,5-dimethylpyrazole and 18.5 g (0.14 mole) 1,1,3,3-tetramethyl-2-thiourea is added to 200 mL of distilled water. This suspension is refluxed for 3 h, during which period the suspended materials dissolve. After the reaction mixture has cooled to room temperature, a solution of 12 g (0.3 mole) sodium hydroxide in 40 mL distilled water is added. The reaction mixture is then further refluxed for 2 h. The aqueous solution is extracted with dichloromethane (5 × 40 mL), in order to remove tetramethylurea, unreacted starting materials, and any neutral side products. The aqueous layer is then acidified to approximately pH 5.5 with concentrated hydrochloric acid, resulting in the separation of the thiol. The acidified solution is extracted with dichloromethane (3 × 40 mL). The yellow organic layer is dried over anhydrous magnesium sulfate. The solvent is removed using a rotary evaporator, resulting in a yellow to orange liquid. The overall yield varies from 40 to 60%.

Properties

The synthesis of N-(2-mercaptoethyl)-3,5-dimethylpyrazole and the workup procedure are performed in air, but the resultant thiol can best be stored under a nitrogen atmosphere. In air, the thiol is resistant to oxidation for several weeks, but the disulfide slowly crystallizes from the oil. According to the NMR spectra, the product is quite pure, but may still contain a small amount of 1,1,3,3-tetramethylurea (up to 5%). The reaction with 1,3-dichloro-2-propanol was carried out in a water/tetrahydrofuran mixture with sodium hydroxide, and appeared to proceed smoothly and in almost quantitative yields.[2] This new reaction path to pyrazole–thioether ligands opens the road to the development of new chelating ligands.

   N-(2-Mercaptoethyl)-3,5-dimethylpyrazole has also been used in the synthesis of coordination compounds and yields a range of mono-, di-, and trinuclear complexes. Complex synthesis is similar to the one described for N-hydroxyethyl-3,5-dimethylpyrazole.[5] ^{1}H NMR (solvent CDCl$_3$, internal standard TMS) of the thiol: 1.37 (t, 1H, SH), 2.20 (s, 3H, CH_3), 2.28 (s, 3H, CH_3), 2.89 (q, 2H, CH_2-SH), 4.08 (t, 2H, pz-CH_2), 5.78 ppm (s, 1H, pz-H); the disulfide is characterized by peaks at 3.05 (t, 2H, CH_2SS) and 4.21 ppm (t, 2H, pzCH_2). ^{13}C NMR (CDCl$_3$): 10.6 and 13.0 (pzCH_3), 24.3 (CH_2SH), 50.5 (pzCH_2), 104.3 (pzCH), 138.7 and 147.2 ppm (pz$C$CH$_3$); (disulfide: 37.34

(CH_2SS) and 46.48 ppm $(pzCH_2)$. IR (neat): 2920 (s), 2860 (m), 2540 (w, br), 1557 (vs), 1460 (s), 1425 (s), 1310 (s), 1272 (m), 1240 (m), 1160 (w), 1120 (m), 1035 (m), 1020 (m), 980 (w), 780 (s, br), 680 (m), 670 (m), 640 (w), 620 (w), 570 (w, br), 480 (w) cm^{-1}.

References

1. W. G. Haanstra, W. L. Driessen, J. Reedijk, U. Turpeinen, and R. Hämäläinen, *J. Chem. Soc., Dalton Trans.*, 2309 (1989).
2. E. Bouwman, P. Evans, R.A.G. de Graaff, H. Kooijman, R. Poinsot, P. Rabu, J. Reedijk, and A. L. Spek, *Inorg. Chem.*, **34**, 6302 (1995).
3. Spiess GmbH, patent Neth. Appl 6,510,637, (1967).
4. E. Bouwman, P. Evans, H. Kooijman, J. Reedijk, and A. L. Spek, *J. Chem. Soc., Chem. Commun.*, 1746 (1993).
5. E. Bouwman and W. L. Driessen, *Inorg. Synth.*, **32**, 82 (1998).

15. 1,5-DIAZACYCLOOCTANE, PENDANT ARM THIOLATO DERIVATIVES AND [N,N'-BIS(2-MERCAPTOETHYL)-1,5-DIAZACYCLOOCTANATO]NICKEL(II)

Submitted by DANIEL K. MILLS,* IVAN FONT,* PATRICK J. FARMER,* YUI-MAY HSIAO,* THAWATCHAI TUNTULANI,* RIZALIA M. BUONOMO,* DAWN C. GOODMAN,* GHEZAI MUSIE,* CRAIG A. GRAPPERHAUS,* MICHAEL J. MAGUIRE,* CHIA-HUEI LAI,* MICHELLE L. HATLEY,* JASON J. SMEE,*, JOHN A. BELLEFEUILLE,* and MARCETTA Y. DARENSBOURG*
Checked by ROBERT D. HANCOCK,† SHEILA ENG,* and ARTHUR E. MARTELL*

The ligand 1,5-diazacyclooctane (daco) has long been known to have interesting steric and binding characteristics.[1] This compound was first synthesized by Buhle and co-workers[2] in an attempt to isolate optically active trivalent nitrogen compounds with steric demands that would eliminate inversion at nitrogen. Since that time, various other synthetic routes have been uncovered, including reaction of the disodium salt of the ditosylate of 1,3-diaminopropane with 1,3-dibromopropane[3] or with the ditosylate of 1,3-hydroxypropane;[4] the reduction and hydrogenation of the eight-membered bicyclic

* Department of Chemistry, Texas A & M University, College Station, TX 77843.
† IBC Advanced Technologies American Fork, UT.

diamide of daco, 1,5-diaza-bicyclo[3.3.0]-octane-4,8-dione;[5] or the use of Ag^+ to template 1,3-diaminopropane before the direct alkylation with 1,3-dibromopropane.[6] None of these are as convenient as the original preparation of Buhle and co-workers. Indeed, one report of the reaction of 1,3-diaminopropane with acrylonitrile followed by reduction with sodium metal to afford daco in ca. 60% overall yield was found after repeated attempts to give only the linear *N*-proplyamino-1,3-diaminopropane.[7] Below is given the optimization of the original synthesis, the neutralization and extraction of the free amine, as well as heretofore unreported spectral characteristics of the free amine and its HBr salt. Synthesis and characterization of two derivatives containing sulfur pendant arms are given. Also presented are the synthesis and characterization of the nickel complexes of these derivatives.

A. 1,5-DIAZACYCLOOCTANE HYDROBROMIDE (DACO · 2HBr)

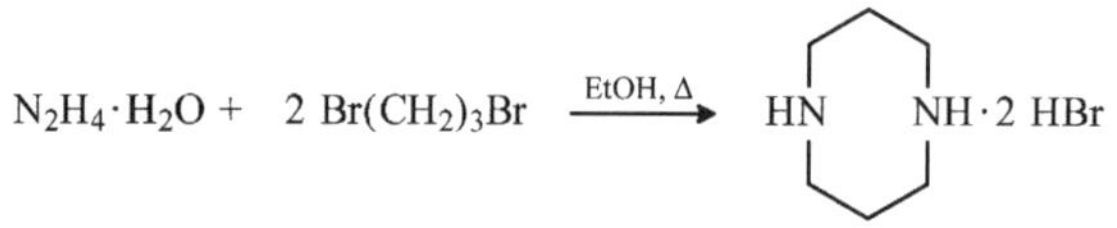

$$N_2H_4 \cdot H_2O + 2 \; Br(CH_2)_3Br \xrightarrow{\text{EtOH, } \Delta}$$

daco · 2 HBr

■ **Caution.** *Hydrazine is a strong reducing agent and is extremely flammable as well as a known carcinogen. The reaction should be carried out in an efficient fume hood, and exposed skin protected against contact with the reagent. The reaction is exothermic and care to maintain gentle reflux only must be taken.*

Procedure

A 1-L, 3-necked, round-bottomed flask, equipped with a stir bar, needle-valve-controlled addition funnel, water-cooled reflux condenser, and nitrogen inlet is charged with 100 mL (2.06 mol) of hydrazine monohydrate* and 400 mL absolute ethanol. The flask is placed in a heating mantle. The mixture is stirred vigorously and brought to gentle reflux under a dry N_2 blanket. Then 100 mL (0.981 mol) 1,3-dibromopropane† is added dropwise from the addition funnel over 4 h, during which time a white precipitate may be observed. During addition, the temperature is controlled to maintain gentle reflux. (If overheating occurs, a rapid release of vapors may occur.) Upon complete addition, the mixture is stirred and refluxed under N_2 for an additional hour. After cooling to room temperature under N_2, the mixture is

---

* Fluka Chemika-BioChemika Buchs, Switzerland.
† Lancaster Synthesis Inc. Windham, NH 03087.

transferred to a 2-L wide mouth Erlenmeyer flask, stoppered, and placed in a refrigerator overnight.

Note: From this point on, the workup must not be interrupted, as prolonged exposure to air greatly reduces product quality by the formation of oxidation byproducts.

The white, crystalline hydrazine monohydrogen bromide which forms on cooling is filtered in air on a Buchner funnel, washed with 3×100 mL absolute EtOH, and discarded. (Because of the explosion hazard of anhydrous hydrazine, recovery of spent hydrazine by neutralization and distillation is strongly discouraged.) A ^{1}H NMR analysis of the hydrazine mononhydrogen bromide indicates the presence of daco $\cdot$ 2HBr, but at levels much too low to warrant recovery. The combined filtrate and washings are returned to the Erlenmeyer flask, 500 mL distilled water is added, and the mixture is acidified with 48% aqueous HBr* to a pH of ca. 3. Acidification will be accompanied by a color shift to yellow. Hydrochloric acid may also be used, but mixed salts will be obtained as HBr is produced from the initial reaction. Benzaldehyde (80 mL, 0.79 mol) is added, resulting in precipitation of a bright yellow solid. If no precipitate is observed, an additional 100 mL of water may be added. The mixture is stoppered and returned to the refrigerator. After cooling for 2 h, the benzalazine that forms is filtered, washed with 3×100 mL H_2O, and discarded. The combined aqueous filtrate and washings are washed with copious amounts (ca. 2 L) of diethyl ether in a separatory funnel and the water layers are reduced to a dark red syrup on a rotary evaporator. Absolute EtOH (500 mL) is added and the mixture is vigorously shaken for 5 min or until a pale yellow precipitate forms. After cooling in the refrigerator several hours or overnight, the product is filtered and washed with 500 mL absolute EtOH and then ether (ca. 5×200 mL portions) until the filtrate is colorless. The daco $\cdot$ 2HBr is sparingly soluble in absolute EtOH (moreso in 95% EtOH); however, insufficient washing will allow an unknown impurity to remain which will air oxidize as the product dries, lowering purity. The solid is placed on a Petri dish and left to air dry. Yield after thorough drying in air: 15–20 g (11–15%).

Properties

The daco $\cdot$ 2HBr is a pale yellow solid that is soluble in water, and sparingly soluble in absolute EtOH. The ^{1}H NMR spectrum in D_2O shows a quintet at 2.2δ, 4H and a triplet at 3.4δ, 8H (H_2O: 4.8δ); mp = 241–243°C.

* EM Science, Gibbstown, NJ 08027.

B. 1,5-DIAZACYCLOOCTANE (DACO-FREE AMINE)

■ **Caution.**   *The daco-free amine, as well as sodium hydroxide, are strong bases and therefore represent contact hazards.*

Procedure

Finely powdered daco · 2HBr (80 g, 0.29 mol) is placed in a 1-L Erlenmeyer flask and dissolved in an absolute minimum of water (ca. 70 mL). With cooling in an ice bath, KOH (80 g, 1.4 mol) is added in small portions with swirling until the solution is saturated in KOH, leaving a few excess pellets on the bottom. The daco-free amine is extremely soluble in H_2O and without complete saturation of the solution with KOH, the subsequent diethyl ether extraction is quite ineffective. At this point, potassium bromide should begin to precipitate. The daco-free amine which begins to separate as a distinct layer is decanted into a 2-L Erlenmeyer flask which is stoppered. Since daco-free amine will react slowly with air to form the amine oxide, air exposure should be minimized by keeping all vessels covered as much as possible. The salt slurry is washed thoroughly with 3×200 mL Et_2O and the ether wash is added to the Erlenmeyer. An additional 80 g KOH is added to the salt mixture and the 3×200 mL ether wash repeated. The ether washings are combined in the Erlenmeyer and shaken to dissolve the daco. The salt mixture is extracted with a final 3×200 mL ether with vigorous shaking and the ether washings transferred to the stoppered Erlenmeyer flask. The combined ether extracts are dried with sodium sulfate and are then transferred in portions to a 2-L 3-necked flask. A slight vacuum assists in solvent removal while maintaining the pot at or below room temperature in a water bath (Fig. 1). The daco-free amine is somewhat volatile. Ether trapped at $-78°C$ before the aspirator can be treated with 100 mL of 2% HBr in absolute EtOH to precipitate any daco lost during solvent removal. The yellow oil which remains is crude daco-free amine. Yield: 31.0 g (94%). Pure daco-free amine can be obtained by adding CaH_2 to the crude oil until no H_2 production is observed and distilling at 60°C, ca. 0.2 torr. The product is a colorless to pale yellow low melting solid. Yield after distillation: 25.6 g (78% based on daco · 2HBr).

Properties

Crude daco-free amine is a yellow oil that is soluble in a variety of solvents including ether, chloroform, benzene, and methanol. The ^1H NMR spectrum in CDCl$_3$ shows a quintet at 1.6δ, 4H, broad singlet (NH) at 2.2d, 2H, and a triplet at 2.9d, 8H.

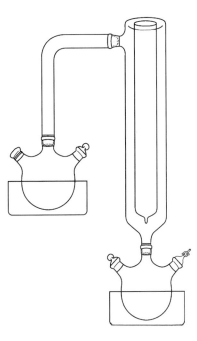

Figure 1. Distilling apparatus for daco-free amine.

Pure daco-free amine is a pale yellow low-melting solid that is soluble in variety of solvents including ether, chloroform, benzene, and methanol. The ^1H NMR spectrum in $CDCl_3$ shows a quintet at 1.6δ, 4H, broad multiplet at 1.98δ, 2H, and a triplet at 2.9δ, 8H; mp $= -23$ to $-18°C$. Density $= 0.978$ g/mL. Mass spectrum: $m/z = 115$.

C. N,N'-BIS(2-MERCAPTOETHYL)-1,5-DIAZACYCLOOCTANE, H_2-BME-DACO

H_2-bme-daco

■ **Caution.** *Several of the reagents in this section are known or suspected carcinogens. All work should be performed in a fume hood, and suitable protective clothing worn.*

Procedure

Using standard anaerobic techniques, a 100-mL Schlenk flask equipped with a stir bar is charged with 50 mL dry benzene and 2 mL (1.96 g, 0.0172 mol) distilled free amine. If crude daco is used, the product will be less pure and of lower yields. The mixture is warmed to 50–60°C under N_2. Three 1-mL (0.05 mol) portions of ethylene sulfide* are added, allowing 20 min reaction time between additions. The mixture is then heated under N_2 for 1 h. Complete reaction is indicated by the formation of a finely divided white precipitate after the final addition. The reaction mixture is filtered anaerobically through a bed of celite in a glass-fritted funnel. Solvent is removed under vacuum while continuing to heat at 50–60°C. The H_2-bme-daco is obtained as a colorless to pale yellow oil. Irrespective of color, this material is of suitable quality to be converted to the nickel complex. If distilled daco is used, the product is quite pure. Attempts at vacuum distillation $(bp_{0.1\,mmHg} = 135°C)$ resulted in partial decomposition. Yield: 3.21 g (80%).

Properties

The H_2-bme-daco is a pale yellow oil (may also be light pink) and is soluble in toluene, benzene, methanol, ether, and chloroform. The ^1H NMR spectrum in $CDCl_3$ shows a quintet at 1.6δ, 4H, broad singlet at 2.1δ, 2H, a broad intense multiplet at 2.7δ, and a multiplet at 2.9δ, 4H. Mass spectrum: $m/z = 235$.

D. ISOBUTYLENE SULFIDE

Procedure

Isobutylene sulfide is prepared according to published procedures from isobutylene oxide and potassium thiocyanate.[8] A 250-mL, round-bottomed flask, fitted with a magnetic stir bar and addition funnel, is charged with 97 g (1.0 mol) potassium thiocyanate and 100 mL water. With vigorous stirring, 72 g (90.0 mL, 1.0 mol) isobutylene oxide[†] is added dropwise over 5 h via the addition funnel. Owing to the exothermicity of the reaction, slow addition is

---

* Aldrich Chemical Company Milwaukee, WI 53233.
[†] TCI America, Portland, OR 97203.

required to maintain the temperature below 40°C and prevent loss of isobutylene oxide. Using a separatory funnel, the top layer is separated and transferred to a 250-mL, round-bottomed flask containing an additional 50 g (0.52 mol) potassium thiocyanate dissolved in 100 mL water. The solution is again stirred for 5 h or overnight. The top layer (ca. 80–90 mL) is separated and dried by careful addition of small (~ 10 mg) portions of CaH_2 until no H_2 evolution is observed. Fractional distillation at ambient pressure first gives unreacted isobutylene oxide at 52–55°C followed by isobutylene sulfide at 84–86°C. Yield: 64 g (73%).

Properties

Isobutylene sulfide is a clear, colorless liquid with a boiling point of 84–86°C and a density of 0.90 g mL. The liquid has a characteristic sulfur stench. The ^1H NMR in C_6D_6 displays singlets at 2.1δ, 2H, and 1.3δ, 6H. During storage a white precipitate may form, but this does not seem to affect the use of isobutylene sulfide in further reactions. Storing the sample in a freezer ($-10°C$) slows the rate of decomposition.

E. N,N'-BIS(2-MERCAPTO-2-METHYLPROPANE)-1,5-DIAZACYCLOOCTANE, H₂-BME*-DACO

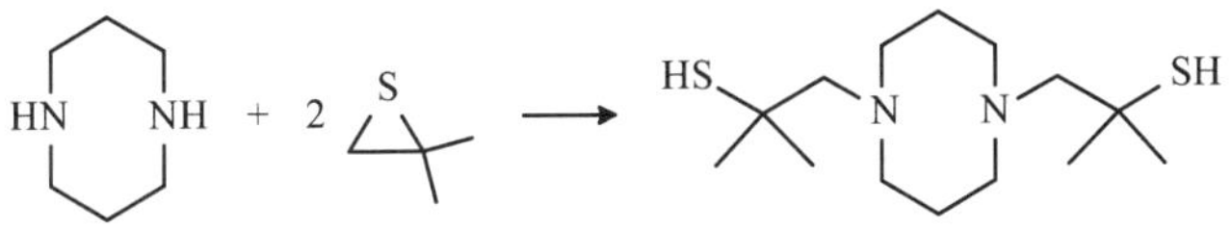

H₂-bme*-daco

■ **Caution.** *Several of the reagents in this section are known or suspect carcinogens. All work should be performed in a fume hood, and suitable protective clothing worn.*

Procedure

A 100-mL Schlenk flask equipped with a magnetic stir bar and attached to an N_2 line with oil bubbler for pressure release is charged with 50 mL dry benzene and 2.0 mL (1.96 g, 0.0172 mol) distilled free amine. Crude daco may be used, but leads to less pure product and lower yields. The mixture is warmed to 50–60°C under N_2 and a single 3.4 mL (0.034 mol) portion of isobutylene sulfide is added. The mixture is then heated under N_2 for 12 h.

* TCI America, Portland, OR 97203.

Solvent is removed under vacuum while continuing to heat at 50–60°C. The H_2-bme*-daco which remains is a colorless to pale yellow oil, and is of suitable quality to be converted to the nickel complex. If distilled daco is used, the product is pure. Yield: 3.3 g (65%).

Properties

The H_2-bme*-daco is a colorless to yellow oil and is soluble in toluene, benzene, methanol, ether, and chloroform. The ^1H NMR spectrum in $CDCl_3$ shows a pair of singlets at 1.362δ and 1.368δ, 12H, a multiplet at 1.56δ, 4H, a pair of singlets at 2.62δ and 2.68δ, 4H, and a multiplet at 2.95δ, 8H. Mass spectrum: $m/z = 291$.

F. [*N,N'*-BIS(2-MERCAPTOETHYL)-1,5-DIAZACYCLOOCTANATO]-NICKEL(II), (BME-DACO)NI^{II}, AND [*N,N'*-BIS(2-MERCAPTO-2-METHYLPROPANE)-1,5-DIAZACYCLOOCTANATO]NICKEL(II), (BME*-DACO)NI^{II}

Procedure

Typically, 1.96 g of daco is converted to H_2-bme-daco or H_2-bme*-daco as described above, taking precaution to avoid air oxidation. Under anaerobic conditions, 3–3.5 g of the ligand is weighed in a tared 500-mL Schlenk flask to which 50-mL distilled toluene is added. Ni(acac)$_2$ (1 equiv. based on daco) is dissolved in 200-mL distilled toluene and the solution carefully transferred dropwise via cannula into the ligand solution over a period of 3 h with vigorous stirring. At this time, a purple precipitate should be observed. If during addition the solution appears to be turning brown, no more nickel should be added. The solution is allowed to stir for 1 h after all of the nickel has been added. Following metallation, the air-sensitive handling techniques may be relaxed. The mixture is filtered on a Buchner funnel in air and washed thoroughly with toluene to obtain the product as a purple solid. The product may be purified further by silica gel column chromatography. The purple product is dissolved in a minimum of MeOH and loaded onto a base-washed silica gel (60–200 mesh) column (12 × 1 in. column, MeOH as eluent). Failure to wash silica with base leads to decreased product yield. Alternately, the purple product may be chromatographed on a neutral alumina column (12 × 1 in.) with MeCN as eluent. Removal of solvent under vacuum yields pure product. Both (bme-daco)Ni^{II} and (bme*-daco)Ni^{II} may be recrystallized from acetonitrile/ether to obtain X-ray quality crystals.

* TCI America, Portland, OR 97203.

Properties

The (bme-daco)Ni^{II} complex is a purple solid that is soluble in a variety of solvents, including methanol, acetonitrile, water, and acetone. The UV-vis in MeCN displays a broad band at 506 nm (640) with a shoulder at 602 nm. The ^{1}H NMR spectrum is complex due to a tetrahedral twist in the N$_2$S$_2$ plane. Mass spectrum: $m/z = 291$.

Anal. Calcd. for (bme-daco)Ni^{II}. (Found): C, 41.3 (40.8); H, 6.92 (6.90); N, 9.62 (9.42).

The (bme*-daco)Ni^{II} complex is a purple/blue solid that is soluble in a variety of solvents including methanol, acetonitrile, water, and acetone. The UV-vis in MeOH displays broad bands at 352 nm (192) and 486 nm (70). The ^{1}H NMR spectrum in CD$_3$OD displays a singlet at 1.42δ, 12H, a multiplet between 1.9 and 2.1δ, 2H, a singlet at 2.55δ, 4H, a multiplet at 2.55–2.70δ, 2H, and a multiplet at 3.27–3.43δ, 8H. Mass spectrum: $m/z = 346$.

Anal. Calcd. for (bme*-daco)Ni^{II}. (Found): C, 48.5 (48.6); H, 8.00 (7.86); N, 8.00 (7.83).

Whereas the (bme-daco)Ni^{II} slowly reacts with O$_2$ in MeCN solution,[9,10] the (bme*-daco)Ni^{II} is air stable in the absence of extensive photolysis or production of ^{1}O$_2$.[11-13] Both, as well as the palladium analog, react with H$_2$O$_2$, yielding S-oxygenates.[9,11,14,15] Other metal derivatives of these ligands have also been reported.[16-18]

References

1. K. Kanamori, W. E. Broderick, R. F. Jordan, R. D. Wilett, and J. J. Legg, *J. Am. Chem. Soc.*, **108**, 7122 (1986) and references therein.
2. E. L. Buhle, A. M. Moore, and F. Y. Wiselogle, *J. Am. Chem. Soc.*, **65**, 29 (1943).
3. R. D. Hancock, personal communication.
4. R. P. Houser, V. G. Young, and W. B. Tolman, *J. Am. Chem. Soc.*, **118**, 2101 (1996).
5. H. Stetter and K. Findeisen, *Chem. Ber.*, **98**, 3228 (1965).
6. D. K. Mills and M. Y. Darensbourg, unpublished results.
7. A. P. Terentev, A. N. Kost, and K. I. Chursina, *C. A.*, **45**, 7008g (1951).
8. H. R. Snyder, J. M. Stewart, and J. B. Ziegler, *J. Am. Chem. Soc.*, **69**, 2672 (1947).
9. P. J. Farmer, T. Solouki, D. K. Mills, T. Soma, D. H. Russell, J. H. Reibenspies, and M. Y. Darensbourg, *J. Am. Chem. Soc.*, **114**, 4601 (1992).
10. P. J. Farmer, T. Solouki, T. Soma, D. H. Russell, and M. Y. Darensbourg, *Inorg. Chem.*, **32**, 4171 (1993).
11. R. M. Buonomo, I. Font, M. J. Maguire, J. H. Reibenspies, T. Tuntulani, and M. Y. Darensbourg, *J. Am. Chem. Soc.*, , **117**, 963 (1995).

* TCI America, Portland, OR 97203.

12. C. A. Grapperhaus, M. Y. Darensbourg, L. W. Sumner, and D. H. Russell, *J. Am. Chem. Soc.*, **118**, 1791 (1996).
13. C. A. Grapperhaus, M. J. Maguire, T. Tuntulani, and M. Y. Darensbourg, *Inorg. Chem.*, **36**, 1860 (1997).
14. I. Font, R. M. Buonomo, J. H. Reibenspies, and M. Y. Darensbourg, *Inorg. Chem.*, **33**, 5897 (1993).
15. T. Tuntulani, G. Musie, J. H. Reibenspies, and M. Y. Darensbourg, *Inorg. Chem.*, **34**, 6279 (1995).
16. T. Tuntulani, J. H. Reibenspies, P. J. Farmer, and M. Y. Darensbourg, *Inorg. Chem.*, **1992**, 31, 3497 (1992).
17. D. K. Mills, Y.-M. Hsiao, P. J. Farmer, E. V. Atnip, J. H. Reibenspies, and M. Y. Darensbourg, *J. Am. Chem. Soc.*, **113**, 1421 (1991).
18. R. P. Houser and W. B. Tolman, *Inorg. Chem.*, **34**, 1632 (1995).

16. POLYDENTATE THIOLATE AND SELENOLATE LIGANDS, RN(CH₂CH₂S(Se)⁻)₂, AND THEIR DIMERIC AND MONONUCLEAR Ni(II) COMPLEXES

Submitted by SURANJAN B. CHOUDHURY,* CHRISTIAN B. ALLAN,*
and MICHAEL J. MARONEY*
Checked by ALDEN D. WOODWARD[†] and C. ROBERT LUCAS[†]

Reactions of Ni^{2+} salts with alkylthiolates and alkylselenolates typically yield polymeric products. The extent of polymerization can be controlled by using polydentate thiolates that incorporate other ligand donor atoms. Here the synthesis of representative complexes of a series of tridentate ligands, $RN[CH_2CH_2S(Se)^-]_2$, that lead to the formation of μ-dichalcogenolato bridged dimers are reported. Two routes are described, one for R = Me that utilizes bis(2-chloroethyl)methylamine as a starting material, and one for $R = CH_2CH_2SMe$ that utilizes the corresponding amine as a starting material. The latter synthesis constitutes a significant improvement over a previously published synthesis of the ligand.[1] The resulting dimeric Ni(II) complexes may be cleaved by the addition of cyanide to form mononuclear *trans*-chalcogenolates. The synthesis of representative mononuclear complexes is also detailed.

* Department of Chemistry, University of Massachusetts, Amherst, MA, 01003-4510.
[†] Department of Chemistry, Memorial University of Newfoundland, St. John's, Newfoundland, Canada A1B 3X7.

General Materials and Procedures

Commercially available reagents are used without further purification. Organic reagents were obtained from Aldrich Chemical Company unless otherwise noted.[‡] Solvents are rigorously dried[2] and distilled under dry dinitrogen. Anhydrous nickel acetate is prepared by heating samples of $Ni(OAc)_2 \cdot 4H_2O$ at 100°C over phosphorus pentoxide in vacuo overnight. All the operations are carried out under an inert atmosphere of dry dinitrogen using standard Schlenk techniques[3] unless otherwise noted.

■ **Caution.** *Handling organic chlorides, thiols, and selenols requires the use of protective clothing and gloves in a well-ventilated hood.*

A. Bis{[(μ-2-MERCAPTOETHYL)(2-MERCAPTOETHYL)-METHYLAMINATO(2-)]NICKEL(II)}[4]

$$MeN(CH_2CH_2OH)_2 + 2SOCl_2 \longrightarrow [MeN(CH_2CH_2Cl)_2 \cdot HCl] + 2SO_2 + HCl \quad (1)$$

$$[MeN(CH_2CH_2Cl)_2 \cdot HCl] + 2SC(NH_2)_2 + 4H_2O \xrightarrow{HCl}$$
$$[MeN(CH_2CH_2SH)_2 \cdot HCl] + 2CO_2 + 4NH_4Cl \quad (2)$$

$$2[MeN(CH_2CH_2SH)_2 \cdot HCl] + 2Et_3N + 2Ni(OAc)_2 \longrightarrow$$

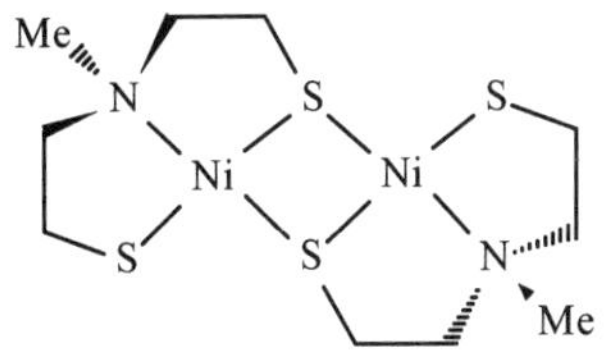

$+ 2[Et_3N \cdot HCl] + 4AcOH \quad (3)$

■ **Caution.** *Bis(2-chloroethyl)methylamine hydrochloride is a vesicant.*

Procedure

Bis(2-chloroethyl)methylamine hydrochloride. *N*-methyldiethanolamine (59.58 g, 0.5 mole) is dissolved in 50 mL of chloroform and added dropwise with stirring to a solution of freshly distilled thionyl chloride (158.17 g, 1.33 mole) in 70 mL of chloroform. After the addition is complete, the solution is refluxed for 3 h and then left to stir overnight at room temperature. The solvent and the excess thionyl chloride are removed by rotary evaporation to yield an off-white solid. This product is recrystallized from acetone and

---

[‡] Aldrich Chemical Company, Inc., 1001 West Saint Paul Avenue, Milwaukee, WI 53233.

washed with a copious amount of ether (5 × 100 mL) to obtain bis(2-chloro-ethyl)methylamine hydrochloride as a white, crystalline solid. Yield = 86.7 g (90%).

Anal. Calcd. for $C_5H_{12}NCl_3$: C, 31.19; H, 6.28; N, 7.28. Found (University of Massachusetts Microanalytical Laboratory): C, 31.08; H, 6.19; N, 7.20.

Properties

Bis(2-chloroethyl)methylamine hydrochloride is a white solid that is soluble in alcohols and very hygroscopic. It is a vesicant and requires careful handling with the use of gloves.

Procedure

Bis(2-mercaptoethyl)methylamine hydrochloride. Bis(2-chloroethyl)methylamine hydrochloride (9.76 g, 50.6 mmole) is dissolved in ethanol (36 mL) and thiourea* (8.10 g, 106.6 mmole) is added. The solution is refluxed for 4.5 h. Upon cooling to room temperature, a gummy solid is formed. Dry hydrogen chloride gas is bubbled through the solution for 30 min and the mixture is stored overnight at 5°C. (Note: it is important to vigorously mix the solid and solution while bubbling HCl.) The liquid phase is decanted and the remaining solid is dissolved in 20% aqueous sodium hydroxide solution (16 mL), heated for 12 min to 90–95°C and then allowed to cool to room temperature. The solution is then extracted with toluene (4 × 10 mL) and the organic layer is dried over anhydrous sodium sulfate for 2.5 h and then filtered. Dry hydro-gen chloride gas is bubbled through the toluene solution for 30 min while holding the temperature at 5°C. The product separates as a white micro-crystalline solid, which is collected by filtration under a nitrogen atmo-sphere, washed several times with ether (5 × 25 mL), and dried in vacuo. Yield = 2.93 g (33%).

Anal. Calcd. for $C_5H_{14}NS_2Cl$: C, 31.99; H, 7.52; N, 7.46. Found (University of Massachusetts Microanalytical Laboratory): C, 33.91; H, 7.55; N, 7.70.

Properties

The solid hydrochloride salt of the ligand is hygroscopic, but not otherwise air sensitive, and may be handled and stored in air as long as it is protected

---

* Acros Organics N.V., Pharmaceuticalaan 3, B-3340 Geel, Belgium.

from moisture. It is soluble in alcohols. IR(cm^{-1}) in KBr: 3005w, 2940m, 2680s, 2620s, 2560s, 2470s, 1475m, 1440m, 960w, 940w, 725w.

Procedure

Preparation of the complex. Anhydrous nickel acetate (0.22 g, 1.2 mmole) is dissolved in warm methanol (3 mL) and added dropwise to a stirred solution of bis(2-mercaptoethyl)methylamine hydrochloride (0.22 g, 1.2 mmole) in warm methanol.

Triethylamine (0.16 mL, 1.2 mmole) is then added to the solution with stirring. The mixture is allowed to stand at room temperature for 24 h, during which time a microcrystalline solid separated. The product is isolated by filtration, washed with ether, and dried in vacuo. Yield = 0.29 g (62%).

Anal. Calcd. for $C_{10}H_{22}N_2S_4Ni_2$: C, 28.88; H, 5.33; N, 6.73. Found (University of Massachusetts Microanalytical Laboratory): C, 28.76; H, 5.17; N, 6.60.

Properties

The compound is diamagnetic and is air stable in solid state. It is soluble in dichloromethane, dimethylformamide, dimethylsulfoxide, and sparingly soluble in acetonitrile. These deep red-brown solutions are mildly air-sensitive due to the oxidation of the thiolate ligands to sulfinates.[5] IR (cm^{-1}) in KBr: 2960m, 2920s, 2860m, 1550w, 1460s, 1315m, 1220m, 1200m, 1045m, 765s. UV-vis (DMF), λ_{max} in nm (ε in cm^{-1} M^{-1}) 552(570), 432(sh), 376(sh), 340(sh), 312(5970), 289(6280).

B. Bis{[(μ-2-(HYDROSELENO)ETHYL)(2-(HYDROSELENO)-ETHYL)METHYLAMINATO(2-)]NICKEL(II)}[4]

$[MeN(CH_2CH_2Cl)_2 \cdot HCl]$ $+$ Et_3N $+$ $2KSeCN$ $\longrightarrow$

$\qquad$ $MeN(CH_2CH_2SeCN)_2$ $+$ $[Et_3N \cdot HCl]$ $+$ 4AcOH   (1)

$MeN(CH_2CH_2SeCN)_2$ $+$ $2NaBH_4$ $\longrightarrow$ $MeN(CH_2CH_2SeH)_2$ $+$ $2NaBH_3CN$ (2)

$2MeN(CH_2CH_2SeH)_2$ $+$ $2[Ni(OAc)_2 \cdot 4H_2O]$ $\longrightarrow$

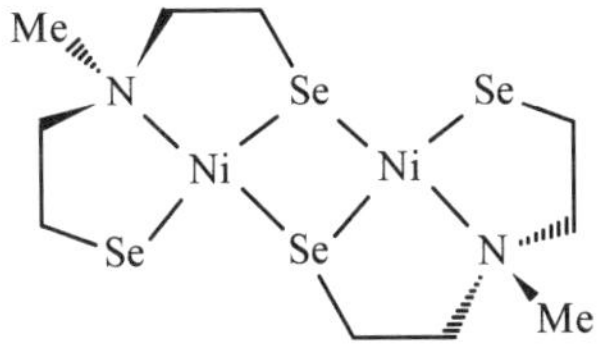

$+$ 4AcOH $+$ 8H$_2$O   (3)

Procedure

Bis(2-(hydroseleno)ethyl)methylamine. Dry triethylamine (1.80 g, 17.8 mmole) is added to a solution of bis(2-chloroethyl)methylamine hydrochloride (see A.1 above) (3.340 g, 17.4 mmole) in methanol (25 mL). This solution is then added dropwise to a warm (ca. 60°C), stirred solution of potassium seleno-cyanate* (5.0 g, 34.7 mmole) in methanol (40 mL). A white precipitate (KCl) immediately forms. The reaction mixture is stirred for an additional 10 h at 60°C, filtered, and concentrated under reduced pressure to 35 mL. After cooling to 15°C, sodium borohydride (2.5 g, 66 mmole) is gradually added to the solution with vigorous stirring. This mixture is allowed to stir for an additional hour and filtered to remove any undissolved material. The filtrate is carefully neutralized with dilute acetic acid (1:1 acetic acid in methanol). Owing to the extreme oxygen-sensitivity of the selenol, this solution, containing bis[(2-(hydroseleno)ethyl]methylamine is used directly in the synthesis of the nickel complex.

Preparation of the complex. The solution of bis[2-(hydroseleno)ethyl]methyl-amine prepared above is gradually added to a solution of nickel acetate tetrahydrate (4.32 g, 17.4 mmole) in ethanol (100 mL) and stirred for 1 h. The solvent is removed under reduced pressure to obtain a dark mass, which is extracted with benzene (10×80 mL). These green extracts are combined, and the benzene is removed under vacuum. The resulting solid is washed with 10 mL of methanol and 50 mL of ether and dried in vacuo to obtain the desired product. Yield = 2.01 g (38.2%).

Anal. Calcd. for $Ni_2C_{10}H_{22}N_2Se_4$: C, 19.89; H, 3.64; N, 4.64. Found (University of Massachusetts Microanalytical Laboratory): C, 20.43; H, 3.55; N, 4.46.

Properties

The compound is diamagnetic and is air-stable in solid state. It is sol-uble in dichloromethane, dimethylformamide, and dimethylsulfoxide and sparingly soluble in acetonitrile, forming green solutions. The DMF solution is air stable. IR (cm^{-1}) in KBr: 2910w, 2860w, 2830w, 1470m, 1450s, 1440s, 1420s, 1410s, 1345w, 1310s, 1260m, 1220s, 1200m, 1170m, 1135w, 1050m, 1030s, 960w, 950w, 910s, 880s, 830s, 750s, 570w, 520m, 460w. UV-vis (DMF), λ_{max} in nm (ε in cm^{-1} M^{-1}) 475(sh), 420(sh), 384(sh), 353(5790), 310(6320), 273(sh).

---

* Aldrich Chemical Company, Inc., 1001 West Saint Paul Avenue, Milwaukee, WI 53233.

C. Bis{[(μ-2-MERCAPTOETHYL)(2-MERCAPTOETHYL)-METHYLTHIOETHYLAMINATO (2-)]NICKEL(II)}[6]

$$[HSCH_2CH_2NH_2 \cdot HCl] \ + \ 2NaOEt \ + \ CH_3I \ \longrightarrow$$
$$CH_3SCH_2CH_2NH_2 \ + \ NaCl \ + \ NaI \ + \ 2EtOH \quad (1)$$

$$CH_3SCH_2CH_2NH_2 \ + \ 2\left[\begin{array}{c}\nabla \\ S\end{array}\right] \ \longrightarrow \ CH_3SCH_2CH_2N(CH_2CH_2SH)_2 \quad (2)$$

$$2[CH_3SCH_2CH_2N(CH_2CH_2SH)_2] \ + \ 2Ni(OAc)_2 \ \longrightarrow$$

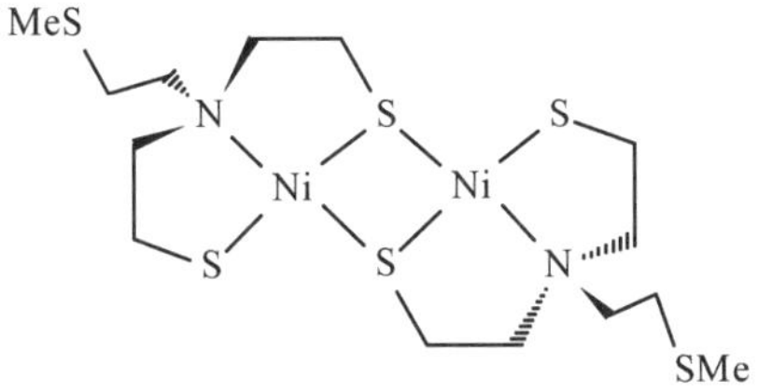

$$+ \ 4AcOH \quad (3)$$

Procedure

2-Methylmercaptoethylamine. At room temperature, 2-aminoethanethiol hydrochloride (5.68 g, 0.05 mole) is dissolved in dry ethanol (60 mL). To this solution, a freshly prepared ethanolic solution of sodium ethoxide [120 mL EtOH + 2.30 g (0.10 mole) of sodium metal] is added dropwise with stirring so that the temperature of the mixture does not rise above 30°C. This mixture is stirred for 30 min and then filtered to remove the NaCl that precipitated. The filtrate is concentrated under reduced pressure to about 80 mL. Freshly distilled methyl iodide* (7.097 g, 0.05 mole) is then added slowly with stirring. After stirring for 6 h at 60–65°C, the solvent is removed under reduced pressure, giving a pale yellow liquid that contained a white solid (NaI). Saturated aqueous sodium hydroxide (25 mL) is added to the yellow liquid at 20°C and the mixture is stirred for 5 min and stored at 10°C. A pale yellow oil separated within 1.5 h. The oil is extracted with ether (2 × 20 mL) and the combined extracts are dried over anhydrous magnesium sulfate overnight, filtered, and the ether is subsequently removed under reduced pressure to obtain 2-methylmercaptoethylamine as a yellow liquid. Yield = 2.65 g, (58%).

* Fisher Scientific, 711 Forbes Avenue, Pittsburgh, PA 15219-4785.

Properties

The liquid has a characteristic pungent smell and undergoes slow decomposition in air at elevated (> 50°C) temperatures. It is stable for months under dry dinitrogen or argon when stored in a refrigerator. ^{1}H NMR (chloroform-d): δ 2.5–2.6(mult), 2.8–2.9(mult), 2.06(s), 1.40(s).

■ **Caution.** *The synthesis of N,N-bis(2-mercaptoethyl)2-methylthioethylamine involves a sealed-tube reaction that should be conducted behind an explosion shield.*

Procedure

N,N-Bis(2-mercaptoethyl)2-methylthioethylamine. In a flame-dried and argon-flushed tube (15 × 3.75 cm) fitted with a teflon needle valve, a solution of ethylene sulfide (4.40 g, 73 mmole) in toluene (13 mL) is added to a solution of 2-methylmercaptoethylamine (2.65 g, 29 mmole) in toluene (10 mL). The valve is then closed to seal the tube and the mixture is allowed to stand for 8 h at room temperature. The sealed tube containing the reaction mixture is then heated to 110–130°C for 30–40 h. After cooling to room temperature, the tube is attached to a Schlenk line and opened under N_2. The product mixture is filtered to remove a white precipitate (ethylene sulfide polymer), and the solvent is removed under vacuum. The residual liquid is fractionally distilled at reduced pressure (114–116°C@0.03 mm) to yield N,N-bis(2-mercaptoethyl)2-methylthioethylamine as a colorless liquid. Yield = 3.56 g (58%).

If the Ni(II) complex, bis{[(μ-2-mercaptoethyl)(2-mercaptoethyl)methyl-thioethylaminato(2-)]nickel(II)}, is the desired product, the crude product obtained prior to fractional distillation may be used in the procedure detailed below.

[Note: If the reaction temperature is too low, or the reaction is not given sufficient time, the secondary amine, N-(2-mercaptoethyl)2-methylthioethyl-amine, is a major product. It is characterized by a ^{1}H-NMR (chloroform-d): 2.75 (A_2B_2 pattern) 2.12 (s), 1.69 (s), and gives an uncharacterized Ni complex that is soluble in methanol.]

Properties

The liquid has strong smell of thiol and undergoes aerial oxidation at room temperature. It should be handled under dry dinitrogen or argon. ^{1}H NMR (chloroform-d): δ 2.5–2.9(mult), 2.12 (s), 1.04 (s).

Procedure

Synthesis of the complex. In a 50-mL Schlenk flask, *N,N*-bis(2-mercapto-ethyl)2-methylthioethylamine (0.1 g, 0.47 mmole) is dissolved in methanol (10 mL) and subsequently cooled to 5°C in an ice bath. To the cooled solution, a solution of anhydrous nickel acetate (84 mg, 0.47 mmole) in methanol (10 mL) is added dropwise with rapid stirring. The product, a red-black, microcrystalline solid, forms immediately and is collected by filtration, washed several times with small amounts of methanol, and dried in vacuo. Yield = 160 mg (65%). The product obtained from purified ligand is pure by elemental analysis. If crude ligand is used, the complex may be purified by recrystallization from CH_2Cl_2 (minimum volume for dissolution) upon addition of petroleum ether or hexane (five times the volume of CH_2Cl_2 used).

Anal. Calcd. for $C_{14}H_{30}N_2S_6Ni_2$: C, 31.36; H, 5.64; N, 5.22. Found (University of Massachusetts Microanalytical Laboratory): C, 31.34; H, 5.03; N, 5.19.

Properties

The solid complex is diamagnetic and not air-sensitive. It may be filtered, handled, and stored in air. It is soluble in dichloromethane, dimethylformamide, and dimethylsulfoxide and sparingly soluble in acetonitrile. These deep red-brown solutions are mildly air-sensitive due to the oxidation of the thiolate ligands to sulfinates.[5] IR (cm^{-1}) in KBr: 1295w, 1258w, 1017m, 968w, 740m, 571m. UV-vis (CH_2Cl_2): λ_{max} in nm (ε in cm^{-1} M^{-1}) 547 (705), 440(970), 384(sh).

D. TETRAETHYLAMMONIUM {*TRANS*-[BIS-(2-MERCAPTOETHYL) METHYTHIOETHYLAMINATO(2-)](CYANATO) NICKELATE(II)}[7]

$\{Ni[RN(CH_2CH_2S)_2]\}_2$ + $2Et_4N(CN)$ $\longrightarrow$ $2(Et_4N)$

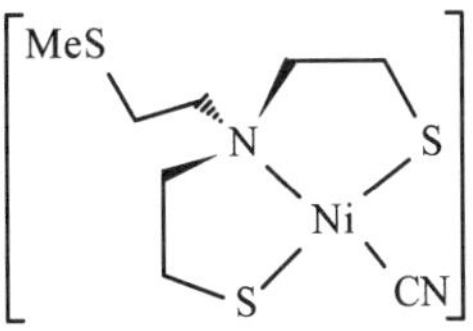

■   **Caution.** *Tetraethylammonium cyanide is a hygroscopic solid that is readily absorbed through skin and is toxic by this route, via inhalation of the dust or if swallowed. Use suitable protective clothing and handle in a well-ventilated hood.*

Procedure

A solution of $Et_4N(CN)$ (0.058 g, 0.37 mmole) in DMF (2 mL) is added to a solution of bis{[(μ-2-mercaptoethyl)(2-mercaptoethyl)methylthioethyl-aminato(2-)]nickel(II)} (see Section C.3) (0.10 g, 0.19 mmole) in DMF (5 mL) which produces a dark green solution. After stirring for 0.5 h, the solvent is removed under vacuum and the green solid is redissolved in CH_3CN (5 mL) and filtered. Toluene (80 mL) is added, and the mixture is stored at $-20°C$ in a freezer. After several days, green crystals of the product form. The crystals are collected by filtration, washed with Et_2O (2 × 5 mL), and dried under vacuum. Yield = 0.15 g (94%).

Anal. Calcd. for $C_{16}H_{35}N_3S_3Ni$: C, 45.29; H, 8.31; N, 9.90. Found (University of Massachusetts Microanalytical Laboratory): C, 45.29; H, 8.14; N, 9.72.

Properties

Tetraethylammonium {*trans*-[bis-(2-mercaptoethyl)methythioethylaminato-(2-)](cyanato)nickelate(II)}[7] is a diamagnetic, green, deliquescent solid that readily reacts with air, forming a monosulfinato complex.[7] The solid should be handled under dry dinitrogen or argon. The complex is soluble in acetone, acetonitrile, DMF, and DMSO. IR (cm^{-1}) in Nujol: 2730w, 2100s ($ν_{CN}$), 1720w, 1590w, 1300m, 1270w, 1210w, 1185m, 1085m, 1040m, 980m, 920w, 900w, 860w, 800s, 750s, 725s.

E. TETRAETHYLAMMONIUM {*TRANS*-[BIS-(2-(HYDROSELENO) ETHYL)METHYLAMINATO-(2-)](CYANATO)NICKELATE(II)}[4]

$\{Ni[RN(CH_2CH_2Se)_2]\}_2$ + $2Et_4N(CN)$ $\longrightarrow$ $2(Et_4N)$

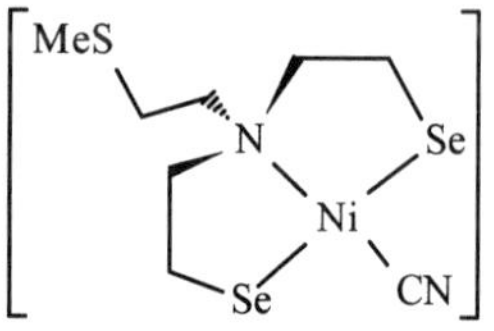

■ **Caution.** *Tetraethylammonium cyanide is a hygroscopic solid that is readily absorbed through skin and is toxic by this route, via inhalation of the dust or if swallowed. Use suitable protective clothing and handle in a well-ventilated hood.*

Procedure

A solution of $Et_4N(CN)$* (62.5 mg, 0.4 mmole) in acetonitrile (2 mL) is added to a solution of bis{[(μ-2-(hydroseleno)ethyl)(2-(hydroseleno)ethyl)methyl-aminato(2-)]nickel(II)} (see Section B.2) (120.6 mg, 0.2 mmole) in acetonitrile (20 mL). The resulting green solution is stirred for 0.5 h prior to the removal of the solvent under reduced pressure. The diamagnetic green solid obtained is redissolved in acetonitrile (5 mL) and precipitated by the addition of ether. The microcrystalline product is collected by filtration, washed with ether, and dried in vacuo over P_4O_{10}. Yield = 118 mg (90%).

Anal. Calcd. for $NiC_{14}H_{31}N_3Se_2$: C, 36.70; H, 6.82; N, 9.17. Found (University of Massachusetts Microanalytical Laboratory): C, 36.99; H, 6.71; N, 9.07.

Properties

Tetraethylammonium {*trans*-[bis-(2-(hydroseleno)ethyl)methylaminato-(2-)](cyanato)nickelate(II)} is a diamagnetic, green, hygroscopic solid that is much less air-sensitive compared to thiolate analogs (e.g., see Section D). The complex is soluble in acetone, acetonitrile, DMF, and DMSO, where it forms modestly air-sensitive solutions.[4] IR (cm^{-1}) in KBr: 2950w, 2840w, 2090s (ν_{CN}), 1485s, 1435m, 1390m, 1365m, 1310m, 1260w, 1220s, 1175s, 1160s, 1050w, 1030m, 1020w, 1000s, 910m, 880m, 840m, 780s, 765s, 610w, 520w, 450s.

References

1. J. L. Corbin, K. F. Miller, N. Pariyadath, S. Wherland, A. E. Bruce, and E. I. Stiefel, *Inorg. Chim. Acta*, **90**, 41 (1984).
2. *Vogel's Textbook of Practical Organic Chemistry*, 5th ed., B. S. Furniss, A. J. Hannaford, P. W. G. Smith, and A. R Tatchell, Eds., Longman Group UK Limited, Harlow, Essex CM20 2JE, England, 1989, p. 395.
3. D. F. Shriver and M. A. Drezdon, *The Manipulation of Air-Sensitive Compounds*, 2nd ed., John Wiley, New York, 1986, p. 326.
4. S. B. Choudhury, M. A. Pressler, S. Mirza, R. O. Day, and M. J. Maroney, *Inorg. Chem.*, **33**, 4831 (1994).
5. S. A. Mirza, R. O. Day, and M. J. Maroney, *Inorg. Chem.*, **35**, 1992 (1996).
6. M. Kumar, R. O. Day, G. J. Colpas, and M. J. Maroney, *J. Am. Chem. Soc.*, **111**, 5974 (1989).
7. S. A. Mirza, M. A. Pressler, M. Kumar, R. O. Day, and M. J. Maroney, *Inorg. Chem.*, **32**, 977 (1993).

* Fluka Chemical Corporation, 980 South 2nd Street, Ronkonkoma, NY 11779-7238.

17. POLY[(METHYLTHIO)METHYL]BORATES AND REPRESENTATIVE METAL DERIVATIVES

Submitted by CHARLES G. RIORDAN,** PINGHUA GE,*
and CARL OHRENBERG*
Checked by WEN-FENG LIAW†

The coordination chemistry of the ligands 1,4,7-trithiacyclononane, 1,4,7-triazacyclononane, and hydridotris(pyrazolyl)borate, which present a face-capping, tridentate donor set to metal ions, has been extensively developed over the last several decades.[1] The ligand field properties of these systems have permitted them to find utility in bioinorganic and organometallic chemistry and in coordination compounds which possess new and unusual magnetic properties.[2] Herein we present a convenient, general synthetic route to the anions, tetrakis[(methylthio)methyl]borate and phenyltris[(methylthio)methyl)]borate, thioether analogs of the versatile poly(pyrazolyl)borates. These ligands provide a monoanionic, S_3 donor set. Additionally, the syntheses of two metal ion derivatives, $[(C_4H_9)_4N][(B(CH_2SCH_3)_4)Mo(CO)_3]$ and $[C_6H_5B(CH_2SCH_3)_3]_2Fe$, are presented.[3] In general, the phenyltris[(methylthio)methyl)]borate metal complexes are more soluble than the corresponding tetrakis[(methylthio)methyl)]borate derivatives.

A. [TETRABUTYLAMMONIUM][TETRAKIS(METHYLTHIO)-METHYL)BORATE]

$$CH_3SCH_3 + C_4H_9Li \xrightarrow{\text{TMEDA}} LiCH_2SCH_3 + C_4H_{10}$$

$$4LiCH_2SCH_3 + BF_3 \cdot Et_2O \longrightarrow LiB(CH_2SCH_3)_4 + 3LiF$$

$$LiB(CH_2SCH_3)_4 + [(C_4H_9)_4N]Cl \longrightarrow [(C_4H_9)_4N][B(CH_2SCH_3)_4] + LiCl$$

Procedure

■ **Caution.** *$(CH_3)_2S$ is a toxic and odiferous flammable liquid.*

*Department of Chemistry, Kansas State University, Manhattan, KS 66506-3701.
†Department of Chemistry, National Changhua University of Education, Changhua, Taiwan 50058.
**Department of Chemistry, University of Delaware, Newark, DE 19716.

A 300-mL, 3-necked, round-bottomed flask is equipped with a magnetic stir bar, fitted with rubber scpta, and filled with N_2. The flask is then charged with $(CH_3)_2S^*$ (15 mL, 200 mmole) and $(CH_3)_2NCH_2CH_2N(CH_3)_2^*$ (TMEDA) (19 mL, 125 mmole). An N_2 inlet needle is inserted through the septum of one side neck and an outflow needle is attached to a bubbler containing commercial bleach solution. [The reaction vessel is continuously purged with N_2 at low flow rates throughout the remainder of the procedure until the reaction is terminated with H_2O.] Then $C_4H_9Li^*$ (40 mL, 2.5 M in hexanes) is added dropwise via syringe to the stirring solution over a period of 5 min.[4] As the C_4H_9Li is added, the mixture becomes viscous and turns yellow. The mixture is allowed to stir for 1 h at 25°C, the solution is then heated in a water bath at 45°C for 30 min to drive off unreacted $(CH_3)_2S$. The solution is cooled to $-78°C$ using a dry ice–acetone bath and the bleach bubbler is replaced with a mineral oil bubbler. With the mixture stirring, $BF_3 \cdot Et_2O^*$ (3.0 mL, 25 mmole) is added slowly via syringe. The mixture is warmed to 25°C and stirred for 30 h. At this point, the reaction can be tested for completion by adding several drops of H_2O to a small aliquot of the reaction mixture removed from the flask via syringe. If the addition of H_2O causes bubbling and an elevation in temperature, the reaction is not complete and should be allowed to continue. When little or no bubbling is seen in this test, the flask is opened and the remainder of the procedure is performed aerobically. The reaction is terminated and any remaining reactive species are quenched by slow addition of H_2O to the stirring mixture until all bubbling ceases. An additional 150 mL of H_2O is added, the mixture is transferred to a 500-mL round bottom flask, and volatile organics are removed by rotary evaporation. The aqueous solution is filtered by vacuum filtration through Celite on a glass funnel frit (medium porosity). The floculent white product is precipitated by addition of aqueous $[(C_4H_9)_4N]Cl^*$ (8 g, 29 mmole) and isolated by vacuum filtration on a glass frit (medium porosity), washed with Et_2O (2 × 30 mL), and dried under vacuum. Yield: 0.50 g (4%). It may be recrystallized from acetone–H_2O.

Properties

[Tetrabutylammonium][tetrakis((methylthio)methyl)borate] is an air-stable, white crystalline solid that is soluble in most common organic solvents. Its ^1H NMR spectrum in $CDCl_3$ displays resonances at δ 3.24 (m, NCH_2, 8H), 2.05 (s, SCH_3, 12H), 17.6 (q, $^2J_{BH} = 4.0$ Hz, BCH_2, 8H), 1.63 (m, CH_2, 8H), 1.45 (m, CH_2, 8H), and 1.01 (t, CH_3, 12H) ppm. The ^{13}C NMR data in $CDCl_3$ are δ 58.9 (s, NCH_2), 35.2 (q, $^1J_{BC} = 40.0$ Hz, BCH_2), 24.0 (s, CH_2),

---

*Aldrich Chemical Company, Inc., Milwaukee, WI 53233.

20.5 (s, SCH_3), 19.8 (s, CH_2), and 13.7 (s, CH_3) ppm. The $^{11}B\{^1H\}$ NMR ($CDCl_3$) displays a single resonance, $\delta - 16.3$ ppm (vs. $BF_3 \cdot Et_2O$); mp: 160°C.

B. [TETRABUTYLAMMONIUM][PHENYLTRIS((METHYLTHIO)-METHYL)BORATE]

$$CH_3SCH_3 + C_4H_9Li \xrightarrow{TMEDA} LiCH_2SCH_3 + C_4H_{10}$$

$$3LiCH_2SCH_3 + C_6H_5BCl_2 \longrightarrow Li[C_6H_5B(CH_2SCH_3)_3] + 2LiCl$$

$$Li[C_6H_5B(CH_2SCH_3)_3] + [(C_4H_9)_4N]Cl \longrightarrow$$

$$[(C_4H_9)_4N][C_6H_5B(CH_2SCH_3)_3] + LiCl$$

See cautionary note in Section A.

Procedure

This compound is prepared in a procedure analogous to that for [tetra-butylammonium]{tetrakis[((methylthio)methyl)borate]} with $C_6H_5BCl_2$* replacing $BF_3 \cdot Et_2O$ as the borane source. $(CH_3)_2S$ (9.2 mL, 125 mmole) and TMEDA (12 mL, 78 mmole) are placed into a 300-mL round-bottomed flask filled with N_2 and fitted and equipped as in Section A. The reaction vessel is purged with N_2 throughout the procedure until the reaction is terminated with H_2O. Then C_4H_9Li (19 mL, 2.5 M in hexanes) is added dropwise via syringe with stirring over a 5-min period, during which time the mixture becomes viscous and turns yellow. After 1 h at 25°C, the solution is heated at 45°C for 30 min to drive off unreacted $(CH_3)_2S$. The solution is cooled to $-78°C$ and $C_6H_5BCl_2$ (2.0 mL, 15 mmole) is added dropwise via syringe. Caution should be taken when adding $C_6H_5BCl_2$ to the reaction mixture as it is very reactive, even at $-78°C$. The mixture is warmed to 25°C and stirred for 48 h. The reaction can be tested for completion in the same manner as in Section A. When the reaction has gone to completion, it is terminated by slow addition of H_2O to the stirring mixture until bubbling ceases. The remaining procedures are performed aerobically. An additional 150 mL of H_2O is added, the mixture is transferred to a 500-mL, round-bottomed flask, and volatile organics are removed by rotary evaporation. The aqueous solution is filtered by vacuum filtration through Celite on a glass funnel frit (medium porosity). The flocculent white product is precipitated by addition of aqueous $[(C_4H_9)_4N]Cl$ (5 g, 18 mmole). The product forms an oil initially, which

*Aldrich Chemical Company, Inc., Milwaukee, WI 53233.

solidifies within an hour upon setting. The solid is isolated by vacuum filtration on a glass frit (medium porosity), washed with Et_2O (2×30 mL), and dried under vacuum. Yield: 3.50 g (42%). It may be recrystallized from acetone–H_2O.

Properties

[Tetrabutylammonium][phenyltris((methylthio)methyl)borate] is an air-stable, white crystalline solid that is soluble in most common organic solvents. Its ^1H NMR spectrum in $CDCl_3$ displays resonances at δ 7.54 (d, CH, 2H), 7.04 (t, CH, 2H), 6.84 (t, CH, 1H), 3 15 (m, NCH_2, 8 H), 2.03 (s, SCH_3, 12 H), 1.97 (q, $^2J_{BH} = 4.4$ Hz, BCH_2, 8 H), 1.35 (m, CH_2, 16 H), and 0.97 (t, CH_3, 12 H) ppm. The ^{13}C NMR data in $CDCl_3$ are δ 163.3 (q, $^1J_{BC} = 50.2$ Hz BC), 133.4 (s, CH), 126.1 (s, CH), 122.4 (s, CH), 58.9 (s, NCH_2), 35.8 (q, $^1J_{BC} = 41.3$ Hz, BCH_2), 24.1 (s, CH_2), 20.7 (s, SCH_3), 19.8 (s, CH_2), and 13.9 (s, CH_3); mp: 74°C.

C. [TETRABUTYLAMMONIUM][(TETRAKIS((METHYLTHIO)MET HYL)-BORATE)TRICARBONYLMOLYBDENUM(0)]

$$[(C_4H_9)_4N][B(CH_2SCH_3)_4] + (C_7H_8)Mo(CO)_3 \rightarrow$$

$$[(C_4H_9)_4N][(B(CH_2SCH_3)_4)Mo(CO)_3] + C_7H_8$$

Procedure

All work is carried out in an Ar-filled glovebox. A 100-mL, round-bottomed flask is charged with [tetrabutylammonium][tetrakis((methylthio)methyl)-borate] (200 mg, 0.40 mmole) and $(C_7H_8)Mo(CO)_3$[5*] (109 mg, 0.40 mmole). A 30-mL portion of THF (freshly distilled from sodium benzophenone ketyl) is added to the flask, yielding an orange solution which turns yellow with stirring over a 30-min period. The solvent volume is reduced to 3 mL in vacuo, and the product is precipitated by addition of Et_2O (30 mL). The yellow solid is isolated by vacuum filtration on a glass frit (medium porosity) and washed with Et_2O (2×15 mL). Yield: 216 mg (80%).

Anal. Calcd. for $C_{27}H_{56}BMoNO_3S_4$: C, 47.78; H, 8.10; N, 2.15; S, 18.68. Found: C, 47.85; H, 8.33; N, 2.07; S, 18.92.

Properties

$[(C_4H_9)_4N][(B(CH_2SCH_3)_4)Mo(CO)_3]$ is an air-sensitive, yellow, crystalline solid soluble in THF, acetone, and chlorinated hydrocarbons. Under an Ar

*Strem Chemical Inc., Newburyport, MA 01950.

atmosphere, solid samples are stable for months. Its IR spectrum in THF displays characteristic ν_{CO} bands: 1899(vs), 1784(vs) cm^{-1}. The ^{1}H NMR spectrum (d_6-acetone) is consistent with a static structure in which the methylthiomethyl arms are not equilibrating on the NMR time scale: δ 3.44 (m, NCH_2, 8H), 2.31 (s, SCH_3, 9H), 1.88 (s, SCH_3, 3H), 1.83 (m, CH_2, 8H), 1.60 (q, $^2J_{BH}$ = 4.0 Hz, BCH_2, 6H), 1.43 (m, CH_2, 8H), 1.29 (q, $^2J_{BH}$ = 4.0 Hz, BCH_2, 2H), and 0.97 (t, CH_3, 12H). The ^{11}B$\{^1$H$\}$ NMR (d_6-acetone) displays a single resonance, δ -17.5 ppm (vs. BF$_3 \cdot$ Et$_2$O).

D. BIS[PHENYLTRIS(METHYLTHIO)METHYL)BORATE]IRON(II)

$$2[(C_4H_9)_4N][C_6H_5B(CH_2SCH_3)_3] + [Fe(H_2O)_6][(BF_4)_2] \rightarrow$$

$$[C_6H_5B(CH_2SCH_3)_3]_2Fe + 2[(C_4H_9)_4N][BF_4]$$

Procedure

The following reaction is performed in an Ar-filled glovebox. A solution of [Fe(H$_2$O)$_6$][BF$_4$]$_2$* (200 mg, 0.59 mmole) in THF (20 mL) is placed in a 100-mL, round-bottomed flask equipped with a magnetic stir bar. To this is added [tetrabutylammonium][phenyltris((methylthio)methyl)borate] (610 mg, 1.2 mmole) in THF (20 mL), yielding an emerald green solution. The flask is fitted with a rubber septum and the solution is stirred for 2 h. The solution is exposed to the atmosphere and the remaining procedures are performed aerobically. The solvent is removed by rotary evaporation and the resulting green solid is washed with acetone (2 × 20 mL) and purified by recrystallization from THF-Et$_2$O (5 mL: 50 mL). Yield: 150 mg (45%).

Anal. Calcd. for C$_{24}$H$_{40}$B$_2$S$_6$Fe: C, 48.17; H, 6.74. Found: C, 48.22; H, 6.70.

Properties

[C$_6$H$_5$B(CH$_2$SCH$_3$)$_3$]$_2$Fe is an emerald green, crystalline solid which is soluble in THF and chlorinated hydrocarbons and is stable to both oxygen and moisture. Its electronic spectrum in THF displays two d–d transitions: 627 (ε_{max} = 33 M^{-1} cm^{-1}) and 439 (ε_{max} = 56 M^{-1} cm^{-1}). Its ^{1}H NMR spectrum in CDCl$_3$ displays broad resonances at δ 7.20 (2H), 7.12 (2H), 6.99 (1H), 4.02 (9H), and 3.89 (6H) ppm. The temperature-dependent magnetic moment (1.8 B.M. at 27°C) is indicative of spin-crossover behavior; mp: 208°C.

---

* Strem Chemical Inc., Newburyport, MA 01950.

References

1. (a) S. Trofimenko, *Chem. Rev.*, **93**, 943 (1993). (b) S. P. Cooper, *Acc. Chem. Res.*, **21**, 141 (1988). (c) P. Chaudhuri and K. Wieghardt, *Prog. Inorg. Chem.*, **35**, 329 (1987).
2. (a) F. Mani, *Coord. Chem. Rev.*, **120**, 325 (1992). (b) P. J. Pérez, P. S. White, M. Brookhart, and J. L. Templeton, *Inorg. Chem.*, **33**, 6050 (1994). (c) R. Hotzelmann and K. Wieghardt, *Inorg. Chem.*, **32**, 114 (1993).
3. (a) P. Ge, B. S. Haggerty, A. L. Rheingold, and C. G. Riordan, *J. Am. Chem. Soc.*, **116**, 8406 (1994). (b) C. Ohrenberg, P. Ge, P. Schebler, C. G. Riordan, G. P. A. Yap, and A. L. Rheingold, *Inorg. Chem.*, **35**, 749 (1996).
4. D. J. Peterson, *J. Org. Chem.*, **32**, 1717 (1967).
5. F. A. Cotton, J. A. McCleverty, and J. E. White, *Inorg. Synth.*, **9**, 121 (1967).

18. POLYAZA BINUCLEATING LIGANDS: OBISTREN AND OBISDIEN

Submitted by ARTHUR E. MARTELL* and DIAN CHEN*
Checked by KRISTEN BOWMAN-JAMES†

Introduction

OBISTREN is an octaaza polyamine which is of interest because it is a dinucleating ligand, which forms mononuclear and dinuclear complexes.[1] The protonated forms of the ligand, and the mononuclear and dinuclear metal complexes, may act as hosts that recognize and bind various anions as guests.[2,3] These guests are bound to the protonated nitrogens of the free ligand, or are coordinated to the metal ions in the mononuclear or dinuclear complexes of OBISTREN. The dinuclear cobalt(II) complex of OBISTREN forms an oxygen adduct which can undergo many oxygenation and deoxygenation cycles without appreciable degradation and is therefore an excellent oxygen carrier.[4,5]

OBISDIEN is a hexaaza macrocyclic ligand that forms dinuclear complexes that are considerably stabilized by bridging groups.[6] These groups are bound to the ligand or its metal complexes by hydrogen bonds, coordinate bonds to the metal ion, or a combination of both.[7–10] Thus the guests are recognized by the host if they have the size, conformation, and functional groups necessary to form hydrogen bonds to the ligand and coordinate bonds to the metal centers. Of special interest is the dinuclear cobalt dioxygen

* Department of Chemistry, Texas A & M University, College Station, Texas 77843-3255.
† Department of Chemistry, University of Kansas, Lawrence, Kansas 66045.

complex of OBISDIEN, which readily oxidized coordinated reducing bifunctional substrates which are bound to the metal centers of the same macrocycle.[11–13]

A. 7,19,30-TRIOXA-1,4,10,13,16,22,27,32-OCTAAZABICYLO-[11.11.11]PENTATRIACONTANE (OBISTREN)

Procedure

The following synthesis of OBISTREN is carried out by the coupling of two tripodal subunits, by triple C—N bond formation without the use of high-dilution conditions. The procedure was first described by Dietrich et al.[14] and was later modified by Motekaitis et al.[3] It had previously been synthesized by a stepwise procedure requiring a high-dilution step.[10] The tripodal synthetic method has been used to synthesize several analogs of OBISTREN.[14] The procedure used here is described in detail by Chen et al.[15]

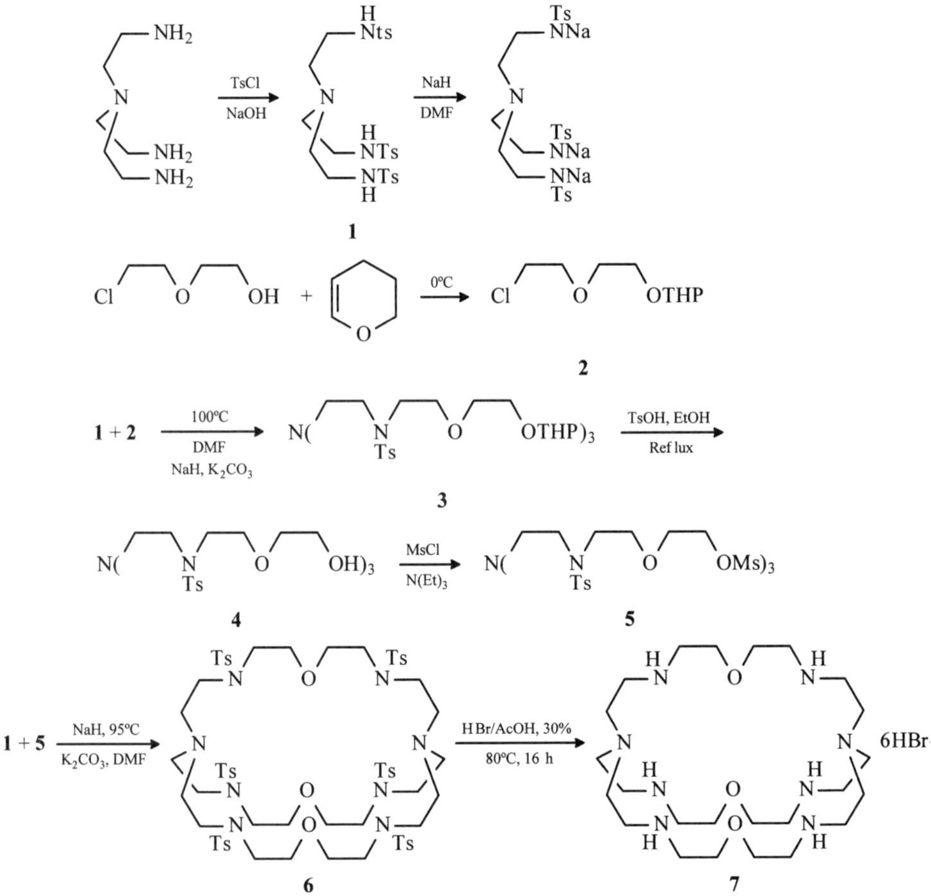

N,N',N''-Tritosyl-2,2',2''-nitrilotriethylamine, **1.** Tris(2-aminoethyl)amine,* 25.2 g, is dissolved in 350 mL of water containing 22 g NaOH in a 1-L, three-necked flask. To this is added dropwise 100 g *p*-toluenesulfonyl chloride* in 300 mL ether with vigorous stirring at room temperature. Stirring is continued for 2 h after the addition and the reaction mixture is allowed to stand for 12 h. The white solid which separates out is collected by filtration, washed with water, and recrystallized from methanol: 88 g, yield 84%. ^{1}H NMR CDCl$_3$ (ppm) 7.7, 7.3 (m, 12H, arom); 6.1 (t, 3H, -NH); 2.9 (br, 6H, NCH$_2$); 2.46 (br, 6H, NCH$_2$); 2.40 (s, 9H, CH$_3$).

1-Chloro-5-(tetrahydro-2H-pyran-2-yloxy)-3-oxapentane, **2.** To monochloro-diethyleneglycol (56 g, 0.45 mol) and CH$_2$Cl$_2$ (150 mL), a solution of freshly distilled 2*H*-dihydropyran (44.8 g, 0.53 mol) in CH$_2$Cl$_2$ (50 mL) is added within 30 min while stirring. After the addition is completed, 12 drops of concentrated HCl are added, and the solution is heated for 1 h at 40°C. After cooling, K$_2$CO$_3$ (10 g) is added, the solvent is evaporated, and the residue is distilled under vacuum, and a fraction, bp 114–115°/3 mm Hg is collected. The yield is 64% and is used directly for the next step. ^{1}H NMR (CDCl$_3$): 1.65 (br, 3CH$_2$); 3.5–4.15 (br, m, ClCH$_2$, 4OCH$_2$); 4.7 (br, OCHO).

N,N',N''-tris(8-tetrahydro-2H-pyran-2-yl)oxy-3-tosyl-6-oxa-3-azaoctyl)amine, **3.** A 30.9-g sample of compound 1 is dissolved in 400 mL dry DMF and 6.1 g NaH (60% dispersion in mineral oil) is added. This reaction mixture is stirred at room temperature for 1.5 h, 10 g of anhydrous K$_2$CO$_3$ is added, and the solution is heated to 100°C. A solution of 33.5 g of compound **2** in 100 mL DMF is then added dropwise over 1 h. Heating is continued for another 18 h. The insoluble material is filtered off, and the solvent is removed from the filtrate by evaporation under reduced pressure. The oily residue is dissolved in 300 mL ether, washed twice with 100 mL water, and dried over anhydrous Na$_2$SO$_4$. Separation from Na$_2$SO$_4$ hydrate and removal of ether leaves a yellow oil (57 g) of compound **3**, which is used directly for the next step.

6,6',6''-Tritosyl-8,8',8''-nitrilo(3-oxa-6-azaoctanol), **4.** Compound **3** (57 g) and *p*-toluenesulfonic acid monohydrate (11 g) are refluxed in 95% EtOH (ca ~400 mL) for 12 h. The solvent is evaporated off and the resulting oil is extracted with 300 mL CH$_2$Cl$_2$, washed with 150 mL water, and dried over K$_2$CO$_3$. After separation of the solid and evaporation of the CH$_2$Cl$_2$, the residue is chromatographed on silica gel (200–400 mesh, 60 Å, 550 g). Compound 4 is eluted with 3% MeOH/CH$_2$Cl as a viscous oil (25 g, 57%). ^{1}H NMR (CDCl$_3$): 7.7, 7.3 (m, 12H, arom); 3.1–3.8 (br, m, 30H, 6CH$_2$NTs,

---

* Aldrich Chemical Co., Milwaukee, WI 53233.

$9CH_2$); 2.82 (br, 6H, $3NCH_2$); 2.40 (s, 9H, CH_3). ^{13}C NMR ($CDCl_3$): 143.5, 136.4, 129.9, 127.2 (arom); 72.6, 70.3, 62.5 (CH_2O); 53.6, 49.1, 47.4 (CH_2N); 21.6 (CH_3).

6,6′,6″-Tritosyl-8,8′,8″-nitrilotri(3-oxa-6-azaoctyl)tris(methanesulfonate), **5.** To a stirred solution of **4** (7.7 g, 8.8 mmole) and Et_3N (22 mL, 160 mmole) in CH_2Cl_2 (300 mL) at 0°C, MsCl (methanesulfonyl chloride, 2.8 mL, 36 mmole) is slowly added. The mixture is stirred for 1 h at 0°C and 2 h at room temperature. The solution is washed successively with 1 M HCl (50 mL), 1 M NaOH (50 mL), and H_2O (50 mL). It is then dried (Na_2SO_4), and evaporated. The residue is dried under vacuum for a few hours. The yellow oil **5** (9.7 g, 98%) is used directly for the cyclization step. ^{13}C NMR ($CDCl_3$): 144.1, 137.1, 130.4, 127.7 (arom. C); 70.5, 69.5, 69.3 (CH_2O); 54.4, 49.4, 48.3 (CH_2N); 38.1 (CH_2Ms); 22.0 ($CH_3C_6H_4$).

4,10,16,22,27,33-Hexatosyl-7,19,30-trioxa-1,4,10,13,16,22,27,33-octa-azabicyclo(11.11.11)pentatriacontane, **6.** This compound is synthesized by a modification of the method of Dietrich et al.[14] The mixture of compound **1** (8.8 g) and NaH (1.7 g) (60% dispersion in mineral oil,* in 400 mL DMF is stirred at room temperature for 1 h; 10 g K_2CO_3 is added, and the solution is heated to 95°C. A solution of compound **5** (16 g) in 120 mL DMF is then added dropwise over 1 h. Heating is continued for another 24 h. The solid is filtered off, and the solvent is removed by evaporation. The residue is dissolved in 300 mL CH_2Cl_2, washed twice with 100 mL water, and dried over Na_2SO_4. It is then separated from Na_2SO_4 hydrate and most of the solvent is removed by evaporation under vacuum. The crude product is chromatographed on silica gel (200 g). Compound **6** is eluted with 5% ether/CH_2Cl_2. The white solid obtained on evaporation of the solvent is further purified one more time with silica gel (same eluent). The pure compound **6** (5.8 g, 28%)† is obtained as a white solid. ^1H NMR ($CDCl_3$): 7.71, 7.33 (m, 24H, arom); 3.47 (br, 12H, CH_2O); 3.24 (br, 24H, CH_2NTs); 2.70 (br, 12H, CH_2N); 2.41 (s, 18H, CH_3); ^{13}C NMR ($CDCl_3$): 143.7, 136.9, 130.1, 127.3 (arom); 70.6 (CH_2O); 53.7, 49.1, 48.1 (CH_2N); 21.5 (CH_3).

■ **Caution.** *30% HBr/HAc is very corrosive. Protective clothing is recommended.*

7,19,30-Trioxa-1,4,10,13,16,22,27,33-octaazabicyclo(11.11.11)pentatriacotane, **7.** This compound is synthesized by a modification of the method of Dietrich

---

* (No. 19923-0) Aldrich Chemical Co., Milwaukee, WI 53233.
† Checker's comment: The checker found that addition of Cs_2CO_3 as a ring-closure template and as the base nearly doubled the yield of this step to 52%.

et al.[6] A mixture of compound 6 (5.8 g), phenol (10.1 g), and 160 mL HBr-HAc (30%) is heated at 80°C for 16 h. After cooling the residue is filtered off and washed with ether and the solid is then dissolved in a minimum of water. The solution is filtered and the EtOH is added to the filtrate until the solution became cloudy. Compound **7** crystallizes out on standing. The product is recrystallized from $H_2O/EtOH$ mixed solvent. The product weighs 2.7 g (68%). ^1H NMR (D_2O) 3.80 (t, 12H, CH_2O); 3.33 (br, 12H, $NHCH_2$); 2.91 (t, 24H, NCH_2CH_2).

Anal. Calcd. for $C_{24}H_{54}N_8O_3 \cdot 6HBr \cdot 2H_2O$: C, 28.14; H, 6.30; N, 10.94. Found: C, 28.34; H, 6.29; N, 10.92.

B. 1,4,7,13,16,19-HEXAAZA-10,22-DIOXACYCLOTETRACOSANE, OR [24]ANE N_6O_2, OBISDIEN

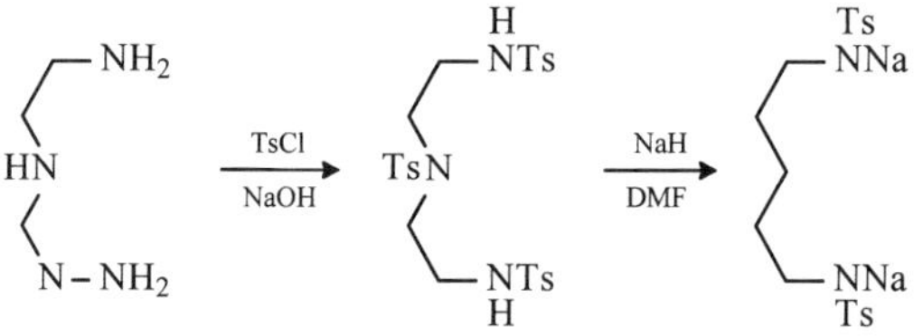

1

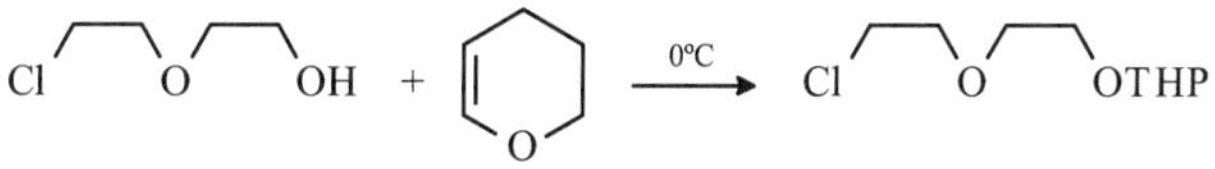

2

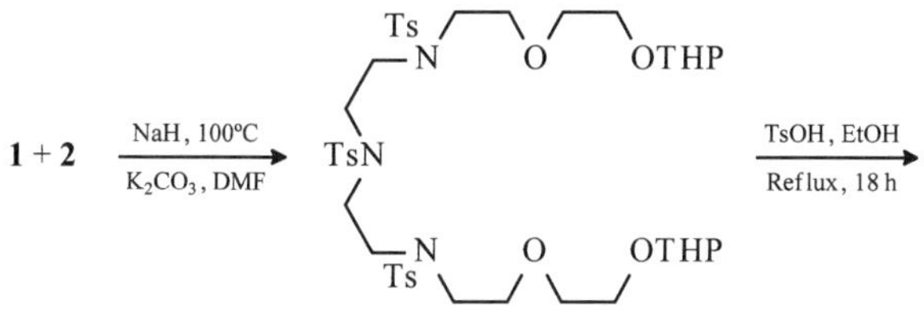

3

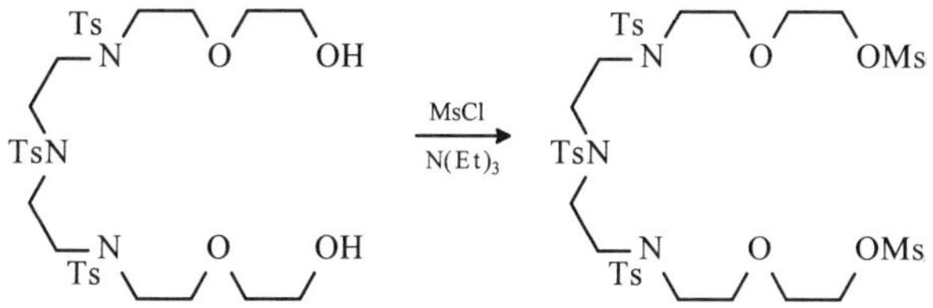

4

5

$$1 + 5 \xrightarrow[\text{K}_2\text{CO}_3, \text{DMF}]{\text{NaH, 100°C}} \mathbf{6} \xrightarrow[\text{80°C, 16 h}]{\text{HBr/HAc, 30\%}} \mathbf{7}$$

The synthesis of **OBISDIEN** was first described by Lehn et al.[7] and an improved procedure was reported subsequently.[8] The procedure used here has been described by Chen et al.[15]

N,N',N''-Tris-(p-tosylsulfonyl)diethylenetriamine, **1**. A 1.25-L aqueous solution of 21.7 g of diethylenetriamine and 76 g $\text{K}_2\text{CO}_3 \cdot 1.5\text{H}_2\text{O}$ is stirred vigorously at 60°C in a 2-L, three-necked flask. Tosyl chloride* (130 g) is added in several small batches over 1 h. Stirring is continued at 60°C for a further 3 h and the mixture is allowed to stand for 16 h at room temperature. The white solid which is formed is filtered off and is washed with water and EtOH and is then refluxed with 500 mL EtOH for 24 h. After the reaction mixture is cooled to room temperature, the white solid which separates out is filtered off and is washed with EtOH. After the product is dried over P_2O_5 under high vacuum at 60°C for 12 h, 98 g of compound **1** is obtained (82%). ^{1}H NMR ($(\text{CD}_3)_2\text{SO}$): 7.5, 7.3 (m, 12H, arom); 2.9, 2.7 (m, 8H, NCH$_2$, NHCH$_2$); 2.26 (s, 9H, CH$_3$). A pure sample which is recrystallized from acetone has a mp of 175–177°C.

1-Chloro-5-(tetrahydro-2H-pyran-2-yloxy)-3-oxopentane, **2**. To monochlorodiethyleneglycol (56 g, 0.45 mole) and CH_2Cl_2 (150 mL), a solution of freshly distilled 2H-dihydropyran (44.8 g, 0.53 mole) in CH_2Cl_2 (50 mL) is added within 30 min while stirring. After the addition was complete, 12 drops of concentrated HCl are added, and the solution is heated for 1 h at 40°C. After cooling, K_2CO_3 (10 g) is added, the solvent is evaporated, the residue is distilled under vacuum, and the fraction boiling at 114–115°C/3 mm Hg is collected. The yield of **2** is 64% and is used directly for the next step. ^{1}H NMR (CDCl$_3$): 1.65 (br, 3CH$_2$); 3.5–4.15 (br, m, ClCH$_2$, 4OCH$_2$); 4.7 (br, OCHO).

6,9,12-Tritosyl-6,9,12-triaza-3,15-dioxa-1,17-bis(tetrahydropyrane-2-yl)oxyheptadecane, **3**. In a 3-necked, round-bottomed flask equipped with reflux condenser, a 30-g sample of compound 1 is dissolved in 350 mL DMF, and

* Aldrich Chemical Co., Milwaukee, WI 53233.

4.5 g of NaH* (60% dispersion in mineral oil) is added, and the solution is then heated to 100°C with stirring. A solution of 23 g compound **2** in 100 mL DMF is added dropwise over 1 h. Heating is continued for another 18 h. The reaction solution is then filtered and the solvent from the filtrate is removed by evaporation. The oily residue is dissolved in 300 mL ether, washed twice with 100 mL water, and dried over Na_2SO_4. The solution is filtered, and evaporation of the ether leaves an oil (48 g), which is used directly in the next step.

6,9,12-Triaza-3,15-dioxa-6,9,12-tritosylheptadecane-1,17-diol, **4.** Compound **3** (48 g) and *p*-toluenesulfonic acid (11 g) are refluxed in 95% EtOH for 14 h. The solvent is evaporated off and the resulting oil is extracted with CH_2Cl_2 (300 mL). The solution is washed with 150 mL 2M NaOH, 150 mL water, and dried over K_2CO_3. After separation of the solid and evaporation of the solvent, the residue is chromatographed on silica gel (200–400 mesh, 60 Å 550 g). Compound **4** is eluted with 3% $MeOH/CH_1Cl_2$. Evaporation of the solvent leaves a viscous oil (22.8 g, 58%). ^1H NMR ($CDCl_3$) 7.75, 7.32 (m, 12H, arom); 3.30–3.70 (br, 24H, NCH_2, OCH_2, CH_2OH); 2.81 (m, 2H, OH). ^{13}C NMR ($CDCl_3$) 143.8, 135.6, 129.8, 127.1 (arom); 72.4, 69.8, 61.4 (CH_2O, CH_2OH); 49.7, 49.2, 48.0 (TsNC); 21.3 (CH_3).

6,9,12-Triaza-3,15-dioxa-6,9,12-tritosylheptadecane-1,17-bis(Ms), **5.** To a stirred solution of compound **10** (15.7 g) and Et_3N (20 mL) in CH_2Cl_2 (270 mL) at 0°C, MsCl (6.6 g) is slowly added. The mixture is stirred for 1 h at 0°C and 2 h at room temperature. The solution is washed successively with 1 M HCl (250 mL), 1 M NaOH (50 mL), and water (50 mL), and is dried over Na_2SO_4. The solid is filtered off and the solution is evaporated. The residue is dried under vacuum for 3 h. The yellow oil obtained (17.1, 90%) is used directly for the cyclization step. ^{13}C NMR ($CDCl_3$) 143.8, 135.6, 129.8, 127.1 (arom); 69.9, 59.2, 68.8 (CH_2O); 49.6, 49.4, 49.1 (CH_2N); 37.6 (CH_3Ms); 21.6 ($CH_3C_6H_4$).

1,13-Dioxa-4,7,10,16,19,22-hexatosyl-1,4,7,16,19,22-hexaazacyclotetracosane, **6.** Compound **18** (9 g) and NaH (1.2 g) (60% dispersion in mineral oil) in 400 mL DMF is stirred at room temperature for 1 h, K_2CO_3 (10 g) is then added and the solution is heated to 95°C. A solution of compound **5** (14.7 g) in 120 mL DMF is then added dropwise over 1 h. Heating and stirring are continued for 24 h. The solution is filtered and the solvent is removed by evaporation. The residue is mixed with 250 mL CH_2Cl_2 and vigorously stirred for 24 h to extract the impurity. Pure compound **6** is obtained by filtration and is washed with CH_2Cl_2 (yield of **6** is 10 g, 50%). ^1H NMR ($CDCl_3$) 7.7, 7.3 (m, 24H, arom); 3.62 (m, 8H, CH_2O); 3.39 (br, 24H, CH_2NTs); 2.41 (s, 18H, CH_3).

---

* Aldrich Chemical Co., Milwaukee, WI 53233.

1,13-Dioxa-4,7,10,16,19,22-hexaazacyclotetracosane, **7.** A mixture of compound **6** (5.7 g), phenol (10 g) and 170 mL HBr-HAc (30%) is heated at 80°C for 16 h. A white solid forms after cooling, the residue is filtered off and is washed with ether, and the solid is then dissolved in a minimum of water. The solution is filtered and the EtOH is added to the filtrate until the solution becomes cloudy. Compound **7** crystallizes out on standing. The product is recrystallized from H_2O/EtOH mixed solvent. The product weighs 2.5 g (67%). ^1H NMR (D_2O): 3.74 (t, 8H, CH_2O); 3.52 (br, 16H, NCH_2); 3.31 (t, 8H, NCH_2). ^{13}C NMR (D_2O): 64.10, 46,74, 43.48, 42.41.

Anal. Calcd. for $C_{16}H_{38}N_6O_2 \cdot 6HBr \cdot 2H_2O$: C, 22.14; H, 5.57; N, 9.68. Found: C, 22.12; H, 5.18; N, 9.57.

References

1. R. J. Motekaitis, A. E. Martell, J. M. Lehn, and E. Watanabe, *Inorg. Chem.*, **21**, 4253 (1982).
2. R. J. Motekaitis, A. E. Martell, B. Dietrich, and J. M. Lehn, *Inorg. Chem.*, **23**, 1588 (1984).
3. R. J. Motekaitis, A. E. Martell, and I. Murase, *Inorg. Chem.*, **25**, 938 (1986).
4. R. J. Motekaitis and A. E. Martell, *J. Chem. Soc. Chem. Commun.*, 1020 (1988).
5. R. J. Motekaitis and A. E. Martell, *J. Am. Chem. Soc.*, **110**, 7715 (1988).
6. R. J. Motekaitis, A. E. Martell, J. P. Lecomte, and J. M. Lehn, *Inorg. Chem.*, **22**, 609 (1983).
7. J. M. Lehn, S. H. Pine, E. I. Watanabe, and A. K. Willard, *J. Am. Chem. Soc.*, **99**, 6766 (1977).
8. J. Comarmond, P. Plumere, J. M. Lehn, Y. Agnus, R. Louis, R. Wiess, O. Kahn, and I. Morgenstern-Badarau, *J. Am. Chem. Soc.*, **104**, 6330 (1982).
9. P. K. Coughlin, J. C. Dewan, S. J. Lippard, E. Watanabe, and J. M. Lehn, *J. Am. Chem. Soc.*, **101**, 265 (1979).
10. P. K. Coughlin and S. J. Lippard, *J. Am. Chem. Soc.*, **103**, 3328 (1981).
11. A. E. Martell and R. J. Motekaitis, *J. Am. Chem. Soc.*, **110**, 8059 (1988).
12. R. J. Motekaitis and A. E. Martell, *Inorg. Chem.*, **30**, 6794 (1991).
13. R. J. Motekaitis and A. E. Martell, *Inorg. Chem.*, **33**, 1032 (1994).
14. B. Dietrich, M. W. Hosseini, J. M. Lehn, and R. B. Sessions, *Helv. Chim. Acta*, **68**, 289 (1985).
15. D. Chen and A. E. Martell, *Tetrahedron*, **51**, 7 (1995).

19. *N,N′*-BIS(2-HYDROXYBENZYL)ETHYLENDIAMINE-*N,N′*- DIACETIC ACID (HBED)

Submitted by ARTHUR E. MARTELL* and ICHIRO MURASE†
Checked by KRISTEN BOWMAN JAMES‡

The ligand *N,N′*-bis(2-hydroxybenzyl)ethylenediamine-*N,N′*-diacetic acid, HBED, H_4L may be considered derived from EDTA, with one of the acetic

* Department of Chemistry, Texas A&M University, College Station, TX 77843-3255.
† 28-14 Umebayashi-4, Jonan-ku, Fukuoka 814-01, Japan.
‡ Department of Chemistry, University of Kansas, Lawrence, KS 66045.

acid groups on each nitrogen replaced by an *o*-hydroxybenzyl group. It forms a very stable Fe(III) complex, and very stable complexes with other small trivalent metal ions, such as those of Al(III) and Ga(III). The synthesis of the ligand is first described by L'Eplattenier et al. in 1967,[1] who also reported the stabilities of several of its metal complexes. Stability constants were reported in a recent publication.[2] The ligand may be used whenever the formation of a very stable iron(III) complex is needed. It has been tested clinically for the removal of iron from the body, with favorable results, and a prodrug, the methyl ester of HBED, seems to offer considerable promise for this purpose.[3] The compound is available commercially from Strem Chemicals, Inc., 7 Mulliken Way, Newburyport, MA, 01950-4098 and from Dijindo and Company, Kumamoto, Japan. In the acid form, the ligand is sparingly soluble in water ($\sim 2.0 \times 10^{-3}$ M), but the alkali metal salts are much more soluble.

An outline of the synthesis of HBED is given by the following sequence of reactions:

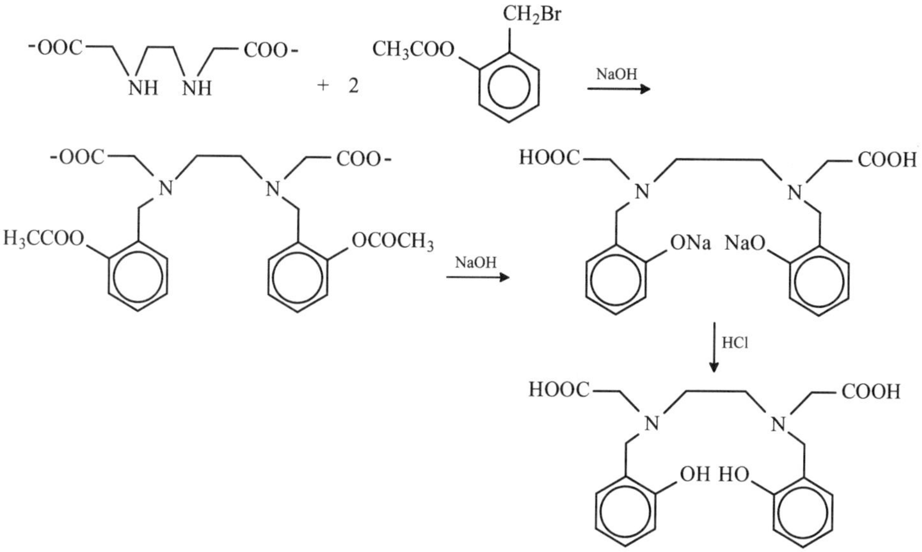

Procedure

In an open beaker containing a magnetic stir bar, 1.8 g (0.010 mole) of ethylenediamine-N,N'-diacetic acid is neutralized with 0.80 g of sodium hydroxide in 10 mL of water and diluted with 25 mL of ethanol. *o*-Acetoxy-benzyl bromide[4] (4.6 g, 0.020 mole) is added. The clear solution is maintained at 35–40°C and stirred while 1.0 g of sodium hydroxide as a 30% aqueous solution is added dropwise at such a rate as to maintain a pH corresponding to the blue-violet color of a phenolphthalein–thymolphthalein mixed indicator. The reaction mixture is then heated at 60°C for 1 h, a second portion of sodium hydroxide (1 g as a 30% aqueous solution) is added, and the solution is further heated at 70°C for 2 h. Ethanol is removed by rotary evaporation and the residue is diluted to 70 mL with water. The pH of the solution is brought to 1–1.5 by the addition of concentrated hydrochloric acid; gummy precipitates are removed by filtration and the filtrate is placed in the refrigerator overnight. The product, *N*,*N′*-di(2-hydroxybenzyl)ethylenediamine-*N*,*N′*-diacetic acid, is obtained as the crystalline dihydrochloride.

The crude product is dissolved in 25 mL of hot 50% (v/v) methanol, filtered, and the filtrate is allowed to stand about 20 h at room temperature. The resulting crystalline product is collected by filtration, washed with cold 50% methanol (a few mL), then with ethanol and dried over calcium chloride in a vacuum desiccator. Yield 2.4 g (30%) mp 143°C (softened at 139°C), sparingly soluble in ethanol and water, and soluble in methanol. The compound is the dihydrochloride, containing 1 mole of water of crystallization, $C_{20}H_{24}N_2O_6 \cdot 2HCl \cdot H_2O$.* ^1H NMR (DMSO): $\delta 7.3$-6.8 (m, 8H, arom); $\delta 4.06$, $\delta 3.66$, $\delta .21$ (s, 12H, NCH_2). ^{13}C NMR (DMSO): $\delta 170$ (COOH); $\delta 156$ (phenol); $\delta 132$, $\delta 130$, $\delta 119$, $\delta 15.5$ (arom); $\delta 52.7$, $\delta 52.1$, $\delta 59.4$ (NCH_2).

Anal. Calcd. for $C_{20}H_{24}N_2O_6 \cdot 2HCl \cdot H_2O$: C, 50.21; H, 5.86; N, 5.86. Found: C, 49.89; H, 5.99; N, 5.78.[5]

References

1. F. L'Eplattenier, I. Murase, and A. E. Martell, *J. Am. Chem. Soc.*, **89**, 837 (1967).
2. R. Ma, R. J. Motekaitis, and A. E. Martell, *Inorg. Chim. Acta*, **224**, 151 (1994).
3. R. W. Grady, A. D. Salbe, M. W. Hilgartner, and P. J. Giardina in *The Development of Iron Chelators for Clinical Use*, R. J. Bergeron and G. M. Brittenham, Eds., CRC Press, Boca Raton, FL, 1994, pp. 395–406.
4. D. L. Fields, J. B. Miller, and D. D. Reynolds, *J. Org. Chem.*, **29**, 2640 (1964).
5. Data reported by checker.

* The HCl and H_2O of crystallization may vary from one to two depending on the solvent used for recrystallization, and the excess HCl present.

20. *N-TERT*-ALKYL-ANILIDES AS BULKY ANCILLARY LIGANDS

Submitted by ADAM R. JOHNSON* and CHRISTOPHER C. CUMMINS*
Checked by SANDRO GAMBAROTTA†

Low-coordinate, low valent transition metal and actinide metal complexes have become readily accessible through the use of bulky ancillary ligands such as silox (t-Bu$_3$SiO$^-$)$^{1-4}$ and bis(trimethylsilyl)amide (N[SiMe$_3$]$_2$).$^{5-10}$ A novel bulky ligand-type, *N-tert*-alkyl anilide, has recently been utilized to stabilize molecules with low coordination numbers and oxidation states.$^{11-25}$ *N-tert*-alkyl anilines were originally prepared by Hunter, in low yield, from phenylacetone imines.^{26} In the following preparations, deuterium is introduced as a spectroscopic handle for the observation of paramagnetic species, as deuterium line widths can be up to 42 times narrower than the corresponding proton line widths in NMR of paramagnetic systems.^{27} More-over, the labeling improved the yield several fold from the original preparation due to the deuterium isotope effect. These ligands are readily prepared in multigram quantities in a few steps, are not susceptible to low-energy decomposition pathways such as β-hydrogen elimination or cyclometallation, and are inherently tunable due to the wide variety of commercially available anilines, further enhancing their utility. Several synthetic methods have been developed and optimized for the preparation of this ligand type, and these are presented herein.

A. *N-d$_6$*-3,5-DIMETHYLPHENYLACETONE IMINE [ArN=C(CD$_3$)$_2$]^{28}

Procedure

■ **Caution.** *Anilines are potentially carcinogenic. Skin contact should be minimized.*

* Department of Chemistry, Massachusetts Institute of Technology, Cambridge, MA 02139.
† Department of Chemistry, University of Ottawa, Ottawa, Ontario, Canada KIN 6N5.

Freshly distilled 3,5-dimethylaniline* (48.24 g, 398.1 mmol), 4 Å molecular sieves (dried under vacuum at $>140°C$ overnight, 90 g), and acetone-d_6[†] (200 mL, 2.72 mol) are added to a 500-mL, round-bottomed flask and placed under a partial vacuum at 4°C for 4 days with occasional stirring. The molecular sieves are removed by filtration of the reaction mixture through a sintered glass frit, and the excess acetone-d_6 is removed by vacuum transfer, leaving the desired imine as a pale yellow oil. More imine is obtained by immersing the sieves in pentane (100 mL) overnight followed by filtration and removal of solvent under vacuum. Yield: 64.60 g (97%). Similarly, N-d_6-2-fluoro-5-methylphenylacetone imine $[Ar_FN=C(CD_3)_2]$ and N-d_6-2-dimethyl-aminophenylacetone imine $[Ar_LN=C(CD_3)_2]$ can be prepared in 73% and 94% yields, respectively.[‡]

Properties

N-d_6-3,5-dimethylphenylacetone imine is an air-stable, pale yellow oil which is sensitive to strong acid. ^{1}H NMR (300 MHz, C_6D_6): $\delta = 6.60$ (s, 1H, para); 6.44 (s, 2H, ortho); 2.15 (s, 6H, aryl CH_3). ^{13}C NMR (75 MHz, C_6D_6): $\delta = 166.66 [C(CD_3)_2]$; 152.65 (ipso); 138.41 (meta); 124.72 (aryl); 117.51 (aryl); 27.60 (CD_3); 21.37 (aryl CH_3); 19.40 (CD_3). High resolution mass spec. Calcd. mass (167.15811), measured mass (167.15828).

$Ar_FN=C(CD_3)_2$: ^{1}H NMR (300 MHz, $CDCl_3$): $\delta = 6.89$ (t, 1H, para), 6.75 (m, 1H, meta), 6.57 (d, 1H, ortho), 2.17 (s, 3H, aryl CH_3). ^{13}C NMR (75 MHz, C_6D_6): $\delta = 21.6$ (q, aryl CH_3), 28.3 (sept, CD_3), 114.6 (d, aryl), 116.7 (d, aryl), 123.9 (t, N $= C[CD_3]_2$), 133.9 (s, aryl), 138.2 (s, aryl CCH_3), 150.1 (d, CF), 172.5 (s, ipso), ^{19}F NMR (282 MHz, benzene): $\delta = -132.8$.

$Ar_LN=C(CD_3)_2$: ^{1}H NMR (300 MHz, $CDCl_3$): $\delta = 6.97$ (dd, 1H, meta), 6.65 (m, 2H), 6.55 (d, 1H, ortho), 2.65 [s, 6H, $N(CH_3)_2$].

B. N-d_6-[t]BUTYL-3,5-DIMETHYLANILINE [HN(R)Ar-d_6][12]

* Aldrich Chemical Company, Milwaukee, WI 53233.
[†] Cambridge Isotope Laboratories, Andover, MA 01810.
[‡] Our checker noted that reaction at atmospheric pressure and room temperature for two days is sufficient.

Procedure

■ **Caution.** *MeLi is spontaneously flammable in air. The reaction should be carried out in a nitrogen filled dry box.*

A solution of MeLi* (1.4 M in diethyl ether, 300 mL, 420 mmol) in a 500-mL Erlenmeyer flask is frozen solid with a liquid nitrogen filled cold well and allowed to thaw until magnetic stirring is just possible. A solution of $ArN = C(CD_3)_2$ (26.00 g, 155.7 mmol) in diethyl ether (ca. 20 mL) at $-35°C$ is added via pipet over 10 min.[†] The reaction mixture is allowed to stir for 18 h, at which point it is quenched by removing the flask from the glove box to a fume hood, and carefully pouring the reaction mixture onto ice (300 mL) in a 1-L Erlenmeyer flask. The mixture is extracted with hexanes (2×150 mL) and flushed through a column of oven-dried alumina (2×25 cm). The column is washed with hexanes to give a total volume of 950 mL. The solvent is removed under vacuum resulting in a canary yellow oil. This oil is shown (by ^1H NMR) to be a 1:1 mixture of $ArN = C(CD_3)(CD_2H)$ and the desired product. Water (200 mL) is added to the oil, and with vigorous stirring, concentrated HCl (25 mL) is added via pipet over a 10-min period. A large quantity of a white flocculent precipitate forms. Stirring is continued for 3 h, at which point the pH of the solution is brought to 12–14 with a concentrated NaOH solution (~ 5 M). The mixture is extracted with hexanes (2×200 mL) and flushed through a column of oven-dried alumina (2×25 cm). Additional hexanes are used to wash the column to give a total volume of 800 mL. The solvent is removed under vacuum, leaving a yellow oil. The oil can be used as prepared or distilled (45°C, $\sim 10^{-1}$ torr). Yield 15.54 g (55%). Mp: $\sim -30°C$.

Anal. Calcd. for $C_{12}H_{13}D_6N$: C, 78.62; H, 10.45; N, 7.64. Found: C, 78.12; H, 10.77; N, 7.16.

High resolution mass spec. Calcd. mass (183.18941), measured mass (183.18948). Similarly, $N-d_6$-^{t}butyl-2-fluoro-5-methylaniline ($HN(R)Ar_F$-d_6) and $N-d_6$-^{t}butyl-2-dimethylaminoaniline [$HN(R)Ar_L$-d_6] can be prepared in 87% and 86% yields respectively, but the acidolysis step can be omitted in these preparations.

Properties

$N-d_6$-^{t}Butyl-3,5-dimethylaniline is an air-stable colorless oil. ^1H NMR (300 MHz, C_6D_6): $\delta = 6.43$ (s, 1H, para); 6.33 (s, 2H, ortho); 3.06 (s, 1H, NH);

---

* Aldrich Chemical Company, Milwaukee, WI 53233.
[†] Our checker noted that the order of addition is extremely important.

2.18 (s, 6H, aryl CH_3); 1.16 [s, 3H, $C(CD_3)_2CH_3$]. ^{13}C NMR (75 MHz, C_6D_6): $\delta = 147.53$ (s, ipso); 138.19 (q, meta); 120.62 (d, aryl); 115.96 (d, aryl); 50.74 [s, $C(CD_3)_2CH_3$]; 30.08 [q, $C(CD_3)_2CH_3$]; 29.56 [m, $C(CD_3)_2CH_3$]; 21.70 (q, aryl CH_3). MS (70 eV): $m/z(\%)$: 183.2(37)$[M^+]$.

HN(R)Ar$_F$-d_6: ^1H NMR (300 MHz, CDCl$_3$): $\delta = 6.89$ (t, 1H, meta), 6.85 (d, 1H, para), 6.50 (s, 1H, ortho), 3.74 (s, 1H, NH), 2.31 (s, 3H, aryl CH_3), 1.38 [s, 3H, $C(CD_3)_2(CH_3)$. ^{13}C NMR (75 MHz, CDCl$_3$): $\delta = 153.4$ (ipso), 150.3 (CF), 134.9 (aryl CCH_3), 133.5 (aryl), 118.4 (aryl), 114.2 (aryl), 51.4 $[C(CD_3)_2(CH_3)]$, 29.9 $[C(CD_3)_2(CH_3)]$, 29.5 $[C(CD_3)_2(CH_3)]$, 21.5 (aryl CH_3). ^{19}F NMR (282 MHz, benzene): $\delta = -138.9$.

HN(R)Ar$_L$-d_6: ^1H NMR (300 MHz, CDCl$_3$): $\delta = 7.05$ (dd, 1H, ortho), 6.97 (dd, 1H, meta), 6.88 (dd, 1H, ortho), 6.64 (m, 1H, meta), 4.91 [br s, 1H, NHC(CD$_3$)$_2$CH$_3$], 2.61 [s, 6H, –N(CH_3)$_2$], 1.40 [s, 3H, –C(CD$_3$)$_2$CH_3]. ^{13}C NMR (75 MHz, CDCl$_3$): $\delta = 142.14$ (aryl), 141.14 (aryl), 124.36 (aryl), 119.62 (aryl), 115.86 (aryl), 112.78 (aryl), 49.93 $[C(CD_3)_2CH_3]$, 44.20 [aryl N(CH_3)$_2$], 29.86 $[C(CD_3)_2CH_3]$, 29.19 [m, $C(CD_3)_2CH_3$].

C. *N*-'BUTYLANILINE [HN('Bu)Ph][29]

Procedure

■ **Caution.** *NH$_3$ and NaNH$_2$ are hazardous. Gases evolved during the reaction and workup should be vented through aqueous acid.*

In a nitrogen-filled glove box, NaNH$_2$* (93.63 g, 2.400 mol, 3 equiv) is placed in a 2-L, two-necked, round-bottomed flask along with a stir bar. A water-cooled condenser topped by a gas adapter is placed in the central neck, and a rubber septum in the other neck. The apparatus is removed from the glove box and taken to a fume hood, where *tert*-butylamine (1 L, freshly distilled under nitrogen) is added to it via cannula. The septum is removed under a counterflow of nitrogen and replaced with a pressure-equalizing addition

* Aldrich Chemical Company, Milwaukee, WI 53233.

funnel containing freshly distilled bromobenzene (84 mL, 0.80 mol). The mixture is stirred at room temperature for 15 min and then cooled in an ice water bath for 15 min. The bromobenzene is then added dropwise over 30 min with stirring. After 90 min, the ice bath is removed and the pink-orange colored reaction mixture is stirred at ambient temperature for 3 h. The vessel is then placed in an ice bath, and water (8 × 5 mL) is added via syringe under nitrogen counterflow over a period of 30 min. Vigorous gas evolution is observed. The flask is removed from the cooling bath and the remaining *tert*-butylamine is removed by distillation under nitrogen (1 atm), leaving a clear, medium orange-brown oil over a white solid. The solid is dissolved with additional water (150 mL) and the aqueous phase extracted with diethyl ether (3 × 50 mL). The organic extracts are combined and placed in a 1-L Erlenmeyer flask in an ice bath. Distilled water (300 mL) is added, followed by the slow addition of concentrated HCl solution (>0.8 mol) until the mixture is at a pH less than 1, and some flocculent, white solid forms. The organic layer is separated and discarded, and the aqueous layer containing the [H_2N(^{t}Bu)Ph]Cl is placed in a large Erlenmeyer flask in an ice bath. Sodium hydroxide solution (ca. 5 M) is added until the pH is at least 13. At this point, the aqueous solution becomes milky white and an orange oil separates from it. The aqueous layer is extracted with diethyl ether (4 × 100 mL), the organic fractions are combined and the solvent is removed under vacuum, yielding 81 g (68%) of a clear, medium brown oil, nearly pure by ^{1}H NMR. The product is purified by vacuum distillation (35°C, ~10^{-2} torr). Yield: 63.1 g (54%).

Properties

N-^{t}Butylaniline is an air-stable, colorless oil. ^{1}H NMR (300 MHz, C_6D_6): $\delta = 7.13$ (dd, 2H, meta), 6.77 (t, 1H, para), 6.62 (dd, 2H, ortho), 3.11 (br s, 1H, -N*H*), 1.08 [s, 9H, NC(CH_3)$_3$]. ^{13}C NMR (126 MHz, CDCl$_3$): $\delta = 146.78$ (t, ipso), 128.79 (dd, ortho), 118.16 (dt, para), 117.33 (m, meta), 51.34 [m, –*C*(CH$_3$)$_3$], 29.98 (m, –C(*C*H$_3$)$_3$).

D. *N*-^{t}BUTYL-4-^{t}BUTYL-ANILINE [HN(R)Ar$_{t\text{-Bu}}$)][30]

Procedure

Into a 1-L, round-bottomed flask with a Teflon stopcock are loaded $Pd_2(dba)_3$ (dba = dibenzylideneacetone)* (0.8589 g, 0.938 mmol, 2 mol%), P(*o*-tolyl)$_3$* (0.8565 g, 2.81 mmol, 6 mol%), *p-tert*-butylbromobenzene (10 g, 46.9 mmol), *tert*-butylamine (4.12 g, 56.3 mmol), $NaOC(CH_3)_3$ (6.31 g, 65.7 mmol), toluene (210 mL), and a stir bar. The reaction mixture is stirred in a 100°C oil bath for 48 h. After this period, the reaction is observed to be near completion by GC/MS. The reaction mixture is cooled and transferred to a 1-L separatory funnel. Diethyl ether (100 mL) is added and the organics are separated and washed with a saturated aqueous solution of sodium chloride (3 × 120 mL). The organic layer is collected and the solvent is removed under vacuum. To the oil is added HCl (80 mL, 1 M). This is stirred for 2 h, during which time the hydrochloride salt of the product precipitates, which is collected on a sintered-glass frit and washed with water and then hexanes. The colorless product is collected and transferred to a 500-mL Erlenmeyer flask. Diethyl ether (100 mL) and NaOH (55 mL of 1 M solution, 1.2 equiv) are added and the solution is stirred until all the hydrochloride salt reacts. The reaction is transferred to a separatory funnel, the organics are collected, and the aqueous layer is extracted with diethyl ether (2 × 100 mL). The ethereal solution is flushed through alumina (2 × 50 cm) and the solvent is removed under vacuum to give a colorless solid. Yield: 4.39 g (46%).

High resolution mass. spec. Calcd. mass: 205.183050. Measured mass: 205.18322. Mp: 31–32.5°C.

Properties

N-'Butyl-4-'butyl-aniline is a colorless, air-stable solid. ^{1}H NMR (300 MHz, CDCl$_3$): $\delta = 7.23$ (dd, 2H, $J_{HH} = 8.1$, $J_{HH'} = 1.8$ Hz), 6.76 (dd, 2H, $J_{FH} = 8.1$, $J_{HH} = 3.3$ Hz), 3.37 (bs, 1H, N*H*), 1.36 (s, 6H), 1.33 (s, 9H). ^{13}C NMR (75 MHz, CDCl$_3$): $\delta = 144.33, 141.21, 125.70, 117.74, 51.48, 33.95, 31.67, 30.38$.

E. N-(1-ADAMANTYL)-3,5-DIMETHYLANILINE [HN(Ad)Ar][31]

* Strem Chemical Company, Newburyport, MA 01950.

Procedure

In a 250-mL, round-bottomed flask equipped with magnetic stir bar, 1-bromoadamantane* (35.0 g, 0.16 moles) is dissolved in 3,5-dimethylaniline[†] (92.0 g, 0.75 moles), yielding a clear, pale yellow solution. This is heated under nitrogen for 24–36 h at 140°C to form a thick brown liquid. The reaction is conveniently monitored by GC/MS. The brown liquid is extracted with ether (200 mL), washed with NaOH (100 mL of a 1 M solution), and vacuum distilled to remove excess aniline, leaving a dark brown solid. The solid is extracted into a minimum volume of hot methanol (ca. 160 mL) and cooled until crystallization occurs. The solid is collected on a sintered-glass frit and washed with cold methanol (2 × 20 mL) to afford white crystals. The filtrate is concentrated again to afford second and third crops. Yield: 29.0 g (71%).

Anal. Calcd. C, 84.16; H, 9.87; N, 5.48. Found: C, 84.16 H, 9.61; N, 5.42.

Mass. spec. Calcd. mass (255.19870). Measured mass: (255.19859).[‡]

Properties

N-(1-Adamantyl)-3,5-dimethylaniline is an air-stable white solid. ^{1}H NMR (300 MHz, CDCl$_3$): δ = 6.50 (s, 1H), 6.48 (s, 2H, aryl), 3.19 (broad, 1H, NH), 2.28 (s, 6H, aryl(CH_3)$_2$, 2.14 (s, 3H, adamantyl methine), 1.91 (d, 6H, adamantyl crown methylene), 1.71 (s, 6H, adamantyl methylene). ^{13}C NMR (75 MHz, CDCl$_3$): δ = 145.90 (aryl), 138.08 (aryl), 121.02 (aryl), 117.13 (aryl), 74.54, 52.01, 43.49, 36.47, 29.71, 21.40.

F. LITHIUM *N-d$_6$-*^{t}BUTYL-3,5-DIMETHYLANILIDE ETHERATE [LiN(R)Ar(OEt$_2$)-d$_6$]

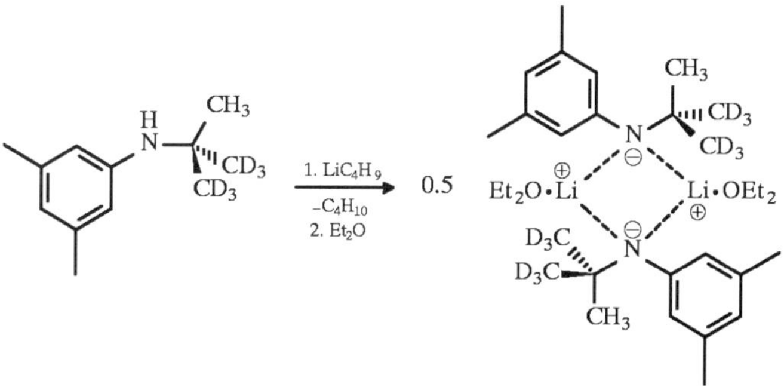

---

* Aldrich Chemical Company, Milwaukee, WI 53233.
[†] Aldrich Chemical Company, Milwaukee, WI 53233.
[‡] Our checker noted that heating the mixture of aniline and 1-Br-Ad for 4 hours at 170°C is sufficient to produce a highly pure white solid in similar yield.

Procedure

- **Caution.** *BuLi is spontaneously flammable in air. The reaction should be carried out in a dry box.*

A solution of HN(R)Ar (13.5 g, 73.7 mmol) in pentane (100 mL) is frozen solid in a liquid nitrogen filled cold well and allowed to thaw until magnetic stirring is just possible. A solution of *n*-BuLi* (49 mL, 1.6 M in hexane, 78 mmol) is slowly added via pipet over a 10-min period. The reaction mixture is stirred for 5 h, the volume is reduced to 60 mL, and diethyl ether (8.30 g, 112 mmol, 1.6 eq.) is added, forming a white precipitate. The volume is further reduced to 30 mL and the product is collected by filtration and dried under vacuum. Yield 13.0 g (67%). Similarly, lithium *N*-'butylanilide etherate [LiN('Bu)Ph(OEt$_2$)] and lithium *N*-d_6-'butyl-2-fluoro-5-methylanilide etherate [LiN(R)Ar$_F$(OEt$_2$)-d_6] can be prepared in 68% and 73% yields respectively. Lithium *N*-'butyl-4-'butyl-anilide [LiN(R)Ar$_{t\text{-Bu}}$], lithium *N*-d_6-'butyl-2-dimethylaminoanilide [LiN(R)Ar$_L$-d_6], and lithium *N*-(1-adamantyl)-3,5-dimethylanilide [LiN(Ad)Ar] can be prepared by following the above procedure (omitting the addition of diethyl ether) in 83%, 92%, and 61% yields respectively.

Properties

Lithium *N*-d_6-'butyl-3,5-dimethylanilide etherate is an air- and moisture-sensitive, white powder soluble in diethyl ether and THF. Some of the derivatives of this complex are only marginally soluble in C$_6$D$_6$; therefore, obtaining good proton NMR spectra is often difficult. ^{1}H NMR (300 MHz, C$_6$D$_6$,): $\delta = 6.54$ (s, 2H, ortho); 6.13 (s, 1H, para); 3.17 (q, 4H, OCH_2CH$_3$); 2.30 (s, 6H, aryl CH_3); 1.60 [s, 3H, C(CD$_3$)$_2$CH_3]; 0.95 (t, 6H, OCH$_2$CH_3). ^{13}C NMR (75 MHz, C$_6$D$_6$): $\delta = 159.48$ (ipso); 138.24 (meta); 116.42 (aryl); 113.47 (aryl); 65.34 (OCH$_2$CH$_3$); 52.13 [*C*(CD$_3$)$_2$CH$_3$]; 31.72 [C(CD$_3$)$_2$*C*H$_3$]; 31.00 [*C*(CD$_3$)$_2$CH$_3$]; 22.14 (aryl *C*H$_3$); 14.87 (OCH$_2$*C*H$_3$).

LiN('Bu)Ph(OEt$_2$): ^{1}H NMR (300 MHz, C$_6$D$_6$): $\delta = 7.19$ (dd, 2H, meta), 6.87 (d, 2H, ortho), 6.47 (t, 1H, para), 3.09 [q, 4H, O(CH_2CH$_3$)$_2$], 1.55 [s, 9H, –C(CH_3)$_3$], 0.86 [t, 6H, O(CH$_2$CH_3)$_2$]. ^{13}C NMR (75 MHz, C$_6$D$_6$): $\delta = 159.26$ (t, ipso), 129.88 (dd, ortho), 118.05 (m, meta), 110.82 (dt, para), 65.06 [tq, O(*C*H$_2$CH$_3$)$_2$], 52.48 [m, –*C*(CH$_3$)$_3$], 31.51 [qh, –C(*C*H$_3$)$_3$], 14.62 [m, O(CH$_2$*C*H$_3$)$_2$].

LiN(R)Ar$_F$(OEt$_2$)-d_6: ^{1}H NMR (300 MHz, C$_6$D$_6$): $\delta = 6.92$ (d, 1H, meta), 6.79 (t, 1H, para), 6.00 (t, 1H, ortho), 3.23 (q, 4H, OCH_2CH$_3$), 2.27 (s, 3H,

---

* Aldrich Chemical Company, Milwaukee, WI 53233.

aryl CH_3), 1.56 [s, 3H,C(CD$_3$)$_2$(CH_3)], 1.01 (t, 6H, OCH_2CH_3). ^{13}C NMR (300 MHz, benzene): $\delta = 158.2$ (C aryl ipso), 155.4 (CF), 149.1 (aryl CCH$_3$), 119.0 (C aryl), 113.3 (C aryl), 107.1 (C aryl), 66.0 (OCH$_2$CH$_3$), 51.9 [C(CD$_3$)$_2$(CH$_3$)], 39.6 [C(CD$_3$)$_2$(CH$_3$)], 31.6 [C(CD$_3$)$_2$(CH$_3$)], 22.6 (aryl CH$_3$), 15.6 (OCH$_2$$CH_3$). ^{19}F NMR (282 MHz): $\delta = -152.1$.

LiN(R)Ar$_{\text{t-Bu}}$: ^{1}H NMR (300 MHz, C$_6$D$_6$): δ 7.23 (dd, 2H, J$_{HH}$ = 8.1, J$_{HH'}$ = 1.8 Hz), 6.76 (dd, 2H, J$_{FH}$ = 8.1, J$_{HH}$ = 3.3 Hz), 3.37 (bs, 1H, NH), 1.36 (s, 6H), 1.33 (s, 9H). ^{13}C NMR (75 MHz, CDCl$_3$): $\delta = 144.33$ (aryl), 141.21 (aryl), 125.70 (aryl), 117.74 (aryl), 51.48 [NC(CH$_3$)$_3$], 33.95 [C(CH$_3$)$_3$], 31.67 (^{t}Bu), 30.38 (^{t}Bu).

LiN(R)Ar$_L$-d_6: ^{1}H NMR (300 MHz, C$_6$D$_6$): $\delta = 6.97$ (ddd, 1H, meta), 6.82 (dd, 1H, ortho), 6.65 (dd, 1H, ortho), 6.40 (ddd, meta), 1.97 [s, 6H, N(CH_3)$_2$], 1.41 [s, 3H, C(CD$_3$)$_2$CH_3]. ^{13}C NMR (75 MHz, C$_6$D$_6$): $\delta = 155.19$ (aryl), 144.32 (aryl), 125.84 (aryl), 119.99 (aryl), 118.74 (aryl), 111.82 (aryl), 51.62 [C(CD$_3$)$_2$CH$_3$], 43.82 [N(CH$_3$)$_2$], 32.79 [C(CD$_3$)$_2$$CH_3$], ca. 32 [m, C(CD$_3$)$_2CH_3$].

LiN(Ad)Ar: ^{1}H NMR (300 MHz, C$_6$D$_6$): $\delta = 6.512$ (s, 3H, aryl), 2.22 (broad, 6H), 2.05 (broad, 3H), 2.02 (broad, 6H), 1.68 (s, 6H).

Acknowledgments

We would like to thank the following people for developing the preparations described in this paper: Brian Budzik, Mike G. Fickes, Marc J. A. Johnson, Catalina E. Laplaza, Aaron L. Odom, Jonas C. Peters, Sheré L. Stokes, and Ryan D. Sutherland. For funding C.C.C. thanks the National Science Foundation (CAREER Award CHE-9501992), DuPont (Young Professor Award), the Packard Foundation (Packard Foundation Fellowship), Union Carbide (Innovation Recognition Award), and 3M (Innovation Fund Award).

References

1. R. E. LaPointe, P. T. Wolczanski, and J. F. Mitchell, *J. Am. Chem. Soc.*, **108**, 6382 (1986).
2. K. J. Covert and P. T. Wolczanski, *Inorg. Chem.*, **28**, 4565 (1989).
3. K. J. Covert, P. T. Wolczanski, S. A. Hill, and P. J. Krusic, *Inorg. Chem.*, **31**, 66 (1992).
4. P. T. Wolczanski, *Polyhedron*, **14**, 3335 (1995).
5. R. A. Andersen, *Inorg. Chem.*, **18**, 2928 (1979).
6. R. A. Andersen, K. Faegri, J. C. Green, A. Haaland, M. F. Lappert, W.-P. Leung, and K. Rypdal, *Inorg. Chem.*, **27**, 1782 (1988).
7. R. A. Andersen and G. Wilkinson, *Inorg. Synth.*, **19**, 262 (1979).
8. D. C. Bradley and M. H. Chisholm, *Accts. Chem. Res.*, **9**, 273 (1976).
9. D. C. Bradley and R. G. Copperthwaite, *Inorg. Synth.*, **18**, 112 (1978).
10. D. C. Bradley, M. B. Hursthouse, K. M. A. Malik, and R. Möseler, *Transition Met. Chem.*, **3**, 253 (1978).

11. A. R. Johnson, P. W. Wanandi, C. C. Cummins, and W. N. Davis, *Organometallics*, **13**, 2907 (1994).

12. M. G. Fickes, W. M. Davis, and C. C. Cummins, *J. Am. Chem. Soc.*, **117**, 6384 (1995).

13. C. E. Laplaza and C. C. Cummins, *Science*, **268**, 861 (1995).

14. C. E. Laplaza, W. M. Davis, and C. C. Cummins, *Angew. Chem. Int. Ed. Engl.*, **34**, 2042 (1995).

15. C. E. Laplaza, W. M. Davis, and C. C. Cummins, *Organometallics*, **14**, 577 (1995).

16. C. E. Laplaza, A. L. Odom, W. M. Davis, C. C. Cummins, and J. D. Protasiewicz, *J. Am. Chem. Soc.*, **117**, 4999 (1995).

17. A. L. Odom and C. C. Cummins, *J. Am. Chem. Soc.*, **117**, 6613 (1995).

18. P. W. Wanandi, W. M. Davis, C. C. Cummins, M. A. Russell, and D. E. Wilcox, *J. Am. Chem. Soc.*, **117**, 2110 (1995).

19. A. R. Johnson, W. M. Davis, and C. C. Cummins, *Organometallics*, **15**, 3825 (1996).

20. C. E. Laplaza, A. R. Johnson, and C. C. Cummins, *J. Am. Chem. Soc.*, **118**, 709 (1996).

21. C. E. Laplaza, M. J. A. Johnson, J. C. Peters, A. L. Odom, E. Kim, C. C. Cummins, G. N. George, and I. J. Pickering, *J. Am. Chem. Soc.*, **118**, 8623 (1996).

22. A. L. Odom and C. C. Cummins, *Organometallics*, **15**, 898 (1996).

23. J. C. Peters, A. R. Johnson, A. L. Odom, P. W. Wanandi, W. M. Davis, and C. C. Cummins, *J. Am. Chem. Soc.*, **118**, 10175 (1996).

24. S. L. Stokes, W. M. Davis, A. L. Odom, and C. C. Cummins, *Organometallics*, **15**, 4521 (1996).

25. M. J. A. Johnson, P. M. Lee, A. L. Odom, W. M. Davis, and C. C. Cummins, *Angew. Chem.*, **109**, 110 (1997).

26. D. H. Hunter, J. S. Racok, A. W. Rey, and Y. Z. Ponce, *J. Org. Chem.*, **53**, 178 (1988).

27. G. N. La Mar, W. D. Horrocks Jr., and R. H. Holm, *NMR of Paramagnetic Molecules*, Academic Press, New York, 1973.

28. D. R. Eaton and P. K. Tong, *Inorg. Chem.*, **19**, 740 (1980).

29. E. R. Biehl, S. M. Smith, and P. C. Reeves, *J. Org. Chem.*, **36**, 1841 (1971).

30. A. S. Guram, R. A. Rennels, and S. L. Buchwald, *Angew. Chem. Int. Ed. Engl.*, **34**, 1348 (1995).

31. P. E. Aldrich, E. C. Hermann, W. E. Meier, M. Paulshock, W. W. Prichard, J. A. Snyder, and J. C. Watts, *J. Med. Chem.*, **14**, 535 (1971).

21. 1,2-BIS(DICHLOROPHOSPHINO)-1,2-DIMETHYLHYDRAZINE AND ALKOXY/ARYLOXY DERIVATIVES

Submitted by V. SREENIVASA REDDY* and KATTESH V. KATTI*
Checked by ROBERT H. NEILSON[†] and JUNMIN JI[†]

1,2-Bis(dichlorophosphino)-1,2-dimethylhydrazine, $Cl_2P\text{-}N(Me)\text{-}N(Me)\text{-}PCl_2$ (**1**) may be produced in a single-step reaction, in high purity and high yields

* Center for Radiological Research and Missouri University Research Reactor, University of Missouri, Columbia, MO 65203.

† Department of Chemistry, Box 298860, Texas Christian University, Fort Worth, Texas 76129.

(>95%), encouraging exploration of the transition metal/organometallic chemistry of hydrazinophosphines. The method described below using 1,2-methyl hydrazine dihydrochloride affords **1** in high purity, even in large-scale operations (e.g., ~500 gm scales).[1] This is an improvement in ease of preparation and in yield over the synthesis of **1** by Gilje et al., and Nöth et al. who produced **1** via (1) the condensation of PCl_3 with highly unstable 1,2-dimethyl hydrazine at $-196°C$,[2,3] and (2) the treatment of heterocyclic-cage compound $P[N(Me)N(Me)]_3P$ with PCl_3,[4] respectively.

Nucleophilic substitutions of **1** result in the formation of a range of alkoxy and aryloxy substituted bisphosphites[5] of the general formula $(RO)_2P$-$N(Me)$-$N(Me)$-$P(OR)_2$, **2**. Detailed investigations of the coordination chemistry of bisphosphanyl hydrazines (e.g., **2**) have demonstrated that they form

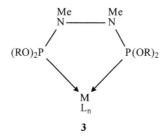

3

well-defined, monoligated, and generally, monomeric complexes of the type **3**.[6-8] As example of these, preparation of alkoxy and aryloxy derivatives, $(RO)_2PN(Me)N(Me)P(OR)_2$ (R=CH_2CF_3, Ph, $C_6H_3Me_2$-2,6) are given along with the parent precursor compound Cl_2P-$N(Me)$-$N(Me)$-PCl_2.

A. 1,2-BIS(DICHLOROPHOSPHINO)-1,2-DIMETHYLHYDRAZINE $Cl_2PN(Me)N(Me)PCl_2$ (1)

$$NMeHNMeH \cdot 2HCl + Excess\ PCl_3 \rightarrow Cl_2P\text{-}N(Me)\text{-}N(Me)\text{-}PCl_2$$

Procedure

1,2-Dimethylhydrazine dihydrochloride* is dried at $80°C/0.1$ torr for 2 h before transferring to the reaction flask under nitrogen atmosphere. A 1-L, three-necked, round-bottomed flask fitted with a mechanical stirrer and water condenser is charged with dry nitrogen. 1,2-dimethylhydrazine dihydrochloride (50 g, 0.375 mol) and phosphorus trichloride (320 mL, 3.78 mol) is

---

* Aldrich Chemical Co., Milwaukee, WI 53233.

added dropwise (30 min) at 25°C. The reaction mixture is heated under reflux with constant stirring until the solution becomes clear ($\sim$ 40 h). The mixture is allowed to cool to 25°C and the excess of PCl_3 is removed in vacuo (0.1 torr at 25°C). The residual PCl_3 is removed by raising the temperature to 60°C at 0.1 torr. The recovered PCl_3 is recycled for use in subsequent reactions. The resulting compound is extracted with dry chloroform and filtered through a sintered glass filtration tube under N_2 atmosphere. Removal of chloroform in vacuo affords the compound **1** in 92% (88.5 g) yield, as a colorless, viscous oil. ^1H NMR: δ 3.15 [t, $^3J(P-H) = 3.7$ Hz, NCH_3]. ^{31}P NMR: δ 160.2(s).

Anal. Calcd. for $C_2H_6Cl_4N_2P_2$: C, 9.2; H, 2.3; N, 10.8; Cl, 53.8%. Found: C, 9.4; H, 2.4; N, 10.6; Cl, 54.5%.

B. $(RO)_2PN(Me)N(Me)P(OR)_2$ $(R = CH_2CF_3, Ph)^5$

$$Cl_2P-N(Me)-N(Me)-PCl_2 + 4ROH + 4Et_3N \rightarrow$$

$$(RO)_2PN(Me)N(Me)P(OR)_2$$

A mixture of 2,2,2-trifluroethanol* or phenol (0.152 mol) and Et_3N (0.160 mol) in *n*-hexane (100 mL) are placed in a pressure-equalized dropping funnel (250 mL). This mixture is added dropwise (30 min) to $Cl_2PN(Me)N(Me)PCl_2$ (10 g, 0.038 mol) also in *n*-hexane (300 mL), contained in a 1L three-necked, round-bottomed flask fitted with a mechanical stirrer and water condenser, at 25°C, with constant stirring. Formation of a white precipitate due to $Et_3N \cdot HCl$ occurs immediately as the reactants are added. The stirring is continued for 4 h, and the $Et_3N \cdot HCl$ is filtered off. Removal of the solvent in vacuo gives analytically pure compound(s) in $\sim$ 85% yield as viscous colorless liquid(s). **R = CH_2CF_3.** Yield: 86% (17.0 g). ^1H NMR: δ 2.85 (t, $^3J_{PH} + {}^4J_{PH} = 2.4$ Hz, 6H, NCH_3), 4.10 (m, br, 8H, OCH_2). ^{31}P NMR: δ 147.3(s).

Anal. Calcd. for $C_{10}H_{14}F_{12}N_2O_4P_2$: C, 23.25; H, 2.73; N, 5.23. Found: C, 23.28; H, 2.68; N, 5.36.

   R = Ph: Yield: 85%(16.0 g). ^1H NMR ($CDCl_3$): δ 2.90 (br, 6H, NCH_3), 7.0-7.3 (m, 20H, C_6H_5); ^{31}P NMR ($CDCl_3$): δ 138.1 (s).

Anal. Calcd for $C_{24}H_{26}N_2O_4P_2$: C, 63.4; H, 5.3; N, 5.7. Found: C, 63.8; H, 5.1; N, 5.4.

---

* Aldrich Chemical Co., Milwaukee, WI 53233.

C. $(RO)_2PN(Me)N(Me)P(OR)_2$ $(R=C_6H_3Me_2-2,6)$

$$Cl_2P-N(Me)-N(Me)-PCl_2 + 4\ HOC_6H_3Me_2-2,6 + 4\ Et_3N \rightarrow$$

$$(RO)_2PN(Me)N(Me)P(OR)_2$$

A mixture of 2,6-dimethylphenol* (18.7 g, 0.152 mol) and Et_3N (0.160 mol) in toluene (100 mL) are placed in a pressure-equalized dropping funnel (250 mL). This mixture is added dropwise (10 min) to $Cl_2PN(Me)N(Me)PCl_2$ (10 g, 0.038 mol) also in toluene (200 mL), contained in a 1-L three-necked, round-bottomed flask fitted with a mechanical stirrer and water condenser, at 25°C, with constant stirring. The reaction mixture is kept at reflux for 6 days (^{31}P NMR indicated that the reaction is incomplete even after 5 days of continuous refluxing) and, after cooling the reaction mixture to room temperature, ~ 50 mL of *n*-hexane is added with stirring and the $Et_3N \cdot HCl$ is filtered off. Removal of the solvent in vacuo gives analytically pure compound in 90% (18.5 g) yield as a viscous liquid. The compound solidifies slowly at room temperature over a few days. ^1H NMR: δ 1.96 (s, 24H, CH_3), 3.13 (t, $^3J_{PH} + ^4J_{PH} = 1.3$ Hz, 6H, NCH_3), 7.0 (m, 12H, phenyl). ^{31}P NMR: δ 141.2(s).

Anal. Calcd for $C_{34}H_{42}N_2O_4P_2$: C, 67.5; H, 7.0; N, 4.6. Found: C, 67.3; H, 7.2; N, 4.5.

Properties

1,2-Bis(dichlorophosphino)-1,2-dimethylhydrazine, $Cl_2P-N(Me)-N(Me)-PCl_2$ (**1**), is nonfuming and undergoes slow hydrolysis in atmospheric moisture. The oxidative stability of alkoxy or aryloxy precursors [e.g., $(RO)_2PN(Me)N(Me)P(OR)_2$] depend on the nature of the substituent on phosphorus. Compounds with bulky substituents (e.g. $OC_6H_3Me_2-2,6$) show higher oxidative and also hydrolytic stabilities as compared to those functionalized with sterically less demanding substituents (e.g., OCH_2CF_3, OEt, or OMe). No products due to Arbuzov-type of rearrangement (see equation) are observed either in the reactions of $Cl_2P-N(Me)-N(Me)-PCl_2$ (**1**), with alcohols or upon storage, under nitrogen atmosphere, for extended periods ($\sim$ one year).

$$(RO)_2PN(Me)N(Me)P(OR)2 \rightarrow (OR)R(O)PN(Me)N(Me)P(O)R(OR)$$

---

* Aldrich Chemical Co., Milwaukee, WI 53233.

Compound **1** can be stored in glass containers, under nitrogen atmosphere, for long periods ($\sim$ one year). The alkoxy and aryloxy derivatives, $(RO)_2PN(Me)N(Me)P(OR)_2$ ($R = CH_2CF_3$, Ph, $C_6H_3Me_2$-2,6) can be stored in glass flasks, under nitrogen atmosphere, for long periods ($\sim$ two years).

References

1. V. S. Reddy and K.V. Katti, *Inorg. Chem.*, **33**, 2695 (1994).
2. T. T. Bopp, M. D. Hevlicek, and J. W. Gilje, *J. Am. Chem. Soc.*, **93**, 3051 (1971).
3. M. D. Havlicek and J. W. Gilje, *Inorg. Chem.*, **11**, 1624 (1972).
4. H. Nöth and R. Ullmann, *Chem. Ber.*, **107**, 1019 (1974).
5. V. S. Reddy, K. V. Katti, and C. L. Barnes, *Inorg. Chem.*, **34**, 5483 (1995).
6. K. V. Katti, V. S. Reddy, and P. R. Singh, *Chem. Soc. Rev.*, **1995**, 97.
7. V. S. Reddy, K. V. Katti, and C. L. Barnes, *Chem. Ber.*, **127**, 1355 (1994).
8. V. S. Reddy and K. V. Katti, *Phosphorus, Sulphur and Silicon and Rel. Ele.*, **109/110**, 161 (1996).

22. TRIS[(*TERT*-BUTYLAMINO)-DIMETHYLSILYL]-METHYLSILANE AND ITS PRECURSORS

Submitted by BERND FINDEIS,* MARTIN SCHUBART,*
and LUTZ H. GADE*
Checked by JOYCE COREY†

Since the pioneering work of Gilman and co-workers in the 1960s,[1-3] several standard preparative methods have been developed for the synthesis of chain and branched-chain organopolysilanes. More recently, the latter have been used in the design of new dendrimeres[4] or cage structures. However, only a few donor-functionalized silanes have found an application in coordination chemistry.

With the aim of developing novel polydentate amido ligands acting as "molecular claws" in the coordination chemistry of the early transition metals, new ligand backbone structures have been synthesised. Tripodal amido ligands containing a silicon-based ligand backbone were found to be particularly versatile in the chemistry of both first row and heavy transition metals.[5] To prepare the ligands on a large scale the established methods of

* Institut für Anorganische Chemie, Universität Würzburg, Am Hubland, 97074 Würzburg, Germany.
† Department of Chemistry, University of Missouri, 8001 Natural Bridge Road, St. Louis, MI 63121-4499.

Si–Si coupling and Si–X functionalization had to be modified and extended. The preparation of $CH_3Si[Si(CH_3)_3]_3$ is based on the procedure reported by Marsmann et al.[6] The functionalization of the branched-chain silane occurs by the method introduced by Ishikawa et al.[7,8] and leads to the desired trichloro-derivative,[9] which can easily be converted to a tripodal amine by aminolysis with a primary amine.[5]

A. TRIS(TRIMETHYLSILYL)METHYLSILANE

$$CH_3SiCl_3 + 3(CH_3)_3SiCl + 6Li \xrightarrow[-10°C \to RT]{THF} CH_3Si[Si(CH_3)_3]_3 + 6LiCl$$

■ **Caution.** *Tetrahydrofuran is extremely flammable and forms explosive peroxides; only freshly distilled, peroxide-free material should be used. Lithium-dispersion is a hazardous material and must be handled in dry conditions and under an inert gas atmosphere. Trimethylchlorosilane and trichloromethylsilane can cause severe skin and eye burns. All manipulations should be carried out in a well-ventilated fume hood; protective gloves and safety glasses should be worn.*

Tetrahydrofuran is dried over sodium benzophenone and is freshly distilled under nitrogen or argon immediately before use. Trimethylchlorosilane and trichloromethylsilane are distilled under nitrogen or argon prior to use. Lithium sand was prepared by the following procedure: Pieces of lithium metal (25 g) and 200 mL of paraffin oil were placed in a 500 mL single-necked Schlenk flask (using argon as inert gas) equipped with a high-speed stirrer (Ultra Turrax, Model TP 18/10, Janke and Kunkel, 79219 Staufen, Germany) and heated to a temperature of 200°C. As soon as the metal began to melt, the mixture was stirred at high speed for ca. 2 min at the end of which a fine Li dispersion was obtained. The highly dispersed Li was isolated by filtration through a Schlenk frit, washed with pentane, and dried in vacuo.

A 1000-mL, two-necked, round-bottomed flask is equipped with a Teflon-coated stirring bar and a 250-mL pressure-equalizing dropping funnel capped with an oil bubbler which is connected to an argon line. The flask is charged with dry THF (300 mL), 7.00 g (1.01 mol) of lithium-sand, and 63.5 mL (54.3 g; 500 mmol) of $(CH_3)_3SiCl$. After cooling the mixture in the flask to $-10°C$ (salt/ice bath), the dropping funnel is charged with a solution of 19.5 mL (24.8 g; 166 mmol) of CH_3SiCl_3 in THF (200 mL). This solution is added dropwise over a period of 60 min to the stirred, cooled mixture in the flask. After completion of the addition, the cooling bath is removed and stirring is continued at room temperature for another 15 h. The mixture is filtered through a Büchner funnel charged with 2 cm Celite, the THF is stripped off by rotary evaporation, and the residues are extracted with hexane

(400 mL). The solid residue is removed by filtration and then washed with hexane (2×100 mL). Removal of the solvent from the combined washings and the filtrate using a rotary evaporator gives a yellow oil. The product is obtained by vacuum distillation (bp 86–94°C/8 Torr). Yield: 24.8 g (56.8%).

Anal. Calcd. for $C_{10}H_{30}Si_4$: C, 45.72; H, 11.51. Found: C, 45.63; H, 11.56.

■ **Caution.** *The residual material in the funnel may include highly activated lithium. It should not come in contact with water and should be destroyed with 2-propanol.*

Properties

The product is a colorless solid (mp 56°C) which is stable to air and water. It is soluble in ethers, aliphatic and aromatic hydrocarbons, and chloroform. The ^1H NMR spectrum ($CDCl_3$) exhibits two singlets at δ 0.03 (3H) and 0.11 (27H) (external tetramethylsilane, TMS). ^{13}C NMR ($CDCl_3$): δ -13.5, 0.35.

B. TRIS(CHLORODIMETHYLSILYL)METHYLSILANE

$$CH_3Si[Si(CH_3)_3]_3 + 3(CH_3)_3SiCl \xrightarrow{AlCl_3} CH_3Si[Si(CH_3)_2Cl]_3 + 3Si(CH_3)_4$$

Procedure

It is essential to use sublimed $AlCl_3$ in this preparation. This may either be obtained commercially (Merck-Schuchardt, Germany) or prepared by sublimation of $AlCl_3$ (99% Aldrich) under nitrogen at atmospheric pressure at 185–195°C (recommendation by the checker). Furthermore, the trisilylsilane starting material should be free of any traces of water (if necessary, drying over Na_2SO_4 is recommended).

A 100-mL, single-necked, round-bottomed flask equipped with a magnetic stirring bar, a 40 cm Vigreux column and a distillation bridge is charged with 42.8 g (394 mmol) of $(CH_3)_3SiCl$, 9.71 g (36.9 mmol) of $CH_3Si[Si(CH_3)_3]_3$ and 1.60 g of aluminum chloride. The bright yellow reaction mixture is heated with an oil bath at 60°C. The temperature of the oil bath is kept between 60 and 80°C for 6–8 h so that the distillation temperature at the head of the column remains below 30°C. The amount of tetramethyl silane collected at the column head (which is slightly contaminated with trimethylchlorosilane) may be used as a rough measure for the conversion achieved (theoretically ca. 12 mL). When no more $Si(CH_3)_4$ is produced, the Vigreux column is replaced by a distillation bridge and excess trimethylchloro-silane is removed by distillation. The $AlCl_3$ catalyst is deactivated by addition of dry acetone (5 mL) to the material in the distillation flask and the product is obtained as a colorless, soft, waxy solid after vacuum distillation (bp 90–98°C/0.1 Torr). Yield: 10.2 g (86%).

Anal. Calcd. for $C_7H_{21}Cl_3Si_4$: C, 25.95; H, 6.53. Found: C, 26.08; H, 6.57.

Properties

The product is a colorless, air- and highly moisture-sensitive solid which has to be handled under an inert gas atmosphere (Schlenk line or glove box). It is soluble in ethers, chloroform, toluene, and acetone. The ^1H NMR spectrum ($CDCl_3$) exhibits two singlets at δ 0.33 (3H) and 0.62 (18H) (external tetramethylsilane, TMS). ^{13}C NMR ($CDCl_3$): δ −13.4, 4.5.

C. TRIS[(*TERT*-BUTYLAMINO)-DIMETHYLSILYL]-METHYLSILANE

$$CH_3Si[Si(CH_3)_2Cl]_3 + 6H_2N^tBu \xrightarrow{-3H_3N^tBuCl} CH_3Si[Si(CH_3)_2NH^tBu]_3$$

Procedure

Diethyl ether is dried over sodium benzophenone and is freshly distilled under argon immediately before use. The reaction product is extremely sensitive to moisture and should be handled under an inert gas atmosphere.

A 250-mL, two-necked, round-bottomed flask is equipped with a stirring bar and a 100-mL pressure-equalizing dropping funnel capped with an oil bubbler which is connected to an argon line. The flask is charged with dry diethyl ether (20 mL) and 15.8 g (261 mmol) of *tert*-butylamine (distilled from NaH prior to use, ca 0.5 g for 75 mL). To the stirred mixture, which is externally cooled with an ice/water bath is added a solution of 5.83 g (18.0 mmol) of $CH_3Si[Si(CH_3)_2Cl]_3$ in diethyl ether (80 mL) over a period of 2 h. After complete addition, the cooling bath is removed and stirring is continued at room temp. for another 12 h. The solvent is removed in vacuo and the residue extracted with 60 mL of pentane. After filtration through a Schlenk frit (porosity 3, 16–40 μm) the filtrate is concentrated to ca. 20 mL and stored at −30°C. The reaction product preciptates as large colorless crystals which are isolated by filtration and dried in vacuo. The excellent solubility of the reaction product requires rapid separation of the solid from the mother liquor in order to avoid partial redissolution. Yield: 4.92 g (63.1%).

Anal. Calcd. for $C_{19}H_{51}N_3Si_4$: C, 52.58; H, 11.84; N, 9.68. Found: C, 52.59; H, 11.91; N, 9.61.

Properties

The colorless crystals (mp 64°C) are air and moisture sensitive. They are highly soluble in ethers, aliphatic and aromatic hydrocarbons, and chloroform. The ^1H NMR spectrum (CDCl$_3$) exhibits three singlets at δ -0.05 (3H), 0.28 (18H), and 1.16 (27H) (external tetramethylsilane, TMS). ^{13}C NMR (CDCl$_3$): δ -13.6, 4.6, 34.0, 51.2.

References

1. H. Gilman, J. M. Holmes, and C. L. Smith, *Chem. Ind.*, 849 (1965).
2. H. Gilman and C. L. Smith, *J. Organomet. Chem.*, **14**, 91 (1968).
3. W. H. Atwell and D. R. Weyenberg, *J. Organomet. Chem.*, **5**, 594 (1966).
4. J. B. Lambert, J. L. Pflug, and C. L. Stern, *Angew. Chem. Int. Ed. Engl.*, **34**, 98 (1995).
5. M. Schubart, B. Findeis, L. H. Gade, W.-S. Li, and M. McPartlin, *Chem. Ber.*, **128**, 329 (1995).
6. H. C. Marsmann, W. Raml, and E. Hengge, *Z. Naturforsch.*, **35b**, 1541 (1980).
7. M. Ishikawa and M. Kumada, *J. Chem. Soc. Chem. Commun.*, 157 (1970).
8. M. Ishikawa, M. Kumada, and H. Sakurai, *J. Organomet. Chem.*, **23**, 63 (1970).
9. G. Kollegger and K. Hassler, *J. Organomet. Chem.*, **485**, 223 (1995).

Chapter Three

TRANSITION METAL COMPLEXES AND PRECURSORS

23. FACILE SYNTHESIS OF ISOMERICALLY PURE cis-DICHLORODIAMMINEPLATINUM(II), *CISPLATIN*

Submitted by VADIM YU. KUKUSHKIN,*,† ÅKE OSKARSSON,*
LARS I. ELDING,* and NICHOLAS FARRELL‡
Checked by STEPHEN DUNHAM§ and STEPHEN J. LIPPARD§

The simple, square-planar complex cis-$[PtCl_2(NH_3)_2]$ (common abbreviations *cisplatin* or cis-DDP) is one of the most potent single anticancer agents in the treatment of solid tumors.[1,2] By displacing the chlorides with, for instance, dicarboxylates, the complex is also the starting material for many "second-generation" antitumor drugs currently in clinical trials.[3] A number of methods are available for preparation of *cisplatin*.[4–9] The standard one is that of Grinberg,[9] adopted for synthesis of *cisplatin* by Dhara.[10] It involves generation in situ of $K_2[PtI_4]$,[11] formation of the intermediate iodo complex

* Inorganic Chemistry 1, Chemical Center, Lund University, P.O. Box 124, S-221 00 Lund, Sweden.
† On leave from Department of Chemistry, St. Petersburg State University, Universitetsky Pr. 2, 198904 Stary Petergof, Russian Federation.
‡ Department of Chemistry, Virginia Commonwealth University, 1001 Main St., Box 2006, Richmond, VA 23284.
§ Department of Chemistry, Massachusetts Institute of Technology, Cambridge, MA 021139.

cis-[PtI$_2$(NH$_3$)$_2$], and subsequent metathesis with a silver salt and chloride:

$$K_2[PtCl_4] \xrightarrow{4KI} K_2[PtI_4] \xrightarrow{2NH_3} cis\text{-}[PtI_2(NH_3)_2] \xrightarrow{2Ag^+/2Cl^-}$$
$$cis\text{-}[PtCl_2(NH_3)_2]$$

This route gives a much better yield and a purer compound than when K$_2$[PtCl$_4$] is treated with ammonia directly. A disadvantage, however, is the necessity to use silver salts (usually nitrate) with overnight stirring, resulting in the possibility of side products formed by hydrolysis of the intermediate aqua species *cis*-[Pt(NH$_3$)$_2$(H$_2$O)$_2$]$^{2+}$.[12] We here present a rapid and facile one-step synthesis of *cisplatin*. The experimental conditions are based on Lebedinsky's method,[8] slightly modified as specified.

■ **Caution.** *Ammonium acetate is an irritant. Both potassium tetrachloroplatinate and cisplatin are known sensitizing agents and cisplatin is a poison. The HCl solution used for recrystallization of cisplatin is highly toxic and corrosive. Contact with the liquid and vapor should be avoided. Appropriate precautions must be taken, and an efficient hood must be used.*

cis-DICHLORODIAMMINEPLATINUM(II)

$$K_2[PtCl_4] + 2NH_4CO_2CH_3 \rightarrow cis\text{-}[PtCl_2(NH_3)_2] + 2CH_3CO_2H + 2KCl$$

A mixture of K$_2$[PtCl$_4$] (Degussa; 2.00 g, 4.8 mmol), ammonium acetate, NH$_4$CO$_2$CH$_3$ (1.60 g, 20.8 mmol), and KCl (2.00 g, 26.8 mmol) in 25 mL of water is refluxed on stirring in a 100-mL round-bottomed flask for 2 h. During this time, the color of the solution turns from dark red to brownish orange and then to greenish yellow; precipitation of a small amount of metallic platinum and Magnus' green salt, [Pt(NH$_3$)$_4$][PtCl$_4$], is also observed. The hot suspension is filtered rapidly through a hot Büchner funnel into a 100-mL Bunsen flask; the filtration set is heated in advance to ca. 80°C in an air thermostat. The yellow filtrate is cooled on a water bath and placed in a refrigerator at 3–5°C for 2 h. The formed yellow precipitate is filtered off while cold, washed on a filter three times with 2 mL of ice-cold water, and dried in air at room temperature. Yield of *cis*-[PtCl$_2$(NH$_3$)$_2$] is 1.08 g, 75% based on Pt.

A pure compound can be obtained by recrystallization from 20 mL of boiling 0.1 *M* aqueous HCl. The recrystallization is performed as described above for isolation of *cis*-[PtCl$_2$(NH$_3$)$_2$] from the reaction mixture. The yield of *cis*-[PtCl$_2$(NH$_3$)$_2$] after recrystallization is 0.65 g (45%, based on initial Pt). Recrystallization as described here or by other good published methods[13,14] will help separate any small amounts of Magnus-like material present in the crude precipitate.

Anal. Calcd. for recrystallized $[PtCl_2(NH_3)_2]$: Cl, 23.6; Pt, 65.0. Found: Cl, 23.5; Pt, 65.0. Cl is analyzed potentiometrically after decomposition of a sample with hydrazine in aqueous alkaline medium followed by filtration of metallic Pt and acidification of the filtrate with HNO_3. The Pt is analyzed gravimetrically after decomposition of a sample by heating with Na_2CO_3.

Observed electron-impact mass spectrum for *cis*-$[PtCl_2(NH_3)_2]$, m/z (relative intensities, fragments) the most abundant mass ion peaks in the isotopic distribution pattern are: 300 (100%, $[PtCl_2(NH_3)_2]^+$), 283 (21%, $[PtCl_2(NH_3)]^+$), 264 (54%, $[PtCl(NH_3)(NH_2)]^+$), 246 (54%, $[PtCl(NH)]^+$), 229 (22%, $[Pt(NH_3)_2]^+$), 211 (64%, $[Pt(NH_2)]^+$), and 195 (35%, $[Pt]^+$).

Properties

The recrystallized *cis*-$[PtCl_2(NH_3)_2]$ is a yellow, crystalline powder which decomposes on heating in a capillary above 270°C. The complex is sparingly soluble in water and easily soluble in *N,N*-dimethylformamide. In water at room temperature it is slowly hydrolyzed. The compound responds to the Kurnakov test[15] on *cis*-geometric configuration: *cis*-$[PtCl_2(NH_3)_2]$ reacts with thiourea in a hot aqueous solution to give a yellow precipitate of $[Pt\{(NH_2)_2CS\}_4]Cl_2$. The ^{195}Pt NMR spectrum in water shows the product to be isomerically pure with one single peak at -2095 ppm relative to external $Na_2[PtCl_6]$, corresponding to a $PtCl_2N_2$ coordination sphere.[16,17] No other peaks are observable in the region from -1400 to -3380 ppm.*

Acknowledgments

We wish to thank J. Champoux for additional checking of the synthesis and M. J. Doedee for recording ^{195}Pt NMR spectra. V. Yu. K. is grateful to the Swedish Royal Academy of Sciences and the Academy of Sciences of Russia for financial support of his stay at the Lund University.

---

* The checkers found a resonance in DMF solution at -2090 ppm for cisplatin and additional resonances at -1775 and -2351 ppm, showing an impurity of 5–10% of $[PtCl_3(NH_3)]$-$[PtCl(NH_3)_3]$. The impurity is reduced to less than 5% by a second recrystallization from 0.1 *M* HCl.

References

1. E. Reed and K. W. Kohn, *Platinum Analogs.* In B. A. Chabner and J. Collins, Eds., *Cancer Chemotherapy—Principles and Practice,* J. B. Lippincott, Philadelphia, PA, 1990, pp. 465–490.
2. N. Farrell, *Transition Metal Complexes as Drugs and Chemotherapeutic Agents.* In B. R. James and R. Ugo, Eds., *Catalysis by Metal Complexes,* Reidel-Kluwer Academic Publishers, Dordrecht, 1989, pp. 46–66.
3. M. C. Christian, *The Current Status of New Platinum Analogs,* Seminars in Oncology, **19**, 720 (1992).
4. G. B. Kauffman and D. O. Cowan, *Inorg. Synth.,* **7**, 239 (1963).
5. C. J. Boreham, J. A. Broomhead, and D. P. Fairlie, *Aust. J. Chem.,* **34**, 659 (1981).
6. M. Chikuma and R. J. Pollock, *J. Magn. Reson.,* **47**, 324 (1982).
7. T. G. Appleton, R. D. Berry, C. A. Davis, J. R. Hall, and H. A. Kimlin, *Inorg. Chem.,* **23** (1984) 3514.
8. V. V. Lebedinsky and V. A. Golovnya, *Izv. Sektora Platiny AN SSSR,* **20**, 95 (1946).
9. A. A. Grinberg and H. I. Gildengerschel, *Izv. Sektora Platiny IONKh AN SSSR,* **26**, 115 (1951).
10. S. C. Dhara, *Indian J. Chem.,* **8**, 193 (1970).
11. Y. Qu, M. Valsecchi, L. de Greco, S. Spinelli, and N. Farrell, *Magnet. Reson. Chem.,* **31**, 920 (1993).
12. R. Faggiani, B. Lippert, C. J. L. Lock, and B. Rosenberg, *Inorg. Chem.,* **17**, 1941 (1978).
13. G. Raudaschl, B. Lippert, and J. D. Hoeshele, *Inorg. Chim. Acta,* **78**, L43 (1983).
14. G. Raudaschl, B. Lippert, J. D. Hoeschele, H. E. Howard-Lock, C. J. L. Lock, and P. Pilon, *Inorg. Chim. Acta,* **106**, 141 (1985).
15. N. S. Kurnakov, *J. Prakt. Chem.,* **50** [2], 480 (1894).
16. P. S. Pregosin, *Ann. Rev. NMR Spectroscopy,* **17**, 285 (1986).
17. T. G. Appleton, J. R. Hall, and S. F. Ralph, *Inorg. Chem.,* **24**, 4685 (1985).

24. *cis*-BIS(BENZENEACETONITRILE)DICHLORO-PLATINUM(II) AND *trans*-BIS(BENZENEACETONITRILE)-DICHLOROPLATINUM(II)

Submitted by VADIM YU. KUKUSHKIN,* VALENTINA M. TKACHUK,[†]
and NIKOLAY V. VOROBIOV-DESIATOVSKY[†]
Checked by STANISLAV CHALOUPKA[‡] and LUIGI M. VENANZI[‡]

The nitrile complexes *cis*- and *trans*-[PtCl$_2$(RCN)$_2$] (R = alkyl or aryl) are convenient starting materials for the synthesis of different platinum

* Department of Chemistry, St. Petersburg State University, University Pr., 2, 198904 Stary Petergof, Russian Federation.
[†] Inorganic Chemistry, St. Petersburg State Technological Institute, Zagorodny Pr., 49, 198013 St. Petersburg, Russian Federation.
[‡] Laboratorium für Anorganische Chemie, ETHZ, Universitätstrasse 6, CH-8092 Zürich, Switzerland.

compounds by substitution reactions of RCN^{1-4} or reactions with coordinated nitriles.[5-7] For preparative purposes $[PtCl_2(CH_3CN)_2]$ and $[PtCl_2(C_6H_5CN)_2]$ are the most widely used.

The complex *cis*-$[PtCl_2(CH_3CN)_2]$ is formed when $K_2[PtCl_4]$ reacts with acetonitrile in water[8,9] and the *cis*-configuration of this compound was confirmed by an X-ray crystallographic study.[10] Chugaev and Lebedinsky[11] have synthesized the *trans*-$[PtCl_2(CH_3CN)_2]$ by careful heating of the *cis*-isomer at a temperature somewhat below its decomposition point. The disadvantage of the *cis*- and *trans*-$[PtCl_2(CH_3CN)_2]$ is their low solubility in water and in the most widely used organic solvents.

The complexes $[PtCl_2(C_6H_5CN)_2]$ were obtained from the interaction of $PtCl_2$ and C_6H_5CN.[3] In this case, benzonitrile is used as both solvent and reagent. According to references 12 and 13, the reaction results in the formation of a *cis*- and *trans*-isomer mixture. The compounds *cis*- and *trans*-$[PtCl_2(C_6H_5CN)_2]$ show relatively good solubility in some organic solvents (e.g., in benzene and chloroform[3]). For synthesis purposes, the mixture of geometric isomers $[PtCl_2(C_6H_5CN)_2]$ is most often used. However, in those cases when syntheses are made specifically from *cis*- or specifically from *trans*-$[PtCl_2(C_6H_5CN)_2]$, the isomers should first be separated by chromatography. The resulting *cis*- and *trans*-$[PtCl_2(C_6H_5CN)_2]$ have been characterized by X-ray analysis.[14]

The title complexes *cis*- and *trans*-$[PtCl_2(C_6H_5CH_2CN)_2]$ have some advantages over their acetonitrile and benzonitrile analogs. The *cis*-isomer is easily formed by interaction of the clathrate $Pt_6Cl_{12} \cdot 0.1C_2H_5Cl \cdot 5.7H_2O$ (obtained from $K_2[PtCl_4]$ and $BF_3 \cdot O(C_2H_5)_2$[15,16] as described below in Section A) and neat benzeneacetonitrile. The isomer *cis*-$[PtCl_2$-$(C_6H_5CH_2CN)_2]$ is isomerized to *trans*-$[PtCl_2(C_6H_5CH_2CN)_2]$ with almost quantitative yield on heating in the solid phase. Both complexes $[PtCl_2$-$(C_6H_5CH_2CN)_2]$ are fairly soluble in a number of organic solvents and can be used for synthesis in a homogeneous liquid phase.

■ **Caution.** *Boron trifluoride etherate is a moisture-sensitive liquid. Small amounts of highly toxic HF and flammable chloroethane and diethyl ether are produced during the synthesis. Benzeneacetonitrile is toxic by inhalation and skin absorption. It also an eye irritant. Appropriate precautions must be taken, and an efficient hood must be used.*

A. CLATHRATE $Pt_6Cl_{12} \cdot 0.1C_2H_5Cl \cdot 5.7H_2O$

$$6K_2[PtCl_4] + 24BF_3 \cdot O(C_2H_5)_2 \rightarrow$$
$$Pt_6Cl_{12} + 12KBF_4 + 12BF_2Cl + 24O(C_2H_5)_2$$

$$BF_2Cl + O(C_2H_5)_2 \rightarrow C_2H_5Cl + F_2BOC_2H_5$$

$$Pt_6Cl_{12} + 0.1C_2H_5Cl + 5.7H_2O^* \rightarrow Pt_6Cl_{12} \cdot 0.1C_2H_5Cl \cdot 5.7H_2O$$

Finely powdered $K_2[PtCl_4]$, 1.000 g (2.41 mmol), is placed in a 50-mL Teflon beaker fitted with Teflon-coated magnetic stirrer bar and Teflon-coated thermometer (if a Pyrex beaker is used instead of a Teflon beaker, the clathrate prepared will contain impurities of SiO_2. The latter does not affect the procedure of synthesis of *cis*-$[PtCl_2(C_6H_5CH_2CN)_2]$, but the yield of this complex will be less). The beaker is filled with 5–10 mL (ca. 40–80 mmol) of freshly distilled $BF_3 \cdot O(C_2H_5)_2$, and a loose lid with a cut for thermometer is put on it.[†] The mixture is heated with stirring from 25 to 100°C for 10–15 min. At about 100°C the suspension becomes brown and gaseous reaction products start to evolve. Within 5–7 min the temperature is raised to 125–130°C and the reaction mixture is kept at this temperature for 5 min. The obtained suspension is cooled to 20–25°C and filtered through paper filter on a Teflon funnel, washed on the filter with 150–175 mL of hot (90°C) water until $K[B(C_6H_5)_4]$ no longer precipitates when aqueous solutions of $Na[B(C_6H_5)_4]$ are added to washing water. Then the precipitate is washed with 20 mL of acetone and 5 mL of diethyl ether, and dried in air at 60°C. Yield of $Pt_6Cl_{12} \cdot 0.1C_2H_5Cl \cdot 5.7H_2O$ is 0.540 g (79%).

Anal. Calcd. for $Pt_6Cl_{12} \cdot 0.1C_2H_5Cl \cdot 5.7H_2O$: Cl, 25.2; Pt, 68.7. Found: Cl, 25.4; Pt, 68.6.

B. *cis*-BIS(BENZENEACETONITRILE)DICHLOROPLATINUM(II)

$$Pt_6Cl_{12} \cdot 0.1C_2H_5Cl \cdot 5.7H_2O + 12C_6H_5CH_2CN \rightarrow$$

$$6\,cis\text{-}[PtCl_2(C_6H_5CH_2CN)_2] + 0.1C_2H_5Cl + 5.7H_2O$$

A 0.516-g (0.30 mmol) portion of $Pt_6Cl_{12} \cdot 0.1C_2H_5Cl$ 5.7H$_2$O is placed into a 30-mL flask thermostated at 100°C and 5 mL (43.32 mmol) of benzene-acetonitrile also heated to 100°C is added. The mixture is kept at this temperature for 2 min (*Attention*: at higher temperatures the resultant product contains the *trans*-isomer as the byproduct and requires recrystallization).

---

* Adsorption during washing of Pt_6Cl_{12} with water.
† The checkers find that the reaction is best carried out under an N_2-atmosphere to avoid significant evolution of noxious fumes. Under their conditions, however, reproducibly high yields (70–73%) of good-quality product required the addition of 60 mg of water to the reaction mixture before heating it up.

Then, the hot solution is filtered, cooled to 20–25°C and, over the course of 1–2 min, 100 mL of diethyl ether is added to the brownish-yellow, filtrate. On stirring, precipitation of pale greenish-yellow, needle-like crystals starts immediately. After 15 min, the precipitate is filtered, washed on a filter with six 5-mL portions of diethyl ether, and dried in air at 20–25°C. Yield of *cis*-[PtCl$_2$(C$_6$H$_5$CH$_2$CN)$_2$] is 0.622 g (69%).* To obtain an analytically pure sample, the compound is recrystallized from a boiling chloroform–acetone mixture (2:1 in volume).

Anal. Calcd. for *cis*-[PtCl$_2$(C$_6$H$_5$CH$_2$CN)$_2$]: Cl, 14.2; Pt, 39.0. Found: Cl, 14.3; Pt, 39.1.

C. *trans*-BIS(BENZENEACETONITRILE)-DICHLOROPLATINUM(II)

$$cis\text{-}[PtCl_2(C_6H_5CH_2CN)_2] \xrightarrow[\text{solid phase}]{\Delta} trans\text{-}[PtCl_2(C_6H_5CH_2CN)_2]$$

The finely powdered *cis*-[PtCl$_2$(C$_6$H$_5$CH$_2$CN)$_2$] complex is spread as a uniform thin layer in a Petri dish† and placed into an oven which is heated from 25 to 125°C for 30–40 min (during this time period about 25% of the *cis*-isomer converts to the *trans*-isomer) and is kept at 125°C for 4 h up to complete conversion of the *cis*-[PtCl$_2$(C$_6$H$_5$CH$_2$CN)$_2$] to the *trans*-[PtCl$_2$(C$_6$H$_5$CH$_2$CN)$_2$]. The process of *cis–trans* isomerization can be monitored by TLC. Yield of the *trans*-[PtCl$_2$(C$_6$H$_5$CH$_2$CN)$_2$] is almost quantitative. The compound is recrystallized from boiling chloroform.

Anal. Calcd. for *trans*-[PtCl$_2$(C$_6$H$_5$CH$_2$CN)$_2$]: Cl, 14.2; Pt 39.0. Found: Cl, 14.4; Pt, 39.0.

Properties

The Pt$_6$Cl$_{12}$·0.1C$_2$H$_5$Cl·5.7H$_2$O clathrate is a fine brown powder. This substance has no characteristic melting point. It decomposes on heating in the solid phase starting from 70°C. IR spectrum (400–200 cm^{-1}, polyethylene pellets) cm^{-1}: 344 s, 337 s, 318 vs, and 204 m.

---

* The checkers obtained a 65% yield.
† The checkers found that carrying out the solid-phase reaction resulted in the formation of some amount of unidentified decomposition products. They state that better results can be obtained by heating a stirred suspension of the *cis*-isomer (0.5 g) in kerosene (3 mL) under nitrogen at 130°C for 4 h. In this case, yields ranged from 75 to 80%.

The complex *cis*-[PtCl$_2$(C$_6$H$_5$CH$_2$CN)$_2$] is crystallized from a chloroform–acetone mixture in the form of lemon-yellow, rod-like crystals. The melting point (Kofler tables, heating rate 25°C/min) is 122–124°C with subsequent crystallization of the *trans*-isomer from the formed melt. The complex *trans*-[PtCl$_2$(C$_6$H$_5$CH$_2$CN)$_2$] is crystallized from chloroform as sabre-like, yellow crystals melting at 174–176°C (dec.). Both compounds are soluble in dichloromethane, acetone, nitromethane, and *N,N*-dimethylformamide and insoluble in water, ethanol, diethyl ether, toluene, and hexane; the *trans*-isomer is soluble in chloroform. Both complexes are nonconducting in acetone. The TLC on SiO$_2$ (Silufol UV 254 plates): R$_f$ (*cis*) = 0.40 and R$_f$(*trans*) = 0.68 (chloroform : acetone = 10 : 1, in volume). In *N,N*-dimethylformamide-d_7 (>99.5 atom % D, CIBA-GEIGY), the ^{1}H-NMR spectra of the compounds show sharp CH$_2$-signals due to coordinated benzeneacetonitrile at 4.71 (J_{PtH} 10.25 Hz) ppm for *cis*-isomer and 4.82 (J_{PtH} 11.6 Hz) ppm for *trans*-isomer. IR spectrum of *cis*-isomer (KBr pellets) cm^{-1}: 2316 m-w n(C≡N), 357 s-m and 348 s-m v(Pt–Cl). IR spectrum of *trans*-isomer (KBr pellets) cm^{-1}: 2360 m and 2315 w v(C≡N), 341 s v(Pt–Cl).

References

1. J. A. Davies, C. M. Hockensmith, V. Yu. Kukushkin, and Yu. N. Kukushkin, *Synthetic Coordination Chemistry: Principles and Practice*, World Scientific, 1996, p. 58.
2. *Comprehensive Coordination Chemistry*, G. Wilkinson, R. D. Gillard, and J. A. McCleverty, Eds., Pergamon Press, 1987, vol. 5, p. 436.
3. F. R. Hartley, *The Chemistry of Platinum and Palladium*, Applied Science, London, 1973, p. 462.
4. V. Yu. Kukushkin, Å. Oskarsson, and L. I. Elding, *Inorg. Synth.*, **31**, 279 (1995).
5. L. Maresca, G. Natile, F. P. Intini, F. Gasparrini, A. Tiripicchio, and M. Tiripicchio-Camellini, *J. Am. Chem. Soc.*, **108**, 1180 (1986).
6. F. P. Fanizzi, F. P. Intini, and G. Natile, *J. Chem. Soc., Dalton Trans.*, 947 (1989).
7. R. A. Michelin, M. Mozzon, and R. Bertani, *Coord. Chem. Rev.*, **147**, 299 (1996).
8. K. A. Hofmann and G. Bugge, *Berichte*, **40**, 1772 (1907).
9. F. P. Fanizzi, F. P. Intini, L. Maresca, and G. Natile, *J. Chem. Soc., Dalton Trans.*, 199 (1990).
10. F. D. Rochon, R. Melanson, H. E. Howard-Lock, C. J. L. Lock, and G. Turner, *Can. J. Chem.*, **62**, 860 (1984).
11. L. Tschugaeff and W. Lebedinski, *Compt. Rend.*, **161**, 563 (1915).
12. T. Uchiyama, Y. Nakamura, T. Miwa, S. Kawaguchi, and S. Okeya, *Chem. Lett.*, 337 (1980).
13. T. Uchiyama, Y. Toshiyasu, Y. Nakamura, T. Miwa, and S. Kawaguchi, *Bull. Chem. Soc. Jpn.*, **54**, 181 (1981).
14. H. H. Eysel, E. Guggolz, M. Kopp, and M. L. Ziegler, *Z. Anorg. Allg. Chem.*, **499**, 31 (1983).
15. N. V. Vorobiov-Desiatovsky, T. P. Smorodina, V. A. Paramonov, E. S. Postnikova, V. V. Sibirskaya, A. I. Marchenko, and Yu. N. Kukushkin, *Zh. Prikl. Khim.*, **58**, 977 (1985); *Chem. Abs.*, **103**: 080805/10.
16. V. V. Sibirskaya, N. V. Vorobiov-Desiatovsky, E. S. Postnikova, and T. P. Smorodina, *Zh. Prikl. Khim.*, **62**, 503 (1989); *Chem. Abs.*, **111**: 108060/12.

25. PLATINUM(II) COMPLEXES OF DIMETHYL SULFIDE

Submitted by GEOFFREY S. HILL,* MICHAEL J. IRWIN,*
CHRISTOPHER J. LEVY,* LOUIS M. RENDINA,* and
RICHARD J. PUDDEPHATT*
Checked by RICHARD A. ANDERSEN† and LUIS McLEAN

Platinum(II) complexes of dimethyl sulfide are convenient and versatile precursors for the preparation of a variety of organoplatinum(II) derivatives.[1-4] The labile S-donor ligands in $[PtMe_2(\mu\text{-}SMe_2)]_2$ and *trans*-$[PtClMe(SMe_2)_2]$ are readily displaced by N-donor ligands to afford complexes of the type $[PtMe_2(NN)]$ and $[PtClMe(NN)]$, respectively, where $NN = 2,2'$-bipyridine, 1,10-phenanthroline, and so on.[1,2] Such complexes have proved very useful in the study of intra- and intermolecular oxidative addition reactions of various substrates including, for example, alkyl and aryl halides.[1]

Herein we present high-yielding syntheses of *cis/trans*-$[PtCl_2(SMe_2)_2]$, $[PtMe_2(\mu\text{-}SMe_2)]_2$ and *trans*-$[PtClMe(SMe_2)_2]$.

■ **Caution.** *Dimethyl sulfide malodorous, toxic, and flammable. All operations should be carried out in a well-ventilated fume hood.*

A. *cis/trans*-DICHLOROBIS(DIMETHYL SULFIDE)PLATINUM(II)

$$K_2[PtCl_4] + 2SMe_2 \xrightarrow{\Delta} cis/trans\text{-}[PtCl_2(SMe_2)_2] + 2KCl$$

Procedure

This is a modified and improved version of that given in the literature.[5] The syntheses of the closely related species, $[PtCl_2(SEt_2)_2]$, are reported elsewhere.[6] To a 500-mL, three-necked, round-bottomed flask equipped with a magnetic stirring bar, a rubber septum, a N_2 inlet, and a gas outlet is added a solution of $K_2[PtCl_4]$ (Digital Specialties, 7.00 g, 16.9 mmol) in distilled water (100 mL). To the stirred solution is added dimethyl sulfide (9.0 mL, 0.123 mol) to afford a mixture of $[Pt(SMe_2)_4][PtCl_4]$ and

* Department of Chemistry, The University of Western Ontario, London, Ontario, Canada N6A 5B7.
† Department of Chemistry, University of California at Berkeley, Berkeley, CA 94720.

cis/trans-[PtCl$_2$(SMe$_2$)$_2$] as a pink/yellow precipitate.[5] The mixture is heated with a heating mantle until the pink precipitate is converted to the bright-yellow *cis/trans*-[PtCl$_2$(SMe$_2$)$_2$] (ca. 30 min). After cooling the mixture to room temperature, the reaction mixture is extracted with CH$_2$Cl$_2$ (3 × 50 mL). The yellow extracts are combined and dried over anhydrous MgSO$_4$. After filtration and evaporation of the solvent in vacuo, the product is obtained as a bright yellow, microcrystalline solid. Yield: 5.23–5.50 g (87.4–92.0%), mp = 159°C.[5] The product is spectroscopically pure and a typical *cis/trans* ratio is 1:2.

Properties

The product is air- and moisture-stable for long periods of time. The chemistry[5] of the complexes is reported elsewhere. The ^{1}H-NMR spectrum of *cis/trans*-[PtCl$_2$(SMe$_2$)$_2$] (300 MHz, CDCl$_3$) shows singlets at $\delta = 2.56$ [^3J(PtH) = 49.5 Hz] and 2.44 [^3J(PtH) = 41.5 Hz] for the protons of the SMe$_2$ ligands of the *cis* and *trans* isomers, respectively.[7]

B. BIS[DIMETHYL(μ-DIMETHYL SULFIDE)PLATINUM(II)]

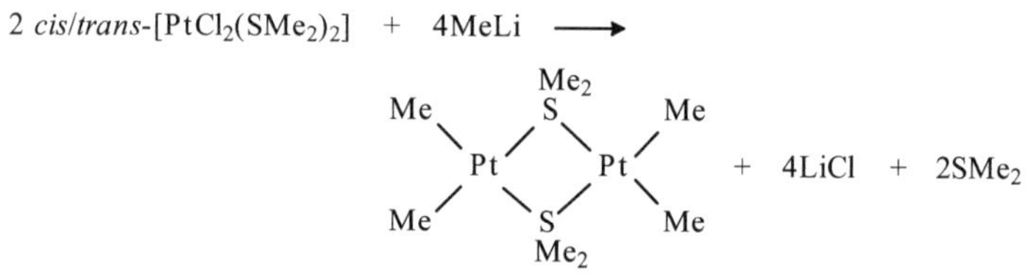

2 *cis/trans*-[PtCl$_2$(SMe$_2$)$_2$] + 4MeLi $\longrightarrow$

+ 4LiCl + 2SMe$_2$

■ **Caution.** *The toxicity of the product is unknown. It should be handled in a fume hood using gloves. Methyllithium is extremely pyrophoric and moisture sensitive.*

Procedure

This is a modified and improved version of that given in the literature.[3] All operations should be performed using standard Schlenk techniques under an atmosphere of dry N$_2$.[8] All Schlenk glassware should be flame dried under vacuum prior to use. All glassware must be cooled to 0°C.

A 500-mL Schlenk tube equipped with a magnetic stirring bar and rubber septum is charged with *finely* powdered *cis/trans*-[PtCl$_2$(SMe$_2$)$_2$] (3.90 g, 11.0 mmol) suspended in dry diethyl ether (160 mL). The ratio of the *cis/trans* isomers is not important. Indeed either pure *cis*- or *trans*-[PtCl$_2$(SMe$_2$)$_2$]

could be used.* The stirred mixture is placed in an ice bath for 30 min, with the ice level at least 3 cm above the level of the mixture. To this is added dropwise, over a period of 10 min, a 1.4 M solution of methyllithium (Aldrich Chemical Co., halide content ca. 0.05 M) in diethyl ether (16.0 mL, 22.4 mmol). The mixture is stirred for 20 min (if a brown color appears during this period, the workup must be commenced immediately). To the resulting white suspension is added a cold solution of saturated, aqueous NH_4Cl (4.0 mL) and distilled water (100 mL, ca. 0°C). The mixture is extracted with ice-cold diethyl ether (3×50 mL). The pale-yellow, organic extracts are combined, cooled to 0°C, and dried over anhydrous $MgSO_4$. A small amount (0.2 g) of decolorizing charcoal is then added to the solution. After 5 min, the black mixture is filtered into a 1-L, round-bottomed flask, and the solvent is evaporated under reduced pressure (15 mmHg) to afford a white powder. The use of a hot water bath must be avoided. Furthermore, prolonged exposure of the product to a vacuum leads to its decomposition. Yield: 2.83–2.89 g (89.6–91.5%), mp 86°C (dec.).† A substantial decrease in the yield of product ensues if the above procedure is scaled up.

Anal. Calcd. for $C_8H_{24}Pt_2S_2$: C, 16.7; H, 4.2%. Found: C, 16.8; H, 4.25%.

Properties

The compound is a white solid that can be handled in air, but should be stored under a N_2 atmosphere at 0°C. The product is stable for several weeks at 0°C, but decomposes within a few hours at room temperature. The compound is very soluble in diethyl ether and CH_2Cl_2, and moderately soluble in acetone. The ^1H NMR spectrum (300 MHz, $CDCl_3$) displays a singlet at $\delta = 0.63$ with quarter intensity ^{195}Pt satellite signals $[^2J(PtH) = 85.2$ Hz] due to Pt–Me, and a $1/8:18:8:1$ multiplet at $\delta = 2.67 [^3J(PtH) = 20.7$ Hz] due to the SMe_2 ligands. The chemistry[9] and X-ray structure[10] of the closely related species $[PtMe_2(\mu\text{-}SEt_2)]_2$ are described elsewhere.

C. CHLOROMETHYLBIS(DIMETHYL SULFIDE)PLATINUM(II)

$cis\text{-}[PtMe_2(SMe_2)_2] + cis/trans\text{-}[PtCl_2(SMe_2)_2] \longrightarrow 2\ trans\text{-}[PtClMe(SMe_2)_2]$

---

* The checkers reported that all free SMe_2 must be absent from $cis/trans$-$[PtCl_2(SMe_2)_2]$ or the complex cis-$[PtMe_2(SMe_2)_2]$ is also obtained.
† The checkers obtained a yield of 59%.

Procedure

This is a modified and improved version of that given in the literature.[3] A 100-mL Schlenk tube equipped with a magnetic stirring bar and rubber septum is charged with $[PtMe_2(\mu\text{-}SMe_2)]_2$ (1.43 g, 2.49 mmol) in CH_2Cl_2 (20 mL). To the stirred solution is added dimethyl sulfide (0.37 mL, 5.0 mmol) to generate *cis*-$[PtMe_2(SMe_2)_2]$ *in situ*. To the clear, colorless solution is added *cis/trans*-$[PtCl_2(SMe_2)_2]$ (1.94 g, 5.47 mmol), and the mixture is stirred for 24 h at room temperature. The ratio of the *cis/trans* isomers is not important. Indeed either pure *cis*- or *trans*-$[PtCl_2(SMe_2)_2]$ could be used. The solvent is removed in vacuo, and the residue is extracted with diethyl ether (3×30 mL) to give a pale-yellow solution. Evaporation of the solvent in vacuo affords a beige solid. Yield: 3.31 g (91.3%), mp 56–60°C.

Anal. Calcd. for $C_3H_9ClPtS_2$: C, 16.2; H, 4.1; S, 17.3%. Found: C, 16.3; H, 4.0; S, 17.1%.

Properties

The product is an air-stable, beige solid which is soluble in most common organic solvents, including diethyl ether and CH_2Cl_2. The ^1H NMR spectrum (300 MHz, $CDCl_3$) displays singlets at $\delta = 0.49[^2J(PtH) = 79.2$ Hz] and $\delta = 2.54$ [$^2J(PtH) = 54.1$ Hz] due to the Pt–Me and SMe_2 ligand protons, respectively.

References

1. (a) P. K. Monaghan and R. J. Puddephatt, *Organometallics*, **3**, 444 (1984); (b) P. K. Monaghan and R. J. Puddephatt, *J. Chem. Soc., Dalton Trans.*, 595 (1988); (c) M. Crespo and R. J. Puddephatt, *Organometallics*, **6**, 2548 (1987); (d) C. M. Anderson, M. Crespo, M. C. Jennings, G. Ferguson, A. J. Lough, and R. J. Puddephatt, *Organometallics*, **10**, 2672 (1991); (e) C. M. Anderson, M. Crespo, G. Ferguson, A. J. Lough, and R. J. Puddephatt, *Organmetallics*, **11**, 1177 (1992); (f) S. Achar, J. D. Scott, and R. J. Puddephatt, *Organometallics*, **11**, 2325 (1992); (g) M. Crespo, M. Martinez, J. Sales, X. Solans, and M. Font-Bardia, *Organometallics*, **11**, 1288 (1992); (h) S. Achar, J. D. Scott, J. J. Vittal, and R. J. Puddephatt, *Organometallics*, **12**, 4592 (1993); (i) M. Crespo, M. Martinez, and J. Sales, *Organometallics*, **12**, 4297 (1993); (j) C. J. Levy, R. J. Puddephatt, and J. J. Vittal, *Organometallics*, **13**, 1559 (1994); (k) K. van Asselt, E. Rijnberg, and C. J. Elsevier, *Organometallics*, **13**, 706 (1994); (l) L. M. Rendina, J. J. Vittal, and R. J. Puddephatt, *Organnometalliics*, **14**, 1030 (1995).
2. (a) R. Bassan, K. H. Bryars, L. Judd, A. W. G. Platt, and P. G. Pringle, *Inorg. Chim. Acta*, **121**, L41 (1986); (b) M. E. Cucciolito, V. De Felice, A. Panunzi, and A. Vitagliano, *Organometallics*, **8**, 1180 (1989); (c) S. Bartolucci, P. Carpinelli, V. De Felice, B. Giovannitti, and A. De Renzi, *Inorg. Chim. Acta*, **197**, 51 (1992).
3. J. D. Scott and R. J. Puddephatt, *Organometallics*, **2**, 1643 (1983).

4. V. Y. Kukushkin, K. Lövqvist, B. Norén, Å. Oskarsson, and L. I. Elding, *Inorg. Chim. Acta*, **219**, 155 (1994).

5. E. G. Cox, H. Saenger, and W. Wardlaw, *J. Chem. Soc.*, 182 (1934).

6. G. B. Kauffman and D. O. Cowan, *Inorg. Synth.*, **6**, 211.

7. R. Roulet and C. Barbey, *Helv. Chim. Acta*, **56**, 2179 (1973).

8. D. F. Shriver, *The Manipulation of Air-Sensitive Compounds*, McGraw-Hill, New York, 1969.

9. J. Kuyper, R. van der Laan, F. Jeanneaus, and K. Vrieze, *Transition Met. Chem.* (*Weinheim, Ger.*), **1**, 199 (1976).

10. D. P. Bancroft, F. A. Cotton, L. R. Falvello, and W. Schwotzer, *Inorg. Chem.*, **25**, 763 (1986).

26. (2,2′:6′,2″-TERPYRIDINE)METHYLPLATINUM(II) CHLORIDE AND (1,10-PHENANTHROLINE)-METHYLCHLOROPLATINUM(II)

Submitted by RAFFAELLO ROMEO* and LUIGI MONSU' SCOLARO*
Checked by VINCENT CATALANO[†] and SUDHIR ACHAR[†]

The synthesis of cationic complexes of the type [Pt(N–N–N)R]X (N–N–N = tridentate nitrogen ligand, R = alkyl group) has been so far restricted to the case of the trimethylsilylmethyl(2,2′:6′,2″-terpyridine)platinum(II) tetrafluoroborate[1] salt. Thus, the interest in the synthesis and characterization of this type of square planar compound is very high and has recently been emphasized by some reports on palladium(II) analogs.[2] In contrast, a number of neutral complexes of the type [Pt(N–N)MeCl] were already prepared by reacting [Pt(COD)MeCl] (COD = 1,5-cyclooctadiene) with the appropriate chelating ligand in refluxing chloroform.[3] Although the cited reaction is a classical one, the synthetic procedure is long and involves delicate steps such as the use of alkyllithium reagents under anaerobic conditions on [Pt(COD)Cl$_2$] and consequently the acidolysis of [Pt(COD)Me$_2$].

Here we report the use of the complex *trans*-[Pt(Me$_2$SO)$_2$MeCl] as a useful synthon to introduce the moieties {PtMe} and {PtMeCl}. The method was applied to the formation of [Pt(terpy)Me]Cl and [Pt(phen)MeCl] that were obtained rapidly in a pure crystalline form and in an almost quantitative

* Dipartimento di Chimica Inorganica, Chimica Analitica e Chimica Fisica, and Istituto di Chimica e Tecnologia dei Prodotti Naturali, (ICTPN-CNR), Sezione di Messina, University of Messina, Messina, Italy.
† Department of Chemistry, University of Nevada, Reno, NV 89557-0020.

yield. The general applicability of the method was checked in the synthesis of analogous complexes containing 1,4,7-triazaheptane (dien),[4] 2,2'-bipyridine, substituted phenanthrolines,[5] 1,2-bis(diphenylphosphino)ethane, and 1,2-bis-(phenylthio)ethane.[6]

A. *CIS*-DICHLOROBIS(DIMETHYLSULFOXIDE)PLATINUM(II)

$$K_2PtCl_4 + 2Me_2SO \rightarrow cis\text{-}[Pt(Me_2SO)_2Cl_2] + 2KCl$$

The synthesis of this starting material has been reported by Price et al.[7] and it takes about 5 h.

- **Caution.** *Dimethylsulfoxide is an irritant and can penetrate the skin very easily. Contact with the skin should be avoided. It is necessary to carry out the reaction in a well-ventilated fume hood.*

Procedure

An aqueous solution of potassium tetrachloroplatinate(II) (1.24 gg, 3 mmol in 10 mL)* is filtered into a 50-mL beaker through paper. This procedure removes impurities due to metallic Pt and/or K_2PtCl_6. Dimethylsulfoxide (0.64 mL, 9 mmol) is added to this solution and, after a gentle hand mixing, the solution is left to stand at room temperature until complete precipitation of yellow needles. The precipitate is filtered, washed with several 5- to 10-mL aliquots of water, ethanol, and diethyl ether, and dried in vacuo for 4 h. The yield is 1.10 g (87% based on K_2PtCl_4).

Anal. Calcd. for *cis*-$[Pt(Me_2SO)_2Cl_2]$: C, 11.38%; H, 2.86%. Found: C, 11.42%; H, 2.82%.

Properties

The complex *cis*-$[Pt(Me_2SO)_2Cl_2]$ is obtained as pale-yellow needle crystals. It is soluble in dimethylsulfoxide and slightly soluble in chloroform and dichloromethane. Two Pt–Cl stretching vibrations occur at 334 and 309 cm^{-1} in nujol mulls. The S–O stretching vibrations are observable at 1157 and 1134 cm^{-1} in nujol mulls.[7]

---

* Strem Chemical Co.

B. TRANS-BIS(DIMETHYLSULFOXIDE)-METHYLCHLOROPLATINUM(II)

$$cis\text{-}[Pt(Me_2SO)_2Cl_2] + Me_4Sn \rightarrow trans\text{-}[Pt(Me_2SO)_2MeCl] + Me_3SnCl$$

This method has been reported by Eaborn et al.[8] The preparation of this compound requires about 36 h.

■ **Caution.** *All operations must be carried out under anhydrous oxygen-free conditions using standard Schlenk techniques. Tetramethyltin is toxic on inhalation and/or other contact and is also flammable. Dimethylsulfoxide is an irritant and can penetrate the skin very easily. Contact with the skin by either reactant should be avoided. It is necessary to carry out the reaction in a well-ventilated fume hood.*

Dimethylsulfoxide is purified by filtration through a 2×25 cm chromatographic column filled with alumina (activated grade I, neutral, 150 mesh),* under a nitrogen atmosphere. This procedure of purification of the solvent is crucial to obtain the indicated yield, since on use of distilled or commercially available anhydrous dimethylsulfoxide,* extensive decomposition takes place.

Cis-dichlorobis(dimethylsulfoxide)platinum(II) (*cis*-$[Pt(Me_2SO)_2Cl_2]$) (1.00 g, 2.37 mmol) is suspended by stirring in 5 mL of dimethylsulfoxide in a 100-mL, round-bottomed, two-necked glass flask equipped with a water-cooled reflux column. Tetramethyltin* (0.6 mL, 0.774 g, d = 1.29 g.mL^{-1}, 4.2 mmol, 99%) is added through a syringe at room temperature. The reaction mixture is heated to 80°C, using a thermostated bath, and allowed to react for 24 h. A dark yellowish color can sometimes develop, indicating partial decomposition. An increase in the reaction time does not improve the yields. The solvent is then rapidly removed from the reaction mixture under vacuum at 80°C. This is a crucial step and the removal of dimethylsulfoxide must be as rapid as possible to avoid decomposition. The residue is then washed several times with 10 mL aliquots of diethyl ether, dissolved in 20 mL of dichloromethane, treated with charcoal until a colorless solution is obtained and then filtered. The volume of the solution is reduced to 5–10 mL and 50 mL of diethyl ether is added to cause precipitation. The yield is 0.54 g (56% based on *cis*-$[Pt(Me_2SO)_2Cl_2]$).

Anal. Calcd. for *trans*-$[Pt(Me_2SO)_2MeCl]$: C, 14.95%; H, 3.76%. Found: C, 14.89%; H, 3.81%.

---

* Aldrich Chemical Co., Milwaukee, WI 53233.

Properties

The complex *trans*-[Pt(Me₂SO)₂MeCl] is obtained as off-white crystals. It is soluble in chloroform, dichloromethane, acetone, alcohols, and water (aquation releases a molecule of sulfoxide). The Pt–Cl stretching vibration occurs at 275 cm^{-1} in nujol mulls. If the solvent is not absolutely anhydrous, in CDCl₃, a partial trans to cis isomerization of *trans*-[Pt(Me₂SO)₂MeCl] takes place. Therefore, the overall ^{1}H NMR pattern can be strongly dependent on the temperature and the water content in the solvent, which controls the distribution of the species in equilibrium. ^{1}H NMR of *trans*-[Pt(Me₂SO)₂MeCl] in CDCl₃: δ 3.42 (s, 12H, $^3J_{PtH}$ = 24 Hz), 0.98 (s, 3H, $^2J_{PtH}$ = 80 Hz). ^{195}Pt NMR in CDCl₃: δ − 4352 (chemical shift relative to [PtCl6]$^{2-}$). ^{1}H NMR of *cis*-[Pt(Me₂SO)₂MeCl] in CDCl₃: δ 3.49 (s, 6H, $^3J_{PtH}$ = 36 Hz), 3.21 (s, 6H, $^3J_{PtH}$ = 8 Hz), 1.17 (s, 3H, $^2J_{PtH}$ = 73 Hz). ^{195}Pt NMR in CDCl₃: δ − 4082 (chemical shift relative to [PtCl₆]$^{2-}$).[9]

C. (2,2′:6′,2″-TERPYRIDINE)METHYLPLATINUM(II) CHLORIDE

trans-[Pt(Me₂SO)₂MeCl] + terpy → [Pt(terpy)Me]Cl + 2Me₂SO

■ **Caution.** *The reaction should be carried out in a well-ventilated fume hood.*

The synthesis of this complex requires about 6 h. A methanolic solution of terpy* (70 mg; 0.3 mmol in 25 mL, 98%) is added dropwise to a stirred solution of *trans*-[Pt(Me₂SO)₂MeCl] (120 mg; 0.3 mmol in 25mL) contained in a 250-mL Erlenmeyer flask. Immediately, the color of the solution turns orange and fine crystals of the same color start to precipitate. After stirring for 60 min, the solid dissolves, giving a clear, orange solution. The precipitation is then forced by adding 100 mL of diethyl ether and cooling at − 30°C. The solid is collected by filtration and washed with several 5- to 10-mL portions of cold dichloromethane and diethyl ether, and finally dried in a desiccator over P₄O₁₀, giving 120 mg of orange-yellow, fine needles (Yield 84%).

Anal. Calcd. for [Pt(terpy)Me]Cl: H, 2.95%, C, 40.13%, N, 8.78%. Found: H, 2.89%, C, 40.3%, N, 8.82%.

Properties

The complex [Pt(terpy)Me]Cl is obtained as an orange to yellow, microcrystalline powder depending on the crystal packing. It is soluble in

---

* Aldrich Chemical Co., Milwaukee, WI 53233.

dimethylsulfoxide, water, and alcoholic solvents and very slightly soluble in chloroform and dichloromethane. The electronic spectrum in water has an absorption maximum at 267 ($\varepsilon = 20\,090$ $dm^3\,mol^{-1}\,cm^{-1}$), 313 (8150), and 331 (7890) nm. ^1H NMR (0.01 M) in dmso-d_6: δ 8.85 (dd, 2H, $^3J_{PtH} = 51.0$ Hz); 8.60 (m, 4H); 8.53 (t, 1H); 8.43 (t, 2H); 7.85 (t, 2H); 1.07 (s, 3H, $^2J_{PtH} = 73.6$ Hz).[4]

D. (1,10-PHENANTHROLINE)METHYLCHLOROPLATINUM(II)

trans-[Pt(Me$_2$SO)$_2$MeCl] + phen → [Pt(phen)MeCl] + 2Me$_2$SO

■ **Caution.** *1,10-phenanthroline is an irritant. Dichloromethane is toxic. Any contact with the skin should be avoided. It is necessary to carry out the reaction in a well-ventilated fume hood.*

The preparation of the complex requires about 3 h. A solution of phen* (0.180 g, 1 mmol, 99%) in 10 mL of dichloromethane is added dropwise to a stirred solution (10 mL) of *trans*-[Pt(Me$_2$SO)$_2$MeCl] (0.401 g, 1 mmol) in the same solvent. The reaction mixture rapidly turns orange and a solid of the same color precipitates. After 20 min the precipitation is completed by adding 20 mL of diethyl ether and cooling at $-30°C$ for 2 h. The precipitate is collected by filtration and washed with 5 mL portions of cold methanol and then with diethyl ether. The yield was 0.42 g (98% based on *trans*-[Pt(Me$_2$SO)$_2$MeCl]).

Anal. Calcd. for [Pt(phen)MeCl]: C, 36.67%; H, 2.60%; N, 6.58%. Found: C, 36.90%; H, 2.45%; N, 6.41%.

Properties

The complex [Pt(phen)MeCl] is obtained as an orange, microcrystalline powder. The color of the solid complex can be yellow or orange as a result of differences in crystal packing depending on the precipitation conditions. It is soluble in chloroform and dichloromethane and very slightly soluble in methanol. In dimethylsulfoxide the solubility is driven by solvolysis of chloride. ^1H NMR in CDCl$_3$ (0.02 M at 298 K): δ 9.78 (dd, 1H, $^3J_{Pt-H} = 13.9$ Hz), 9.46 (dd, 1H, 3J$_{Pt-H} = 60.7$ Hz), 8.65 (dd, 1H), 8.57 (dd, 1H), 7.96 (s, 2H), 7.94 (dd, 1H), 7.66 (dd, 1H), 1.35 (s, 3H, $^2J_{Pt-H} = 78.6$ Hz). The NMR data are slightly concentration and temperature dependent.

---

* Aldrich Chemical Co., Milwaukee, WI 53233.

References

1. S. K. Thomson and G. B. Young, *Polyhedron*, 7, 1953 (1988).
2. R. E. Rulke, I. M. Han, C. J. Elsevier, K. Vrieze, P. W. N. M. van Leeuwen, C. F. Roobeek, M. C. Zoutberg, Y. F. Wang, and C. H. Stam, *Inorg. Chim. Acta*, 169, 5 (1990). B. A. Markies, P. Wijkens, J. Boersma, H. Kooijman, N. Veldman, A. L. Spek, and G. van Koten, *Organometallics*, 13, 3244 (1994).
3. H. C. Clark and L. E. Manzer, *J.Organomet. Chem.*, 59, 411 (1973).
4. G. Arena, L. Monsù Scolaro, R. F. Pasternack, and R. Romeo, *Inorg.Chem.*, 34, 2994 (1995).
5. R. Romeo, L. Monsù Scolaro, N. Nastasi, and G. Arena, *Inorg.Chem.*, 35, 5087 (1996).
6. R. Romeo and L. Monsù Scolaro, unpublished results.
7. J. H. Price, A. N. Williamson, R. F. Schramm, and B. B. Wayland, *Inorg. Chem.*, 11, 1280 (1972).
8. C. Eaborn, K. Kundu, and A. J. Pidcock, *J. Chem. Soc. Dalton Trans.*, 933 (1981).
9. R. Romeo, L. Monsù Scolaro, N. Nastasi, B. E. Mann, G. Bruno, and F. Nicolo, *Inorg. Chem.*, 35, 7691 (1996).

27. (N,N-CHELATE)(OLEFIN)PLATINUM(0) COMPLEXES

Submitted by FRANCESCO RUFFO* AUGUSTO DE RENZI,*
and ACHILLE PANUNZI*
Checked by ROBERT G. BERGMAN[†] and WILLIAM H. HOWARD[†]

Phosphines and phosphites have been typically preferred as ligands in the synthesis of platinum group derivatives, whose preparations have been reviewed.[1] Recent results[2] in the field of metal-mediated synthesis indicate that the nitrogen-chelating ligands are a valid alternative to phosphines and phosphites. Here we describe the synthesis of (N,N-chelate)(olefin)platinum(0) complexes, which are suitable precursors for the preparation of several types of 18e⁻ Pt(II) derivatives[3,4] and are involved in rare examples of reversible oxidative-addition/reductive-elimination equilibria.[5] The synthesis of (dmphen)(ethene)platinum (dmphen = 2,9-dimethyl-1,10-phenanthroline) herein described[6] uses bis(1,5-cyclooctadiene)platinum[7] as starting material. Previous syntheses of (N,N-chelate)(olefin)platinum(0) species are an extension of that achieved for the corresponding Pd(0) compounds.[8] The method reported here, which involves the readily available tris(2-norbornene)platinum,[7] appears to be the most convenient[9] in comparison with synthetic approaches analogous to those of Pt(0) phosphine compounds.[10-11]

* Dipartimento di Chimica, Università di Napoli "Federico II", Napoli, Italy
† Department of Chemistry, University of California, Berkeley, CA 94720-1460.

A. 2,9-DIMETHYL-1,10-PHENANTHROLINE

$$[Pt(C_8H_{12})_2] + 3C_2H_4 \rightarrow [Pt(C_2H_4)_3] + 2C_8H_{12}$$

$$[Pt(C_2H_4)_3] + C_{14}H_{12}N_2 \rightarrow [Pt(C_{14}H_{12}N_2)(C_2H_4)] + 2C_2H_4$$

Procedure

Bis(1,5-cyclooctadiene)platinum[7] (0.411 g, 1.0 mmol) is suspended in dry toluene (5 mL) in a 10-mL, two-necked, round-bottomed flask containing a magnetic stir bar and flushed continuously with ethene* at 0°C (ice bath). The complex dissolves within 10 min, giving rise to a yellow solution containing (ethene)₃platinum.[7] Anhydrous dmphen† (0.250 g, 1.2 mmol) is added in 0.050-g portions. When the addition is complete, the ice-bath is removed and the stream of ethene is stopped. Stirring is maintained vigorously while (dmphen)(ethene)platinum separates from the solution in the form of a sparkly, red microcrystalline solid. After 15 min, ethene is readmitted and stirring is stopped. The supernatant liquid is decanted with a syringe. The crystals are washed with toluene (2 × 3 mL) and dried under vacuum (1 torr, 15 min). Yield: 0.35 g (81%).

Analogous Complexes

A toluene solution of (dmphen)(propene)platinum could be obtained according to the above procedure. Although its isolation at room temperature was not achieved owing to decomposition, satisfactory yields of reaction products with electrophiles could be obtained.[12-13]

Properties

(Dmphen)(ethene)platinum can be stored in ethene atmosphere at − 20°C without appreciable decomposition for several weeks. It is soluble in chloroform, which reacts with the complex within a few seconds, affording the five-coordinate product [PtCl(CHCl₂)(dmphen)(ethene)].[14] Methylene

---

* Ethene (99.5% purity) was purchased from SON s.p.a., Napoli, Italy.
† Commercially available as its hydrate (Aldrich). Dmphen was dried as follows: a 3.0-g sample was dissolved in 20 mL of dry methylene chloride containing anhydrous sodium sulphate. After 24 h stirring, filtration under nitrogen and removal of the solvent under vacuum afforded dmphen with a sufficient degree of dryness. The checkers dried dmphen by heating a solid sample to 80–90°C under reduced pressure (60 mtorr) for 6 h. The dryness was checked by recording an NMR spectrum in the presence of a water scavenger i.e., [(η⁵-C₅H₅)₂ZrMe₂]. The checkers also report that significant impurities result if dmphen is not dried before use.

chloride provides a lower solubility, but the oxidative addition of the solvent, which gives rise to $[PtCl(CH_2Cl)(dmphen)(ethene)]$,[14] takes longer. Thus, methylene chloride can be used as solvent for reactions which are complete within a few seconds. Alternatively, (dmphen)(ethene)platinum can be reacted in a suspension of an anydrous nonprotic solvent (e.g., toluene) under ethene atmosphere.

B. (DIMETHYLFUMARATE)(2,9-DIMETHYL-1,10-PHENANTHROLINE)PLATINUM(0)

$$[Pt(C_7H_{10})_3] + E\text{-}MeO_2CCH{=}CHCO_2Me + C_{14}H_{12}N_2 \rightarrow$$
$$[Pt(C_{14}H_{12}N_2)(E\text{-}MeO_2CCH{=}CHCO_2Me)] + 3C_7H_{10}$$

Procedure*

A solution of dimethylfumarate (0.144 g, 1.0 mmol) in dry diethyl ether (10 mL) is transferred under nitrogen with a syringe into a 25-mL, two-necked, round-bottomed flask which contains tris(2-norbornene)platinum[7] (0.477 g, 1.0 mmol) and magnetically stirred. The resulting solution is added with a syringe into another magnetically stirred 25-mL, two-necked, round-bottomed flask which contains anhydrous dmphen (0.208 g, 1.0 mmol) suspended in dry diethyl ether (3–4 mL). The crystallization of (dimethylfumarate)(dmphen)platinum starts at once. After 24 h stirring, the suspension is centrifuged and the supernatant is decanted. The yellow crystals are washed with diethyl ether (3 × 10 mL) and dried under vacuum (1 torr, 15 min). Yield: 0.500 g (90%).

Analogous Complexes

This procedure is useful for the preparation of the complexes listed in Table I (with the exception of the ethene derivative) as well as many others where the olefin bears electron-withdrawing substituents. The yields are higher than 80%.

Properties

The color of the (*N*,*N*-chelate)(olefin)platinum(0) complexes described in this section varies from light to dark yellow. These complexes are air stable and are fairly soluble in chloroform or methylene chloride. The stability in solution increases as the electron-withdrawing properties of the alkene increase. For example, the fumarate or maleate derivatives decompose very

---

* Exclusion of air and use of dry ether is not strictly necessary. Satisfactory yields can be obtained by carrying out reaction in analytical degree ether in air.

TABLE I. Selected ^{1}H NMR^a and Analytical Data for the Platinum(0) Complexes

	NMR			Analysis					
	Olefin^b	Me(H)-CN	OMe	Carbon Calcd.	Found	Hydrogen Calcd.	Found	Nitrogen Calcd.	Found
[Pt(dmphen)(CH$_2$=CH$_2$)]^c	2.20(87,4H)	3.33(6H)		44.55	44.31	3.74	3.89	6.49	6.34
[Pt(dmphen)(CH$_2$=CHCO$_2$Me)]^d	3.2(88,m,1H); 2.65(74,dd,1H); 2.38(68,dd,1H)	3.22(3H); 3.18(3H)	3.63(3H)	44.17	44.16	3.71	3.88	5.72	5.87
[Pt(dmphen)(CH$_2$=CHCN)]	2.5(98,m,1H); 2.25(m,2H)	3.29(3H); 3.07(3H)		44.74	44.65	3.31	3.18	9.21	9.46
[Pt(dmphen)-(E-MeO$_2$CCH=CHCO$_2$Me)]	3.75(87,2H)	3.18(6H)	3.62(6H)	43.88	43.84	3.68	3.65	5.12	5.30
[Pt(dmphen)-(Z-MeO$_2$CCH=CHCO$_2$Me)]	3.45(85,2H)	3.21(6H)	3.65(6H)	43.88	44.04	3.68	3.87	5.12	4.81
[Pt(phen)(CH$_2$=CHCO$_2$Me)]^d	3.43(98,dd,1H); 2.53(78,dd,1H); 2.31(68,dd,1H)	9.73(d,1H); 9.61(d,1H)	3.67(3H)	41.65	41.23	3.06	2.92	6.07	6.11
[Pt(phen)(CH$_2$=CHCN)]	2.61(96,t,1H); 2.30(m,2H)	9.67(d,1H); 9.61(d,1H)		42.06	42.37	2.59	2.49	9.81	9.29
[Pt(phen)-(E-MeO$_2$CCH=CHCO$_2$Me)]	3.90(95,2H)	9.50(d,2H)	3.67(6H)	41.62	41.68	3.10	3.03	5.39	5.30
[Pt(phen)-(Z-MeO$_2$CCH=CHCO$_2$Me)]	3.46(82,2H)	9.53(d,2H)	3.70(6H)	41.62	41.75	3.10	3.22	5.39	5.45

^{a} At 200 or 270 MHz and 25°C; CDCl$_3$ ($\delta = 7.26$ ppm) as solvent [CHCl$_3$ ($\delta = 7.26$ ppm) as internal standard]. Abbreviations: d, doublet; dd, double doublet; m, multiplet; no attribute, singlet; t, triplet.
^{b} $^2J_{Pt-H}$ (Hz) in parentheses (when measurable).
^{c} Recorded at 30°C with a 60-MHz CW spectrometer which allowed the spectrum to be recorded within 20 sec from dissolution. The checkers recorded the spectrum under more leisurely conditions in C$_6$D$_6$, where the complex, although poorly soluble, is more stable.
^{d} Recorded within 2 min from dissolution. On longer standing, the signals of decomposition products become detectable.

slowly in chloroform, while (dmphen)(methyl acrylate)platinum undergoes oxidative addition of $CHCl_3$ within a few hours. The NMR data of some (N,N-chelate)(olefin)platinum(0) complexes have been reported.[6]

References

1. Chapter 2 of *Inorg. Synth.*, **28** (1990) is dedicated to the synthesis of low-valent complexes of Rh, Ir, Ni, Pd, and Pt.
2. A. Togni and L. M. Venanzi, *Angew. Chem. Int. Ed. Engl.*, **33**, 497 (1994).
3. V. G. Albano, G. Natile, and A. Panunzi, *Coord. Chem. Rev.*, **133**, 67 (1994).
4. V. G. Albano, C. Castellari, M. Monari, V. De Felice, M. L. Ferrara, and F. Ruffo, *Organometallics*, **14**, 4213 (1995) and references therein.
5. V. De Felice, A. Panunzi, F. Ruffo, and B. Åkermark, *Acta Chem. Scand.*, **46**, 499 (1992).
6. V. De Felice, A. De Renzi, F. Ruffo, and D. Tesauro, *Inorg. Chim. Acta*, **219**, 169 (1994).
7. L. E. Crascall and J. L. Spencer, *Inorg. Synth.*, **28**, 126 (1990).
8. N. Ito, T. Saji, and S. Aoyagui, *Bull. Chem. Soc. Jpn.*, **58**, 2323 (1985). More recent syntheses also used the same approach, which involves bis(dibenzylideneacetone)platinum(0) as starting material: R. van Asselt, C. J. Elsevier, W. J. J. Smeets, and A. L. Spek, *Inorg. Chem.*, **33**, 1521 (1994).
9. M. Bisbiglia, M. E. Cucciolito, V. De Felice, and F. Ruffo, *Rend. Accad. Sci. Fis. Mat. Soc. Naz. Sci. Lett. Arti in Napoli*, Ed. Liguori, **62** (1995).
10. For the use of ethanolic KOH, see section 33 of ref. 1.
11. For the use of sodium naphthalide, see section 35 of ref. 1.
12. V. G. Albano, C. Castellari, V. De Felice, A. Panunzi, and F. Ruffo, *J. Organomet. Chem.*, **425**, 177 (1992).
13. V. De Felice, M. Funicello, A. Panunzi, and F. Ruffo, *J. Organomet. Chem.*, **403**, 243 (1991).
14. V. De Felice, B. Giovannitti, A. Panunzi, F. Ruffo, and D. Tesauro, *Gazz. Chim. Ital.*, **123**, 65 (1993).

28. DIMETHYLPALLADIUM(II) AND MONOMETHYLPALLADIUM(II) REAGENTS AND COMPLEXES

Submitted by PETER K. BYERS,* ALLAN J. CANTY,[†] HONG JIN,[†] DENNIS KRUIS,[‡] BERTUS A. MARKIES,[‡] JAAP BOERSMA,[‡] and GERARD VAN KOTEN[‡]
Checked by GEOFFREY S. HILL,[§] MICHAEL J. IRWIN,[§] LOUIS M. RENDINA,[§] and RICHARD J. PUDDEPHATT[§]

The displacement of a weak ligand by a stronger donor ligand is a widely used strategy in the synthesis of organometallic complexes. This strategy is

* School of Chemistry, University of Birmingham, Edgbaston, Birmingham B15 2TT, Great Britain.
[†] Chemistry Department, University of Tasmania, Hobart, Tasmania 7001, Australia.
[‡] Department of Metal-Mediated Synthesis, Debye Institute, Utrecht University, Padualaan 8, 3584 CH Utrecht, The Netherlands.
[§] Department of Chemistry, University of Western Ontario, London, Ontario N6A 5B7, Canada.

ideal when the labile reagent complex is stable in air in the solid state at convenient temperatures. Several suitable organopalladium(II) reagents have been developed recently, in particular $[PdMe_2(tmeda)]$ (tmeda = N,N,N',N'-tetramethylethylenediamine)[1,2] and $[PdMe_2(pyridazine)]_n$[3,4] for the synthesis of complexes of 2,2'-bipyridyl (bpy), tertiary phosphines, and 1,4,7-tri-thiacyclononane. The reagents PdIMe(tmeda)[5] and $[PdMe(SMe_2)(\mu\text{-}X)]_2$ (X = Cl, Br, I)[3,4,6-9] are also being used for the synthesis of a wide range of monomethylpalladium(II) complexes. The dimethylpalladium(II) reagents may also be used for the synthesis of a wide range of monomethyl-palladium(II) complexes. The dimethylpalladium(II) reagents may also be used for the synthesis of organopalladium(IV) complexes.[1,7,10-16]

The synthesis of the tmeda, pyridazine, and dimethylsulfide complexes is described here, followed by representative syntheses of bpy, PPh_3, and bis(diphenylphosphino)ferrocene (dppf) complexes. Anaerobic techniques are used for organolithium and oxidative addition reactions, and are recommended, although not essential, for the subsequent ligand-exchange reactions. Products were stored at ambient temperature, preferably under nitrogen, except for $PdMe_2(tmeda)$, $PdMe_2(pyridazine)$, and $[PdMe(SMe_2)$-$(\mu\text{-}X)]_2$ (X = Cl, Br, I), which were stored at $-20°C$.

A. *TRANS*-DICHLOROBIS(DIMETHYLSULFIDE)PALLADIUM(II)

$$PdCl_2 + 2SMe_2 \rightarrow trans\text{-}[PdCl_2(SMe_2)_2]$$

■ **Caution.** *Dimethylsulfide is toxic and has an unpleasant odor. The synthesis must be carried out in a well-ventilated hood.*

Dimethylsulfide (0.40 mL, 10.5 mmol) is added at ambient temperature to a magnetically stirred suspension of $PdCl_2$ (0.46 g, 4.3 mmol) in dichloromethane (30 mL) under a nitrogen atmosphere in a 100-mL, round-bottomed flask. The suspension is stirred for 1 h, giving a clear orange solution. The solvent and the excess of dimethylsulfide are evaporated in a vacuum, the resulting orange solid is redissolved in dichloromethane (15 mL) and filtered through filter-aid (Celite). The filter was rinsed with dichloromethane (2 × 10 mL). The combined fractions were collected in a 50-mL, round-bottomed flask and evaporated to dryness in a vacuum to give a microcrystalline, orange solid. The yield is 0.74 g (100%). ^1H NMR in $CDCl_3$: δ 2.39 (s, SMe_2). ^{13}C NMR in $CDCl_3$: δ 22.93 (s, SMe_2).

Anal. Calcd. for C, 15.9; H, 4.0. Found: C, 15.9; H, 4.0.

Properties

The orange solid may be stored at ambient temperature and is soluble in a range of organic solvents.

B. *TRANS*-DIBROMOBIS(DIMETHYLSULFIDE)PALLADIUM(II)

$$trans\text{-}[PdCl_2(SMe_2)_2] + 2KBr \rightarrow trans\text{-}[PdBr_2(SMe_2)_2] + 2KCl$$

Potassium bromide (0.60 g, 5.0 mmol) in water (2 mL) is added at ambient temperature to a magnetically stirred orange solution of *trans*-$[PdCl_2(SMe_2)_2]$ (0.50 g, 1.66 mmol) in acetone (15 mL) in a 50-mL, round-bottomed flask. A yellow precipitate and red solution formed immediately, and after stirring for 1 h the solvent is evaporated in a vacuum. The red-orange residue is extracted with benzene (2×2 mL) and the suspension filtered through filter aid (Celite) to remove a yellow-brown solid. The volume of the filtrate is reduced to 5 mL in a vacuum and petroleum ether (10 mL, bp 60–80°C) is added to precipitate the product as a red-orange solid (0.55 g, 85%). ^1H NMR in $CDCl_3$: δ 2.53 (s, SMe_2). ^{13}C NMR in $CDCl_3$: δ 24.3 (s, SMe_2).

Anal. Calcd. for C, 12.3; H, 3.1; S, 16.4. Found: C, 12.5; H, 3.2; S, 16.6.

Properties

The red-orange solid may be stored at ambient temperature and is soluble in a range of organic solvents.

C. DIMETHYL(PYRIDAZINE)PALLADIUM(II)

$$trans\text{-}[PdCl_2(SMe_2)_2] + 2LiMe + C_4H_4N_2 \rightarrow$$
$$1/n[PdMe_2(C_4H_4N_2)]_n + 2LiCl + 2SMe_2$$

■ **Caution.** *Solutions of LiMe are flammable, and SMe$_2$ is toxic and has an unpleasant odor.*

Methyllithium (0.4% LiCl; 3.33 mL, 3.50 mmol) is added to a cooled ($-78°C$) and stirred suspension of *trans*-$[PdCl_2(SMe_2)_2]$ (0.50 g, 1.67 mmol) in anhydrous diethyl ether (100 mL) under a nitrogen atmosphere in a three-necked, 250-mL, round-bottomed flask fitted with a nitrogen inlet, pressure-equalizing dropping funnel, and a septum. The suspension is stirred for 1 h at

− 78°C, followed by stirring at ca. − 40°C until a clear, colorless solution is obtained free from unreacted *trans*-[PdCl₂(SMe₂)₂]. Pyridazine* (1,2-diazine) (0.13 mL, 1.81 mmol) in anhydrous diethyl ether (10 mL) is added dropwise to yield a yellow-orange precipitate, followed by water (2 mL) at − 10°C and rapid filtration in air on a sintered glass filter. This solid is washed with several portions of water (3 × 10 mL) and diethyl ether (4 × 10 mL) and dried immediately under high vacuum at ambient temperature. The yield is 0.26 g (72%). ¹H NMR in (CD₃)₂CO: δ 9.23 m (H3,6), 7.98 m (H4,5), 0.06 s (PdMe).

Properties

The yellow solid may be stored at − 20°C; it slowly darkens over a period of weeks but is still useful in syntheses. The complex is soluble in a range of organic solvents and pyridazine/ligand exchange reactions have been satisfactorily accomplished in acetone, benzene, dichloromethane, and chloroform, but solutions of the complex are unstable in the absence of added ligand.

D. DIMETHYLBIS(DIMETHYLSULFIDE)DI(μ-CHLORO)-DIPALLADIUM(II)

trans-[PdCl₂(SMe₂)₂] + LiMe →

$$1/2\ [PdMe(SMe_2)(\mu\text{-}Cl)]_2 + 2\,LiCl + 2\,SMe_2$$

Methyllithium (0.4% LiCl; 1.71 mL, 1.69 mmol) is added to a cooled (− 78°C) and stirred suspension of *trans*-[PdCl₂(SMe₂)₂] (0.51 g, 1.68 mmol) in anhydrous diethyl ether (100 mL) under a nitrogen atmosphere in a three-necked, 250-mL, round-bottomed flask fitted with a nitrogen inlet and a septum. The suspension is stirred for 1 h at − 78°C, giving a clear colorless solution with some unreacted *trans*-[PdCl₂(SMe₂)₂]. Gradual warming of the mixture to − 15°C gives an orange solution with only traces of *trans*-[PdCl₂(SMe₂)₂] evident. Hydrolysis with 0.2 mL of water and filtration at − 10°C, followed by evaporation of solvent in a vacuum at 0°C gives a black solid. The solid is extracted with dry acetone (5 × 5 mL) and filtered. Dry hexane (30 mL) is added and slow evaporation of solvents in vacuum at 0°C gives the product as a yellow, microcrystalline solid (0.16 g, 44%). ¹H NMR in CDCl₃: δ 2.34 (SMe₂), 0.78 (PdMe).

---

* Aldrich Chemical Co., Milwaukee, WI 53233.

Anal. Calcd. for C, 16.5; H, 4.1. Found: C, 16.6; H, 3.7.

E. DIMETHYLBIS(DIMETHYLSULFIDE)DI(μ-BROMO)-DIPALLADIUM(II)

$trans$-$[PdBr_2(SMe_2)_2]$ + LiMe →

$$1/2 \, [PdMe(SMe_2)(\mu\text{-}Br)]_2 + 2 \, LiBr + 2 \, SMe_2$$

The procedure outlined above is followed, using $trans$-$[PdBr_2(SMe_2)_2]$ (0.50 g, 1.28 mmol) and LiMe (1.22 mL, 1.28 mmol). The yield of the orange-tan product is 0.23 g (68%). ^{1}H NMR in $CDCl_3$: δ 2.43 (SMe_2), 0.87 (PdMe).

Anal. Calcd. for C, 13.7; H, 3.4. Found: C, 13.8; H, 3.4.

F. DIMETHYLBIS(DIMETHYLSULFIDE)DI(μ-IODO)-DIPALLADIUM(II)

$trans$-$[PdCl_2(SMe_2)_2]$ + 2 LiMe + MeI →

$$1/2 \, [PdMe(SMe_2)(\mu\text{-}I)]_2 + 2 \, LiCl + SMe_2$$

The reaction given above is assumed to occur via the formation of dimethyl-palladium(II) species which react with iodomethane to give ethane and the monomethylpalladium(II) product. The general procedure outlined for the synthesis of $[PdMe_2(\text{pyridazine})]_n$ is employed, using the same quantities of palladium(II) reagent, methyllithium, and diethyl ether solvent, but iodo-methane (1.0 mL, 16.5 mmol) in diethyl ether (10 mL) is added in place of pyridazine. The resulting mixture is stirred at $-40°$C (dry ice/acetone bath) for 1 h, followed by gradual warming to $-15°$C to give a clear, yellow solution. Hydrolysis with 2 mL of water at $-15°$C, filtration, and evapor-ation of diethyl ether solvent in a vacuum at $0°$C gives the product as a yellow-green solid (0.41 g, 80%). Recrystallization is not necessary, al-though acetone/hexane is a suitable solvent system. The yield is 0.23 g (68%). ^{1}H NMR in $CDCl_3$: δ 2.39 (SMe_2), 0.93 (PdMe).

Anal. Calcd. for C, 11.6; H, 2.9. Found: C, 11.9; H, 2.8.

Properties of the Complexes $[PdMe(SMe_2)(\mu\text{-}X)]_2$

The complexes may be stored at $-20°$C for many weeks without any visible sign of decomposition or deterioration of synthetic utility. They are soluble in dichloromethane and chloroform, and partially soluble in acetone. Exchange reactions have been successfully accomplished in these solvents and in diethyl ether and benzene.

G. DIMETHYL (*N,N,N',N'*-TETRAMETHYLETHYLENEDIAMINE)- PALLADIUM(II)

$$1/n[PdCl_2]_n + tmeda \rightarrow PdCl_2(tmeda)$$

$$PdCl_2(tmeda) + 2MeLi \rightarrow PdMe_2(tmeda) + 2LiCl$$

In a round-bottomed flask containing 200 mL of acetonitrile, 6.00 g (33.8 mmol) of palladium(II) chloride* is dissolved at reflux. The solution is cooled to ambient temperature and tetramethylenediamine, tmeda (6 mL, 40 mmol), is added. The yellow precipitate formed is collected by filtration, washed with diethyl ether (3 × 25 mL), and dried in vacuo. Yield: 9.61 g (97%). A suspension of $PdCl_2(tmeda)$ (3.00 g, 10.2 mmol) in diethyl ether (50 mL) is cooled to $-40°C$ under a nitrogen atmosphere in a 100-mL Schlenk tube. Methyllithium (19 mL, 1.6 M in diethyl ether) is added and the mixture is allowed to warm slowly to 0°C after which stirring is continued for 1 h at this temperature. Water (20 mL) is added, upon which the brown suspension becomes black. The water layer is frozen (using liquid nitrogen), allowing decantation of the clear ether layer; conventional separating funnel techniques are also satisfactory. The water layer is extracted with diethyl ether (3 × 50 mL), the combined ether layers are evaporated to dryness, and the residue is dissolved in benzene (30 mL). The solution is filtered through Celite and the filter is rinsed with benzene (2 × 10 mL). Evaporation of the filtrate gives a white, crystalline solid, which was suspended in pentane (50 mL), collected on a sintered-glass filter, and dried in vacuo. Yield: 2.11 g (81%) of pure $PdMe_2(tmeda)$.

If required, the product may be recrystallized by dissolution in diethyl ether (40 mL) and acetone (7 mL) with subsequent filtration of the solution. Addition of pentane (10 mL) and storage overnight at $-20°C$ gives colorless crystals (1.10 g, 43%). ^1H NMR in CD_3COCD_3: δ 2.57, (S, CH_2), 2.40 (s, NMe_2), -0.34 (s, PdMe).

Anal. Calcd. for C, 38.0; H, 8.8; N, 11.1. Found: C, 38.0; H, 8.6; N, 11.0.

Properties

The white solid may be kept at $-20°C$ for months without apparent decomposition. The complex is very soluble in acetone, benzene,

---

* Aldrich Chemical Co., Milwaukee, WI 53233.

dichloromethane, tetrahydrofuran, and chloroform and slightly soluble in diethyl ether. Ligand-exchange reactions are best performed in benzene or acetone. If not freshly prepared, solutions of $PdMe_2$(tmeda) should be filtered through Celite to remove traces of palladium metal.

H. CHLORO(METHYL)(N,N,N',N'-TETRAMETHYLETHYLENE-DIAMINE)PALLADIUM(II)

$$PdMe_2(\text{tmeda}) + MeCOCl \rightarrow PdClMe(\text{tmeda}) + MeCOMe$$

Acetyl chloride (0.25 mL, 3.5 mmol) is added to solution of $PdMe_2$(tmeda) (0.75 g, 3.0 mmol) in benzene (50 mL) at $\sim 5°C$ in a 100-mL, round-bottomed flask. The solution is stirred at this temperature for 1 h, after which the solvent is evaporated. The residue is stirred with pentane (50 mL) and subsequently collected on a sintered-glass filter. The crude product is washed on the filter with pentane (2 × 10 mL) and diethyl ether (2 × 10 mL) and dried in vacuo, giving a slightly yellow power (0.73 g, 90%).

Although it is not necessary, the product may be recrystallized via dissolution in dichloromethane (12 mL), filtration and addition of diethyl ether (15 mL) to induce slow precipitation, collection by filtration, and washing with diethyl ether (2 × 10 mL) (0.50 g, 62%). ^{1}H NMR in $CDCl_3$: δ 2.80–2.45 (m, CH_2), 2.67 (s, NMe_2), 2.57 (s, NMe_2), 0.48 (s, PdMe).

Anal. Calcd. for C, 30.8; H, 7.0; N, 10.3. Found: C, 30.5; H, 7.0; N, 10.3. (N.B.: The acetyl chloride (and the acetyl bromide used in the next preparation) should be checked for acetic acid contamination as the presence of this will severely lower the yield.)

I. BROMO(METHYL)(N,N,N',N-TETRAMETHYLETHYLENE-DIAMINE)PALLADIUM(II)

$$PdMe_2(\text{tmeda}) + MeCOBr \rightarrow PdBrMe(\text{tmeda}) + MeCOMe$$

This complex is prepared in a manner similar to the chloro complex and obtained as a slightly yellow power (0.91 g, 97%). The product may be recrystallized by stirring with dichloromethane (50 mL), filtration, and addition of hexane (50 mL) to induce slow precipitation. The suspension is concentrated to ca. 10 mL and the product collected by filtration and washed with diethyl ether (2 × 10 mL) and dried in vacuo (0.78 g, 83%). ^{1}H NMR in $CDCl_3$: δ 2.80–2.48 (m, CH_2), 2.65 (s, NMe_2), 2.59 (s, NMe_2), 0.50 (s, PdMe).

Anal. Calcd. for C, 26.5; H, 6.0; N, 8.8. Found: C, 26.4; H, 6.1; N, 8.7.

J. IODO(METHYL)(N,N,N',N'-TETRAMETHYLETHYLENE-DIAMINE)PALLADIUM(II)

$$PdMe_2(tmeda) + MeI \rightarrow PdIMe(tmeda) + MeMe$$

Methyl iodide (0.2 mL, 3.2 mmol) is added to an ice-cold solution of $PdMe_2$(tmeda) (0.67 g, 2.7 mmol) in acetone (5 mL) in a 100-mL, round-bottomed flask. A yellow precipitate is immediately formed, accompanied by ethane evolution. The suspension is stirred at this temperature for 15 min, after which hexane (80 mL) is added. The bright yellow powder is collected by filtration, washed with hexane (2×10 mL), and dried in vacuo (0.91 g, 94%). The product may be recrystallized via dissolution in dichloromethane (3.5 mL), filtration, and slow addition of hexane (25 mL) to induce slow crystallization. The suspension is then concentrated to 5 mL and the product collected and washed with hexane (10 mL) (0.59 g, 79%). ^1H NMR in $CDCl_3$: $\delta 2.80-2.45$ (m, CH_2), 2.63 (s, NMe_2), 2.62 (s, NMe_2), 0.45 (s, PdMe).

Anal. Calcd. for C, 23.1; H, 5.2; N, 7.7. Found: C, 23.1; H, 5.1; N, 7.5.

K. SYNTHESIS OF DIMETHYL{BIS(1,1'-DIPHENYLPHOSPHINO)-FERROCENE}PALLADIUM(II)

$$1/n[PdMe_2(C_4H_4N_2)]_n + Fe(C_5H_4PPh_2)_2 \rightarrow PdMe_2(dppf) + C_4H_4N_2$$

Acetone (20 mL) in a 50-mL, round-bottomed flask is deaerated by nitrogen purge, bis(1,1'-diphenylphosphino)ferrocene (dppf)* (0.25 g, 0.45 mmol) is added and $[PdMe_2(pyridazine)]_n$ is added to the stirred suspension. The mixture is stirred at ambient temperature for 15 min and filtered, and the solvent is removed from the filtrate in a vacuum to give crude $[PdMe_2(dppf)]$ as an orange solid in near quantitative yield. Recrystallization is achieved by dissolution in acetone/dichloromethane (15 mL/5 mL), filtration, addition of hexane (15 mL), and slow removal of solvent in a vacuum to ca. 10 mL to give the product as a yellow powder (0.26 g, 68%). ^1H NMR in $CDCl_3$: δ 7.64–7.71 (m, 4H) and 7.31–7.45 (m, 6H) $[C_6H_5]$, 4.22 and 4.08 ('t', 4H) $[C_5H_4]$, 0.13 (d, 2H, J_{H-P} 3 Hz) $[PdCH_3]$. $^{31}P\{^1H\}$ NMR in $CDCl_3$: δ 20.7. ^{13}C NMR: 127.8–134.7 (m) $[C_6H_5]$, 71.6 and 74.8 (s) $[C_5H_4]$, 8.52 (dd, J_{C-H} 15 Hz, J_{C-P} 109 Hz) $[PdCH_3]$.

* Aldrich Chemical Co., Milwaukee, WI 53233.

Anal. Calcd. for C, 62.6; H, 5.0. Found: C, 62.3; H, 5.1.

L. SYNTHESIS OF OTHER COMPLEXES FROM [PdMe₂(pyridazine)]ₙ OR [PdMe(SMe₂)(μ-X)]₂

The following complexes may be obtained similarly to PdMe$_2$(dppf), with yields given in parentheses: cis-PdMe$_2$(PPh$_3$)$_2$] (55%), [PdMe$_2$(bipy)] (82%), *trans*-[PdXMe(PPh$_3$)$_2$] [X = Cl (75%), Br (79%), I (82%)], [PdXMe(bipy)] [X = Cl (73%), Br (75%), I (79%)], [PdXMe(dppf)] [X = Cl (71%), Br (72%), I (70%)].

M. *CIS*-DIMETHYLBIS(TRIPHENYLPHOSPHINE)PALLADIUM(II)

$$PdMe_2(tmeda) + 2\,PPh_3 \rightarrow \textit{cis}\text{-}PdMe_2(PPh_3)_2 + tmeda$$

In a 25-mL, round-bottomed flask, PPh$_3$ (0.22 g, 0.84 mmol) is added to a solution of PdMe$_2$(tmeda) (0.10 g, 0.40 mmol) in benzene (5 mL). A precipitate is formed in a short time. After stirring for 45 min the solvent is removed in vacuo. The product is washed with hexane (2 × 20 mL) and dried in vacuo. The solid is redissolved in a mixture of dichloromethane (14 mL) and acetone (6 mL). Hexane (15 mL) is added and the solution is concentrated in vacuo, giving a white solid. The solid is washed with hexane (2 × 30 mL) and dried in vacuo (0.22 g, 85%). ^{1}H NMR in CDCl$_3$: δ 7.0–7.4 (m, 15H) [C$_6$H$_5$], 0.22 (dd, 3H, $J_{H\text{-}P}$ 4 and 7 Hz) [PdMe]. ^{31}P{^{1}H} NMR in CDCl$_3$: δ 24.0.

N. *TRANS*-CHLORO(METHYL)BIS(TRIPHENYLPHOSPHINE)-PALLADIUM(II)

$$PdClMe(tmeda) + 2\,PPh_3 \rightarrow \textit{trans}\text{-}PdClMe(PPh_3)_2 + tmeda$$

In a 25-mL, round-bottomed flask, PPh$_3$ (0.25 g, 0.95 mmol) is added to a solution of PdClMe(tmeda) (0.12 g, 0.44 mmol) in dichloromethane (10 mL). A precipitate is formed within a few minutes, and after stirring for 45 min the solvent is removed in a vacuum and the white product is washed with hexane (3 × 10 mL) and dried in vacuo (0.29 g, 97%). ^{1}H NMR in CDCl$_3$: δ 7.70 (m, 12H), 7.39 (m, 18H) [C$_6$H$_5$], − 0.01 (t, 3H, $J_{H\text{-}P}$ 6 Hz) [PdMe]. ^{31}P{^{1}H} NMR in CDCl$_3$: δ 30.9.

O. SYNTHESIS OF PPh₃ AND dppf COMPLEXES FROM PdXMe(tmeda)

Using methods K and L, the following complexes can be obtained, with yields in parentheses: PdMe$_2$(bpy) (86%), PdMe$_2$(dppf) (85%), *trans*-[PdXMe(PPh$_3$)$_2$] [X = Br (91%), I (92%)], and [PdXMe(dppf)] [X = Cl (70%), Br (87%), I (70%)]. The yields reported for the dppf complexes are following recrystallization from dichloromethane/hexane. Acetone was used as a solvent for the synthesis of bromo and iodo complexes.

P. CHLORO(METHYL)(2,2′-BIPYRIDYL)PALLADIUM(II)

$$PdClMe(tmeda) + bpy \rightarrow PdClMe(bpy) + tmeda$$

In a 25-mL, round-bottomed flask, bpy (0.065 g, 0.42 mmol) is added to a solution of PdClMe(tmeda) (0.11 g, 0.40 mmol) in dichloromethane (10 mL). After stirring for 45 min hexane (2 mL) is added. The last step is repeated three times and during this period a yellow precipitate is formed. The solvent is evaporated in vacuo and the resulting solid washed with hexane (2 × 10 mL) and dried in vacuo. The product is a mixture of starting material and the desired complex. To ensure complete conversion, the entire procedure is repeated four times, after which the pure, pale yellow product can be obtained by repeated recrystallization from dichloromethane/hexane (0.08 g, 63%). ^{1}H NMR in CDCl$_3$: δ 9.17 (d, 1H), 8.65 (d, 1H), 7.91–8.12 (m, 4H), 7.52 (m, 2H) [bpy], 1.00 (s, 3H) [PdMe].

Anal. Calcd. for C, 42.4; H, 3.5; N, 8.9. Found: C, 41.4; H, 3.5; N, 8.6.

The analogous complexes PdBrMe(bpy) and PdIMe(bpy) can be prepared in a similar way to give yields of 61 and 41%, respectively.

References

1. W. de Graaf, J. Boersma, W. J. J. Smeets, A. L. Spek, and G. van Koten, *Organometallics*, **8**, 2907 (1989).
2. M. A. Bennett, A. J. Canty, J. K. Felixberger, L. M. Rendina, C. S. Sutherland, and A. C. Willis, *Inorg. Chem.*, **32**, 1951 (1993).
3. P. K. Byers and A. J. Canty, *Organometallics*, **9**, 210 (1990).
4. P. K. Byers, A. J. Canty, and R. T. Honeyman, *J. Organomet. Chem.*, **385**, 417 (1990).
5. B. A. Markies, M. H. P. Rietveld, J. Boersma, A. L. Spek, and G. van Koten, *J. Organomet. Chem.*, **424**, C12 (1992).
6. P. K. Byers, A. J. Canty, B. W. Skelton, P. R. Traill, A. A. Watson, and A. H. White, *Organometallics*, **9**, 3080 (1990).
7. P. K. Byers, A. J. Canty, B. W. Skelton, and A. H. White, *J. Organomet. Chem.*, **393**, 299 (1990).
8. V. G. Albano, C. Castellari, M. E. Cucciolito, A. Panunzi, and A. Vitagliano, *Organometallics*, **9**, 1269 (1990).
9. A. J. Canty, H. Jin, A. S. Roberts, P. R. Traill, B. W. Skelton, and A. H. White, *J. Organomet. Chem.*, **489**, 153 (1995).

10. W. de Graaf, J. Boersma, and G. van Koten, *Organometallics*, **9**, 1479 (1990).
11. P. K. Byers, A. J. Canty, B. W. Skelton, and A. H. White, *Organometallics*, **9**, 826 (1990).
12. D. G. Brown, P. K. Byers, and A. J. Canty, *Organometallics*, **9**, 1231 (1990).
13. A. J. Canty, R. T. Honeyman, A. S. Roberts, P. R. Traill, R. Colton, B. W. Skelton, and A. H. White, *J. Organomet. Chem.*, **471**, C8 (1994).
14. W. Klaui, M. Glaum, T. Wagner, and M. A. Bennett, *J. Organomet. Chem.*, **472**, 355 (1994).
15. A. J. Canty, J. Jin, A. S. Roberts, B. W. Skelton, P. R. Traill, and A. H. White, *Organometallics*, **14**, 199 (1995).
16. A. J. Canty, S. D. Fritsche, H. Jin, B. W. Skelton, and A. H. White, *J. Organomet. Chem.*, **490**, C18 (1995).

29. 2,4-PENTANEDIONATOGOLD(I) COMPLEXES AND 2,4-PENTANEDIONATOTHALLIUM

Submitted by JOSE VICENTE* and MARIA-TERESA CHICOTE*
Checked by ROSA M. DÁVILA,* MATTHEW S. ADRIAN,*
and JOHN P. FACKLER, JR.*

The 2,4-pentanedionato or acetylacetonato (acac) complexes are useful synthetic intermediates because they react with protic acids through the reaction (1):[1]

$$[M](acac) + BH \rightarrow [M]B + Hacac \tag{1}$$

In addition, the byproduct acetylacetone is very easily separated from the other reaction product. We have shown the utility of [Au(acac)PPh₃] and of PPN[Au(acac)₂] [PPN = bis(triphenylphosphoranylidene)ammonium = N(PPh₃)₂] in the synthesis of a great variety of organometallic and coordination gold(I) complexes (see Scheme 1).[2] Laguna et al. have also found interesting applications for these complexes.[3]

The preparation of [Au(acac)PPh₃] was first reported by Lewis et al.[4] by the reaction of [AuCl(PPh₃)] with Tl(acac), but few experimental details were given. We report here a simpler method that improves the yield and uses less Tl(acac) than that reported later by Ingold et al.[5]

Although Tl(acac) is a very useful thallium compound and is commercially available (Strem) it is very easily obtained by reacting the cheaper Tl₂CO₃ and acetylacetone as described below. We also report the synthesis of PPN[AuCl₂], which is used to prepare PPN[Au(acac)₂].

* Grupo de Química Organometálica, Departamento de Química Inorgánica, Facultad de Química, Universidad de Murcia, Aptdo. 4021, E-30071 Murcia, Spain.

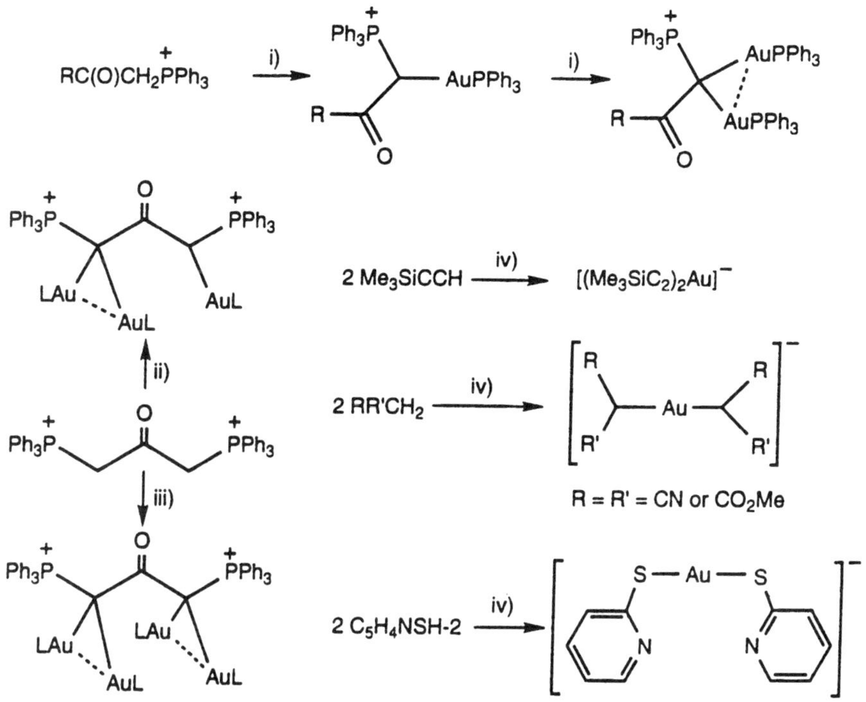

Scheme 1

A. (2,4-PENTANEDIONATO-*O-O'*)THALLIUM

$$Tl_2CO_3 + 2acacH \rightarrow 2Tl(acac) + CO_2 + H_2O$$

■ **Caution.** *Thallium compounds are extremely toxic. Avoid inhalation and contact with skin. Carry out this procedure in a well-ventilated hood and use neoprene gloves. Residues containing thallium compounds must carefully be disposed in accordance with its toxic nature.*

Procedure

A 250-mL, round-bottomed flask equipped with a reflux condenser is charged with a Teflon-coated magnetic stirring bar, anhydrous ethanol (100 mL, commercial 98%), acetylacetone (5 mL, 48.6 mmol, Fluka) and solid Tl_2CO_3 (4 g, 8.53 mmol, Fluka). The resulting suspension is refluxed with vigorous stirring until a pale yellow solution is obtained (usually 4–6 h). If the Tl_2CO_3 is impure, a suspension of a few milligram of a pale cream solid is produced. In this case, filtration of the hot suspension through a Celite pad is required. The resulting solution is concentrated under vacuum (30 mL) to precipitate a first crop of Tl(acac) (3.5 g, 11.5 mmol) which is collected by filtration, washed with diethyl ether (15 mL), and air dried. Concentration of the mother liquor (10 mL) and addition of diethyl ether (50 mL) yields a second crop (1 g, 3.29 mmol). Yield: 87%.

Anal. Calcd. for $C_5H_7O_2Tl$: C, 19.79; H, 2.32. Found: C, 19.80; H, 2.10.

Properties

The compound Tl(acac) is a white, crystalline solid which is air and moisture stable at room temperature. It decomposes at 136°C and is soluble in refluxing ethanol, slightly soluble in acetone and dichloromethane, and insoluble in diethyl ether and *n*-hexane. Its IR spectrum shows $\nu(CO)$ as a strong broad band at 1600 cm^{-1}.

B. BIS(TRIPHENYLPHOSPHORANYLIDENE)AMMONIUM DICHLOROAURATE(I)

$$[AuCl(SC_4H_8)] + PPNCl \rightarrow PPN[AuCl_2] + SC_4H_8$$

Procedure

A 100-mL, round-bottomed flask is charged with a Teflon-coated magnetic stirring bar, dichloromethane (30 mL), $[AuCl(SC_4H_8)]^6$ (SC_4H_8 = tetrahydrothiophene) (552 mg, 1.72 mmol), and PPNCl* (998.4 mg, 1.72 mmol). The resulting solution is stirred for 45 min, then concentrated under vacuum (2 mL), and diethyl ether (30 mL) is added to precipitate $PPN[AuCl_2]$ (1330 mg, 1.64 mmol) as a white solid which is recrystallized from dichloromethane–diethyl ether. Yield: 90%.

Anal. Calcd. for $C_{36}H_{30}AuCl_2NP$: C, 53.61; H, 3.75; N, 1.74; Au, 24.42. Found: C, 53.91; H, 4.05; N, 1.93; Au, 24.86.

Properties

The compound $PPN[AuCl_2]$ is a white, crystalline solid which is air and moisture stable at room temperature. It melts at 210°C and is soluble in

* Aldrich Chemical Co., Milwaukee, WI 53233.

acetone and dichloromethane and insoluble in diethyl ether and *n*-hexane. Its IR spectrum shows $v(\text{AuCl})$ as a strong band at 350 cm^{-1}.

C. 2,4-PENTANEDIONATO(TRIPHENYLPHOSPHINE)GOLD(I)

$$[\text{AuCl}(\text{PPh}_3)] + \text{Tl}(\text{acac}) \rightarrow [\text{Au}(\text{acac})\text{PPh}_3] + \text{TlCl}$$

A 100-mL, 3-necked, round-bottomed flask is charged with a Teflon-coated magnetic stirring bar, acetone (20 mL), $[\text{AuCl}(\text{PPh}_3)]$[7] (750 mg, 1.52 mmol), and Tl(acac) (506 mg, 1.67 mmol) and the resulting suspension is stirred under a nitrogen atmosphere for 5 h. The suspension is concentrated to dryness and the residue is extracted four times (4 × 32 mL) with a solvent mixture of acetone (2 mL) and diethyl ether (30 mL). The combined extracts are filtered through Celite. The resulting solution (concentrated to 3 mL) and *n*-hexane (20 mL), is added to precipitate $[\text{Au}(\text{acac})\text{PPh}_3]$ (711 mg, 1.27 mmol) as a white solid which is filtered off and dried in the air. Yield: 84% based on gold.

Anal. Calcd. for $\text{C}_{23}\text{H}_{22}\text{AuO}_2\text{P}$: C, 49.48; H, 3.97; Au, 35.28. Found: C, 49.26; H, 4.03; Au, 34.86.

Properties

The compound $[\text{Au}(\text{acac})\text{PPh}_3]$ is a white solid which is air and moisture stable at room temperature for 2 or 3 months. However, after longer periods of time it decomposes, becoming a white grayish solid. It can then be recrystallized by dissolving in acetone, filtering over Celite, and adding *n*-hexane. However, it is indefinitely stable when stored in a stoppered flask under nitrogen at 0°C. It melts with decomposition at 192°C. It is very soluble in acetone and dichloromethane, moderately soluble in diethyl ether, and insoluble in *n*-hexane. Its IR spectrum shows two very strong bands assignable to $v(\text{CO})$ at 1646 and 1662 cm^{-1}. NMR data: ^1H (CDCl$_3$, TMS), δ, 2.31 (s, 6H, CH$_3$), 4.69(s, br, 1H, CH), 7.40–7.56 (m, 15H, Ph); $^{31}\text{P}\{^1\text{H}\}$ (CDCl$_3$, H$_3$PO$_4$), δ, 39.61 (s) ppm.

D. BIS(TRIPHENYLPHOSPHORANYLIDENE)AMMONIUM BIS(2,4-PENTANEDIONATO)AURATE(I)

$$\text{PPN}[\text{AuCl}_2] + 2\text{Tl}(\text{acac}) \rightarrow \text{PPN}[\text{Au}(\text{acac})_2] + 2\text{TlCl}$$

A 50-mL, 3-necked, round-bottomed flask fitted with a nitrogen inlet and an outlet to an oil bubbler is charged with a Teflon-coated magnetic stirring bar, freshly distilled dichloromethane (10 mL), PPN[AuCl$_2$] (310 mg, 0.38 mmol), and Tl(acac) (234 mg, 0.77 mmol). The resulting suspension is stirred under a nitrogen atmosphere for 45 min and then filtered through Celite, to remove a small amount of colloidal gold. The filtrate is collected into a 50-mL, round-bottomed flask containing stirred diethyl ether (30 mL). Then PPN[Au(acac)$_2$] (270 mg, 0.29 mmol) precipitates as a white solid which is collected by filtration, washed with diethyl ether (5 mL), and dried in the air. Yield: 76%.

Anal. Calcd. for C$_{46}$H$_{44}$NAuO$_4$P$_2$: C, 59.17; H, 4.75; N, 1.50; Au, 21.09. Found: C, 59.64; H, 4.90; N, 1.50; Au, 21.73.

Properties

The white solid PPN[Au(acac)$_2$] decomposes slowly in the solid state at room temperature, but is stable at low temperature (4–10°C). When heated in a Reichert melting-point apparatus, it becomes grey at 90–94°C without melting. A 3×10^{-4} mol L^{-1} solution in acetone gives a molar conductivity of 97 Ω^{-1} cm^2 mol^{-1} which is in accordance with that of a 1 : 1 electrolyte.[8] It is soluble in acetone and dichloromethane and insoluble in diethyl ether and *n*-hexane. The IR spectrum shows a very broad and strong band assignable to ν(CO) at 1640 cm^{-1}. NMR data: ^{1}H (CDCl$_3$, TMS), δ, 2.23 (s, 12H, CH$_3$), 4.35 (s, 2H, CH), 7.40–7.71 (m, 30H, Ph); ^{31}P{^{1}H} (CDCl$_3$, H$_3$PO$_4$), δ, 21.7 (s) ppm.

References

1. See, for example, R. Usón, J. Vicente, and M. T. Chicote, *J. Organomet. Chem.*, **209**, 271 (1981); J. Vicente, M. D. Bermúdez, M. T. Chicote, and M. J. Sánchez-Santano, *J. Chem. Soc., Chem. Commun.*, 141 (1989); K. Dedeian, P. I. Djurovich, F. O. Garces, G. Carlson, and R. J. Watts, *Inorg. Chem.*, **30**, 1685 (1991); D. Matt and A. Van Dorsselaer, *Polyhedron*, **10**, 1521 (1991); S. A. Westcott, H. P. Blom, T. B. Marder, and R. T. Baker, *J. Am. Chem. Soc.*, **114**, 8863 (1992); M. Kita and M. Nonoyama, *Polyhedron*, **12**, 1027 (1993); J. Forniés, F. Martinez, R. Navarro, E. P. Urriolabeitia, and A. J. Welch, *J. Chem. Soc., Dalton Trans.*, 2147 (1993); M. A. Calvo, A. M. Manotti Lanfredi, L. A. Oro, M. T. Pinillos, C. Tejel, A. Tiripicchio, and F. Ugozzoli, *Inorg. Chem.*, **32**, 1147 (1993).
2. J. Vicente, M. T. Chicote, J. A. Cayuelas, J. Fernandez-Baeza, P. G. Jones, G. M. Sheldrick, and P. Espinet, *J. Chem. Soc., Dalton Trans.*, 1163 (1985); J. Vicente, M. T. Chicote, I. Saura-Llamas, J. Turpin, and J. Fernandez-Baeza, *J. Organomet. Chem.*, **333**, 129 (1987); J. Vicente, M. T. Chicote, I. Saura-Llamas, P. G. Jones, K. Meyer-Bäse, and C. F. Erdbrügger, *Organometallics*, **7**, 997 (1988); J. Vicente, M. T. Chicote, and I. Saura-Llamas, *J. Chem. Soc., Dalton Trans.*, 1941 (1990); J. Vicente, M. T. Chicote, M. C. Lagunas, and P. G. Jones, *J. Chem. Soc., Dalton Trans.*, 2579 (1991); J. Vicente, M. T. Chicote, M. C. Lagunas, and P. G. Jones, *J. Chem. Soc., Chem. Commun.*, 1730 (1991); J. Vicente, M. T. Chicote, I. Saura-Llamas,

and M. C. Lagunas, *J. Chem. Soc., Chem. Commun.*, 915 (1992); J. Vicente, M. T. Chicote, and M. C. Lagunas, *Inorg. Chem.*, **32**, 3748 (1993); J. Vicente, M. T. Chicote, and P. G. Jones, *Inorg. Chem.*, **32**, 4960 (1993); J. Vicente, M. T. Chicote, P. González-Herrero, P. G. Jones, and B. Ahrens, *Angew. Chem., Int. Ed. Engl.*, **33**, 1852 (1994); J. Vicente, M. T. Chicote, P. González-Herrero, and P. G. Jones, *J. Chem. Soc., Dalton Trans.*, 3183 (1994); J. Vicente, M. T. Chicote, and M. D. Abrisqueta, *J. Chem. Soc., Dalton Trans.*, 497 (1995); J. Vicente, M. T. Chicote, R. Guerrero, and P. G. Jones, *J. Chem. Soc., Dalton Trans.*, 1251 (1995); J. Vicente, M. T. Chicote, P. González-Herrero, and P. G. Jones, *J. Chem. Soc., Chem. Commun.*, 745 (1995); J. Vicente, M. T. Chicote, R. Guerrero, and P. G. Jones, *J. Am. Chem. Soc.*, **118**, 699 (1996); J. Vicente, M. T. Chicote, and C. Rubio, *Chem. Ber.*, **129**, 327 (1996).

3. E. J. Fernández, M. C. Gimeno, P. G. Jones, A. Laguna, M. Laguna, and J. M. López-de-Luzuriaga, *J. Chem. Soc., Dalton Trans.*, 3365 (1992); M. C. Gimeno, A. Laguna, M. Laguna, F. Sanmartin, and P. G. Jones, *Organometallics*, **12**, 3984 (1993); *Ibid.*, **13**, 1538 (1994); E. J. Fernández, M. C. Gimeno, P. G. Jones, A. Laguna, M. Laguna, and J. M. López de Luzuriaga, *Angew. Chem. Int. Ed. Engl.*, **33**, 87 (1994).
4. D. Gibson, B. F. G. Johnson, J. Lewis, and C. Oldham, *Chem. Ind.*, 342 (1966); D. Gibson, B. F. G. Johnson, and J. Lewis, *J. Chem. Soc. A*, 367 (1970).
5. B. J. Gregory and C. K. Ingold, J. Chem. Soc. B, 276 (1969).
6. R. Usón, A. Laguna, and M. Laguna, *Inorg. Synth.*, **26**, 85 (1989).
7. R. Usón and A. Laguna in *Organometallic Syntheses*, Vol. 3, R. B. King and J. J. Eisch, Eds., Elsevier, Amsterdam, 1986, p. 325.
8. W. J. Geary, *Coord. Chem. Rev.*, **7**, 81 (1981).

30. TETRAKIS(PYRIDINE)SILVER(2+)PEROXYDISULFATE

$$2AgNO_3 + 8C_5H_5N + 3K_2S_2S_8 \rightarrow 2[Ag(C_5H_5N)_4]S_2O_8 + 2K_2SO_4 + 2KNO_3$$

**Submitted by GEORGE B. KAUFFMAN,* RICHARD A. HOUGHTEN,†
ROBERT E. LIKINS,* PHILIP L. POSSON,* and R. K. RAY‡
Checked by JOHN P. FACKLER, JR.§ and R. THERON STUBBS§**

The dipositive silver ion is unstable because of its powerful oxidizing nature in solution. Because of the high potential required to oxidize Ag(I) to Ag(II), only a few simple compounds of dipositive silver are known. However, inasmuch as this potential can be radically changed by coordination, a number of complexes of dipositive silver have been prepared. The most stable

* Department of Chemistry, California State University, Fresno, CA 93740-0070.
† Torrey Pines Institute for Molecular Studies, 3550 General Atomics Court, San Diego, CA 92121.
‡ Department of Chemistry, Rama Krishna Mission Vivekananda Centenary College, Rahara 743186, 24-Parganas (North), West Bengal, India.
§ Department of Chemistry, Texas A & M University, College Station, TX 77843-3255.

and easily prepared of such coordination compounds contain as ligands bases with nitrogen atoms, especially heterochelates such as pyridines, poly-pyridines, and macrocycles, which are capable of acting as donors in the formation of coordinate covalent bonds.[1-6]

These compounds are usually highly colored substances of general formula $[AgA_4]X_2$ (in which A = one molecule of a monoamine or 1/2 molecule of a diamine), with a square planar tetracoordinate structure,[2,3,6] but a few are of the octahedral trisbidentate variety.[6-8] Most silver(II) complexes are insol-uble in polar and nonpolar solvents, which facilitates their isolation. Some are soluble in aqueous media, but the resulting solutions decompose rapidly. The peroxydisulfates are generally only sparingly soluble in water but dissolve in nitric acid without reduction.[9] The presence of dipositive silver in these compounds has been proven by their paramagnetism, which corres-ponds closely to the value expected for the presence of one unpaired electron.[2,6,10-12]

The first Ag(II) compound to be isolated in the solid state was tetra-kis(pyridine)silver(2 +) peroxydisulfate.[13,14] Its stability can be attributed to its insolubility, the coordination of the Ag(II) ion, and the presence of an anion with an element in a high oxidation state. It can be prepared conve-niently, rapidly, and in high yield by reaction of a solution containing silver nitrate and pyridine with a solution of potassium peroxydisulfate. The corres-ponding, but less stable, nitrate can be prepared by the electrolytic oxidation of silver nitrate in concentrated aqueous pyridine.[15]

Procedure

■ **Caution.** *All operations involving pyridine should be carried out in the hood. Inasmuch as pyridine is flammable, open flames should be absent.*

Twenty grams (0.074 mol) of potassium peroxydisulfate* is dissolved in 1800 mL of ice-cold water contained in a 3-L beaker. Eight milliliters of pyridine (0.1 mol) is added to a solution of 2.00 g (0.0118 mol) of silver nitrate in 40 mL of water, and the resulting solution is slowly added with stirring to the potassium peroxydisulfate solution.

The last traces of the silver–pyridine complex solution may be rinsed into the large beaker with distilled water. The resulting solution soon becomes yellow, then orange, and a precipitate of fine, orange crystals soon begins to deposit. The beaker and contents are allowed to stand in an ice bath until precipitation appears complete (about 15 min). The product is collected

---

* Reagent Grade. Code 2049, Baker & Adamson Products, General Chemical Division. Allied Chemical Corporation, Morristown, NJ 07960.

**TABLE I. Spin Hamiltonian Parameters Obtained from ESR Studies of
$[Ag(C_5H_5N)_4]S_2O_8$ at 77 K[9a]**

Medium	g_x	g_y	$g_z = g_\parallel$	$A\parallel^{Ag}$	$A\perp^{Ag}$	$A\parallel^{N}$	$A\perp^{N}$	Ref.
Solid	2.048	2.100	2.150					27
	2.049	2.098	2.148					21
	2.048	2.100	2.150					9
	2.048	2.100	2.150					26
Diluted with								
$[Cd(C_5H_5N)_4]S_2O_8$		2.04	2.18	34	22	17	22	22
		2.042	2.204	18.0	34.5	21.0	19.6	9
		2.042	2.204	18.0	34.5	21.0	19.6	26

[a] g values $= \pm 0.005$. A values $= \pm 1 \times 10^{-4}$ cm^{-1}.

on a 5-cm, medium-porosity, sintered-glass funnel, washed with five 10-mL portions each of ice-cold water, 95% ethanol, and diethyl ether, and then air-dried. The yield is 6.90 g (95.5%). The product can be stored for several days in a tightly stoppered dark-glass vial, but it decomposes and becomes colorless if allowed to stand in air for several days. When dried in vacuo for several hours over potassium hydroxide, it is stable for several months.

Anal. A freshly dried sample of product is added to water, aqueous ammonia is added, and the suspension is boiled for 15 min to destroy the complex and to reduce the Ag(II) to Ag(I) and the peroxydisulfate to sulfate. The resulting solution is acidified with dilute hydrochloric acid, and the silver is determined as silver chloride. Peroxydisulfate is determined by precipitation of the sulfate in the filtrate with barium chloride. Calcd. for $Ag(C_5H_5N)_4S_2O_8$: Ag, 17.50; $S_2O_8^{2-}$, 31.17. Found: Ag, 17.90; $S_2O_8^{2-}$, 31.32.

Properties

Tetrakis(pyridine)silver(2+) peroxydisulfate consists of orange microscopic needles which are virtually insoluble in water and decompose at 137°C. It dissolves in nitric acid without reduction.[9] The compound shows the typical oxidizing properties of dipositive silver; it oxidizes manganese(II) to permanganate ion, ammonia to nitrogen, iodide ion to iodine, and hydrogen peroxide to oxygen. It has also found use in organic chemistry as a reagent for oxidative transformations such as the oxidation of alcohols, aldehydes, and amines;[16,17] the decarbonylation of acids;[18] the oxidation of benzylic alcohols to carbonyl compounds, aromatic thiols and allylaryl thioethers to arylsulfonic acids, and benzylic carbon–hydrogen bonds to carbonyl groups.[19] Sodium hydroxide solution converts it to black silver(II) oxide.* As d^9 species, silver(II) compounds should be paramagnetic. The magnetic moment of $[Ag(C_5H_5N)_4]S_2O_8$ is 1.71 B.M. at room temperature in close agreement with the "spin only" value. Moreover, the compound follows the Curie law down to 1.6 K.[20] The ESR studies (Table I) reveal a relatively small deviation of $g\parallel$ values from the free-spin value, indicating that the unpaired d-electron is not localized on the silver ion.[6,21,22] Diffused reflectance and single-crystal studies show d–d transitions at 22 and 20.4 kK, respectively, consistent with a square-planar configuration.[2,3,23,24] The presence of the peroxydisulfate ion is confirmed by a broad band at $1230-1330\ cm^{-1}$ and a sharp band at $1060\ cm^{-1}$.[25]

References

1. G. T. Morgan and F. H. Burstall, *J. Chem. Soc.*, **1930**, 2594.
2. H. N. Po, *Coord. Chem. Rev.*, **20**, 171 (1976).
3. W. Levason and M. D. Spicer, *Coord. Chem. Rev.*, **76**, 45 (1987).
4. J. A. McMillan, *Chem. Rev.*, **62**, 65 (1962).
5. P. Rây and D. Sen, *Chemistry of Bi- and Tripositive Silver*, National Institute of Sciences of India, New Delhi, 1960.
6. R. J. Lancashire in *Comprehensive Coordination Chemistry*, Vol. 5, G. Wilkinson, R. D. Gillard, and J. A. McCleverty, Eds., Pergamon, Oxford, England, 1987, pp. 839–849.
7. J. A. Arce Sagüés, R. D. Gillard, and P. A. Williams, *Inorg. Chim. Acta*, **36**, L411 (1979).
8. D. P. Murtha and R. A. Walton, *Inorg. Nucl. Chem. Lett.*, **11**, 301 (1975).
9. J. C. Evans, R. D. Gillard, R. J. Lancashire, and P. H. Morgan, *J. Chem. Soc., Dalton Trans.*, **1980**, 1277.
10. L. Capatas and N. Perakis, *C.R. Acad. Sci.*, **202**, 1773 (1936).
11. W. Klemm, *Z. Anorg. Allgem. Chem.*, **201**, 32 (1931).
12. S. Sugden, *J. Chem. Soc.*, **1932**, 161.
13. G. A. Barbieri, *Gazz. Chim. Ital.*, **42**(II), 7 (1912).
14. G. A. Barbieri, *Atti Accad. Lincei*, **21**(I), 560 (1912).
15. G. A. Barbieri, *Berichte*, **60B**(I), 2424 (1927).
16. T. G. Clarke, N. A. Hampson, J. B. Lee, J. R. Morley, and B. Scanlon, *Can. J. Chem.*, **47**, 1649 (1969).
17. J. B. Lee, C. Parkin, M. J. Shaw, N. A. Hampson, and K. I. MacDonald, *Tetrahedron*, **29**, 751 (1973).
18. J. M. Anderson and J. K. Kochi, *J. Org. Chem.*, **35**, 986 (1970).
19. H. Firouzabadi, P. Salehi, and I. Mohammadpour-Baltork, *Bull. Chem. Soc. Jpn.*, **68**, 2878 (1992).
20. H. M. Gijsman, H. J. Gerritsen, and J. van Handel, *Physica*, **20**, 15 (1954).
21. J. A. McMillan and B. Smaller, *J. Chem. Phys.*, **35**, 1698 (1961).
22. T. Buch, *J. Chem. Phys.*, **43**, 761 (1965).

* Although these types of compounds have been shown to be shock-sensitive as well as explosive on heating, the checker found $[Ag(C_5H_5N)_4]S_2O_8$ does not explode when struck and on heating in an open flame merely melted, then burned, and finally decomposed. However, caution should be exercised when handling these types of compounds.

23. H. G. Hecht and J. B. Frazier, III, *J. Inorg. Nucl. Chem.*, **29**, 613 (1967).
24. R. S. Banerjee and S. Basu, *J. Inorg. Nucl. Chem.*, **26**, 821 (1964).
25. A. Simon and H. Richter, *Naturwissen.*, **44**, 178 (1957).
26. R. K. Ray and G. B. Kauffman, unpublished results.
27. T. S. Johnson and H. G. Hecht, *J. Mol. Spectroscopy*, **17**, 98 (1965).

31. A ONE-POT SYNTHESIS OF TETRAHYDRONIUM TRIS(4,4'-DICARBOXYLATO-2,2'-BIPYRIDINE)-RUTHENIUM(II) DIHYDRATE

Submitted by MD. K. NAZEERUDDIN,*[†] K. KALYANASUNDARAM[†]
and M. GRÄTZEL[†]
Checked by B. PATRICK SULLIVAN[‡] and KEVIN MORRIS[‡]

Tris(2,2'-bipyridine)ruthenium(II) and its substituted analogs have been extensively studied during the last two decades, because of their useful photophysical and redox properties.[1,2] Amongst various polypyridyl ligands, 4,4'-dicarboxy-2,2'-bipyridine (dcbpyH$_2$) is unique, owing to its anchoring capability to metal oxide surfaces.[3] Tris-chelated complexes of Ru(II) containing 4,4'-dicarboxy-2,2'-bipyridine ligand have been used in visible light sensitization of polycrystalline TiO$_2$ electrodes[4] and homogenous catalysis of water oxidation.[5] The light-induced electron injection from the photoexcited state onto the conduction band of the TiO$_2$ is nearly quantitative. The high efficiency could be due to the overlap of the π-orbitals of bipyridine and the carboxylate group, which is in contact with TiO$_2$. The reported synthetic route for this interesting complex suffers from long reaction time (4 days) and several steps.[6] Here we describe a one-pot synthesis, which takes 8 h to get an analytically pure sample of the title compound. The key for the facile synthesis and shorter reaction time is the addition of base to the reaction mixture after formation of the bis complex. The base in the reaction mixture deprotonates the carboxyl groups, causing the nitrogens of the third bipyridine ligand to be more basic. We also report a modified synthesis of the required ligand 4,4'-dicarboxy-2,2'-bipyridine (dcbpyH$_2$),[7] which is free from the often found monocarboxy byproduct 4-methyl-4'-carboxy 2,2'-bipyridine.

* e-mail: Md Khaja. Nazeeruddin@icp.dc.epfl.ch.
[†] Laboratory for Photonics and Interfaces, Swiss Federal Institute of Technology, CH-1015 Lausanne, Switzerland.
[‡] University of Wyoming, Department of Chemistry, Laramie, WY 82071.

Materials. The ligand 4,4′-dimethyl-2,2′-bipyridine (dmbpy) and $RuCl_3 \cdot 3H_2O$ are obtained from Fluka and Johnson Matthey, respectively. All reagents and solvents purchased are reagent grade and are used without further purification.

A. THE LIGAND 4,4′-DICARBOXY-2,2′-BIPYRIDINE (DCBPYH₂)

The ligand 4,4′-dicarboxy-2,2′-bipyridine (1), is synthesized by a modification of the Sasse method.[7] Fifteen grams of 4,4′-dimethyl-2,2′-bipyridine is dissolved in 800 mL of 4 M H_2SO_4 cooled in an ice-salt bath to $-5°C$. To this solution, 32 g of solid $KMnO_4$ is added in 6 portions over a period of 10 min while stirring. After an additional half hour of stirring at the same temperature, another 32 g of $KMnO_4$ is added in a similar manner. The cooling bath is removed and the reaction mixture is heated under reflux on a silicone oil bath for 18 h with stirring. The mixture is then cooled to room temperature and filtered by suction using a Buchner funnel. The black solid is dissolved in 800–1000 mL of 1 M Na_2CO_3 solution and filtered at atmospheric pressure, through a fine-grained filter paper supported on a Buchner funnel. The pH of this filtrate was lowered to 2 by slow addition of a 1:1 mixture (by volume) of CH_3COOH and 4 M HCl. The precipitated white solid is collected on a fine-porosity sintered-glass funnel (G-4) and dried under vacuum at room temperature. Yield: 9–9.5 g (45–48% of crude ligand). The ^{1}H and ^{13}C NMR spectra (Table I) show that the product contains less than 2% of the monocarboxy derivative, 4-methyl-4′-carboxy 2,2′-bipyridine (based on integration of the methyl resonance peak at δ 2.28 against an aromatic doublet at δ 8.78 ppm). The formation of this byproduct can be reduced to < 0.01% by further refluxing the crude compound in 500 mL of 6 M HNO_3 solution for 12 h. The solution is poured into a 1-L beaker containing 600 g of ice cubes and the resulting dcbpyH₂ ligand, in the form of a white solid, is collected on a sintered-glass crucible (G-4) and dried under vacuum at room temperature. Yield: 8.8 g. (44% of the crude ligand).

TABLE I. NMR Data of the Ligand 4,4′-Dicarboxy-2,-bipyridine and Complex 2 in D₂O Solution Containing 0.05 M NaOD[a]

Complex/Ligand	COOH	C-H (6)	C-H (5)	C-H (4)	C-H (2)	C-H (3)
^{1}H NMR of **1**		8.78(d)	7.87(dd)			8.4(d)
^{13}C NMR of **1**	173.64	150.44	124.08	146.99	156.33	121.94
^{1}H NMR of **2**		7.93(d)	7.73(d)			8.94(s)
^{13}C NMR of **2**	166.39	152.94	127.39	140.26	157.23	124.56

[a] δ in ppm from TMS.

B. TETRASODIUM TRIS(4,4′-DICARBOXYLATO-2,2′-BIPYRIDINE)-RU(II)

$$RuCl_2 \cdot 3H_2O + 3LL \xrightarrow[\text{0.5 M NaOH}]{\text{DMF, N}_2} [Na]_4[Ru(LL)_3] \cdot 13H_2O$$

$$LL = 4,4'\text{-dicarboxylato-2,2'-bpy}$$

Procedure

Reagent grade N,N' dimethylformamide (DMF, 60 mL) is added to a three-necked, round-bottomed flask fitted with a reflux condenser, a magnetic stir bar and a gas inlet. The DMF solvent is degassed with a N_2 purge for 10 min. Ruthenium(III) chloride trihydrate (0.261 g, 1 mmol) is dissolved in DMF and this is followed by the addition of the ligand 4,4′-dicarboxyl-2,2′-bpy (0.754 g, 3.09 mmol). The flask is placed in an oil bath and the reaction mixture is refluxed for 3 h. During this reflux time, the solution is stirred magnetically and the temperature of the oil bath is kept at 150°C. After the reflux (3 h), the UV-vis absorption spectrum of the reaction mixture in water shows three bands at 310, 386, and 526 nm. At this stage, 12 mL of 0.5 M of NaOH or $NaHCO_3$ solution in water is added into the reaction flask and the mixture is refluxed for another 4 h. The initial blue-green color of the solution turns red and then orange. The solution is concentrated to 10 mL on a rota-vap.* The precipitated dark orange material is then collected on a medium-porosity sintered-glass filter and washed with 3×15 mL DMF. The resulting solid is recrystallized from methanol–diethylether to give reddish-orange microcrystalline material. Yield: 0.763 g (82%).

Anal. Calcd. for $C_{36}H_{18}N_6O_{12}Na_4Ru$. $13H_2O$: C, 37.5; H, 3.84; N, 7.28. Found: C, 36.56, H, 3.55, N, 7.26.

C. TETRAHYDRONIUM TRIS(4,4′-DICARBOXYLATO-2,2′-BIPYRIDINE)-RUTHENIUM(II)DIHYDRATE

$$[Na]_4[Ru(LL)_3] \cdot 13H_2O \xrightarrow[\text{HCl, pH = 2.5}]{H_2O} [H]_4[Ru(LL)_3] \cdot 2H_2O$$

---

* Checkers' comment: Since our rotovaps are incapable of removing DMF, high-vacuum distillation is utilized. During reduction of the volume of the solvent, all the solvent is removed and then 40 mL of DMF is added. The solid is scraped from the walls of the flask and vigorously stirred for 2 h to homogenize the solid in the DMF. The solid filtered washed 4×10–15 mL DMF and 2×4 mL diethyl ether to remove "trapped" DMF.

Procedure

Tetrasodium tris(4,4'-dicarboxylato-2,2'-bipyridine)ruthenium(II)hydrate (**3**) (0.2 g) is taken in a 50-mL flask containing 20 mL of distilled water. The pH of this solution is taken to 2.5 by the addition of 0.1 M HCl or HNO_3, and the mixture is allowed to stand for 24 h at 0°C. The resulting precipitate is collected on a medium-porosity sintered-glass fritted filter and washed 2–3 times with pH 2.5 water. The yield is 0.13 g (88%).

Anal. Calcd. for $C_{36}H_{22}N_6O_{12}Ru. 2H_2O$: C, 49.87; H, 3.02; N, 9.69. Found: C, 50.02, H, 3.25, N, 10.0.

Properties

The ^1H NMR spectrum of complex **2** in D_2O shows three resonances in the aromatic region. The singlet at 8.94, and two doublets centered at 7.93 and 7.73 ppm (Table I), with respect to TMS, are assigned to C-3H, C-6H, and C-5H, respectively (Fig. 1). Table II presents the absorption and emission spectral properties of complex **2**.[8] The absorption spectrum of this complex

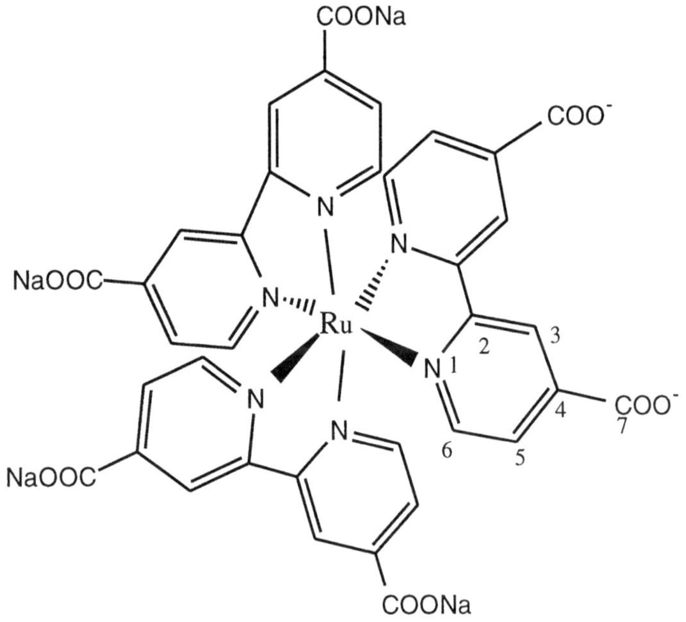

Figure 1. Complex 2.

TABLE II. Absorption and Emission Spectral Properties of Complex 2 in Solution at Room Temperature

Medium	Absorption Maximum $\lambda_{(nm)}$	$\varepsilon(M^{-1} cm^{-1})$	Emission Maximum $\lambda_{(nm)}$	τ (ns)
H_2O, pH $= 9$	466	2.09×10^4	640	700
	442 sh	1.75×10^4		
	342	1.54×10^4		
	304	7.37×10^4		
1M HCl				
	470	2.16×10^4	645	755
	446	1.79×10^4		
	354	1.64×10^4		
	308	6.7×10^4		

shows a strong visible band at 466 nm, due to a charge transfer transition from the metal t_{2g} orbital to the π^* orbital of the ligand. Upon protonation the visible and UV bands are red shifted by 4 and 10 nm, respectively. The emission spectrum of complex **2** in water (pH $= 9$) shows a maximum at 618 nm. Upon lowering the pH to zero, the emission maximum shows a red shift up to pH 3.5 and then blue shifts as the pH is lowered to zero. Complex **2** is much more soluble in H_2O than is complex **3**.

References

1. K. Kalyanasundaram, *Photochemistry of Polypyridine and Porphyrin Complexes*, Academic Press, London, 1992.
2. A. Juris, V. Balzani, F. Barigelleti, S. Campagna, P. Belser, and A. von Zelewsky, *Coord. Chem. Rev.*, **84**, 85 (1988).
3. K. Murakoshi, G. Kano, Y. Wada, S. Yanagida, H. Miyazaki, M. Matsumoto, and S. Murasawa, *J. Electroanal. Chem.*, **396**, 27 (1995); R. Argazzi, C. A. Bignozzi, T. A. Heimer, F. N. Castellano, and G. J. Meyer, *Inorg. Chem.*, **33**, 5741 (1994); M. K. Nazeeruddin, A. Kay, I. Rodicio, R. Humphry, E. Muller, P. Liska, N. Vlachopoulos, and M. Grätzel, *J. Am. Chem. Soc.*, **115**, 6382 (1993).
4. S. J. Valenty, *Anal. Chem.*, **50**, 669 (1978).
5. J. Desilvestro, D. Duonghong, M. Kleijn, and M. Grätzel, *Chimia*, **39**, 102 (1985); J. Desilvestro, M. Grätzel, L. Kavan, J. Moser, and J. Augustynski, *J. Am. Chem. Soc.*, **107**, 2988 (1985).
6. P. J. Delaive, J. T. Lee, H. Abruna, H. W. Sprintschnik, T. J. Meyer, and D. G. Whitten, *Adv. Chem. Ser.*, **168**, 28 (1978); G. Sprintschnik, H. W. Sprintschnik, P. P. Kirsc, and D. G. Whitten, *J. Am. Chem. Soc.*, **99**, 4947 (1977); M. J. Cook, A. P. Lewis, G. S. G. McAuliffe, V. Skarda, A. J. Thomson, J. L. Glasper, and D. J. Robbins, *J. Chem. Soc. Perkin Trans. II*, 1293 (1984).

7. K. D. Bos, J. G. Kraaijkamp, and J. G. Noltes, *Synth. Comm.*, **9**(6), 497 (1979); A. Launikonis, P. A. Lay, A. W. H. Mau, A. M. Sargeson, and W. H. F. Sasse, *Aust. J. Chem.*, **39**, 1053 (1986).
8. Md. K. Nazeeruddin and K. Kalyanasundaram, *Inorg. Chem.*, **28**, 4251 (1989).

32. TRICHLORO[2,2′:6′,2″-TERPYRIDINE]RUTHENIUM(III) AND PHOSPHINE LIGAND DERIVATIVES

Submitted by CAROL A. BESSEL,* RANDOLPH A. LEISING,[†]
LISA F. SZCZEPURA,[‡] WILLIE J. PEREZ,[§]
MY HANG VO HUYHN,[§] and KENNETH J. TAKEUCHI[§]
Checked by KAREN J. BREWER[§§] and SUMNER W. JONES[§§]

Ruthenium(II) chemistry typically involves inert ligand substitution and octahedral geometry about the metal center. These two properties can be exploited to enable the preparation of ruthenium(II) complexes which possess specific and unique ligand ensembles. Herein are described the general synthetic routes in which one or two tertiary phosphine ligands can be added to, or rearranged about, the ruthenium(II) center in specific ligand substitution patterns.[1-4] Phosphine ligands of greatly varying size [PMe_3, cone angle $118°$ and PBz_3 ($Bz = CH_2C_6H_5$), cone angle $148°$] are chosen to demonstrate the generality of the synthetic method. Notably, while phosphine ligands are studied extensively in regard to their steric and electronic ligand effects,[5] most of these phosphine ligand studies involve transition metal complexes which contain only one type of phosphine ligand per metal center. Thus, interest in the preparation of mixed-phosphine complexes stems from the use of these complexes to study the combined and competitive phosphine steric and electronic ligand effects of two different tertiary phosphine ligands.

A. TRICHLORO[2,2′:6′,2″-TERPYRIDINE]RUTHENIUM(III)[1,2]

$$RuCl_3 \cdot xH_2O + trpy \rightarrow RuCl_3(trpy) + xH_2O$$

* Department of Chemistry, Villanova University, Villanova, PA 19085–1688.
† Wilson Greatbatch Ltd., 10,000 Wehrle Drive, Clarence, NY, 14031.
‡ Department of Chemistry, Illinois State University, Normal, IL 61790-4160.
§ Department of Chemistry, SUNY at Buffalo, Buffalo, NY, 14260.
§§ Department of Chemistry, Virginia Polytechnic Institute and State University, Blacksburg, VA 24061.

Procedure

■ **Caution.** *$RuCl_3 \cdot xH_2O$ is corrosive and hygroscopic; store in a vacuum desiccator.*

A 200-mL, round-bottomed flask containing a teflon-coated magnetic stir bar and equipped with a reflux condenser and gas inlet tube is charged with $RuCl_3 \cdot xH_2O^*$ (assumed to be $RuCl_3 \cdot 3H_2O$; 0.653 g, 2.50 mmol), 2,2':6',2''-terpyridine[†] (0.583 g, 2.50 mmol, 1 equiv. trpy), and methanol (50 mL, previously dried by distillation over Mg). While stirring vigorously, the dark brown-black mixture is outgassed with $N_2(g)$ for 5 min. [This is accomplished by flowing a steady stream of $N_2(g)$ through the gas-inlet tube and condenser while the condenser is disconnected and is slightly separated from the round-bottomed flask.] Once the condenser is reattached to the round-bottomed flask, the mixture is heated to reflux under $N_2(g)$ for 3 h. After this time, the reaction mixture is cooled to room temperature and then to 0°C for 0.5 h. The product, a fine brown powder, is vacuum filtered from the red solution. The product is washed with cold methanol (3 × 30 mL) followed by Et_2O (4 × 30 mL) and air dried. The yield of $RuCl_3(trpy)$, **1**, is 1.011 g (2.29 mmol, 92%).[‡]

Properties

The complex $RuCl_3(trpy)$, **1**, is a fine brown solid which is air stable. It is generally insoluble and therefore difficult to characterize. It is used without further purification in the following procedures.

B. *trans*-DICHLORO[2,2':6',2''-TERPYRIDINE](TRIMETHYL OR TRIBENZYLPHOSPHINE)RUTHENIUM(II)[1–3]

$$RuCl_3(trpy) + PR_3 + NEt_3 \rightarrow \textit{trans-}[RuCl_2(trpy)(PR_3)] + Cl^- + NEt_3$$

$$\text{oxidation products}^6$$

$$[\text{where } PR_3 = PMe_3 \ (\textbf{2}) \text{ or } PBz_3 \ (\textbf{3})]$$

---

* $RuCl_3 \cdot xH_2O$ (99.9%) obtained from Johnson-Matthey Catalog Company, Inc., 30 Bond Street, Ward Hill, MA 01835-8099.
† Aldrich Chemical Company, Inc., 1001 West Saint Paul Avenue, Milwaukee, WI 53233.
‡ The checkers obtained a 76% yield of $RuCl_3(trpy)$, **1**.

Procedure

■ **Caution.** *NEt_3 is flammable and corrosive. It should be used in a well-ventilated hood with proper safety equipment; exposure to water should be avoided. The material is outgassed with $N_2(g)$ after each use.*

■ **Caution.** *PMe_3 is an oxygen-sensitive, flammable liquid; PBz_3 is an irritant. These phosphines are best stored in an inert atmosphere glovebox. The 1.0 M PMe_3 solution in toluene is recommended.*

In an inert-atmosphere glovebox, a 100-mL round-bottomed flask containing a 2.5-cm teflon-coated magnetic stir bar and fitted with a reflux condenser and gas inlet tube is charged with $RuCl_3$(trpy) (0.200 g, 0.454 mmol of **1**), PMe_3[7] (0.681 mL of 1.0 M PMe_3 in toluene, 0.681 mmol, 1.5 equiv.),* or PBz_3† (0.276 g, 0.908 mmol, 2.0 equiv.) and N_2-outgassed CH_2Cl_2 (60 mL). Once the reagents are combined, the round-bottomed flask, reflux condenser, and gas inlet tube (closed stopcock) are assembled with Keck clamps and removed from the glovebox. Immediately after removing from the glovebox, the gas inlet tube is connected and opened to a nitrogen manifold in a fume hood. While purging with a strong flow of $N_2(g)$, the reflux condenser is briefly opened to add a sample of $N_2(g)$-outgassed NEt_3 (3 mL, 40.8 mmol, 90 equiv.) as a reductant which converts the metal center from ruthenium(III) to ruthenium(II). This mixture is outgassed with $N_2(g)$ for 5 min and then the condenser is reattached to the round-bottomed flask. The vigorously stirring mixture is heated at reflux under $N_2(g)$ for 1.5 h. The color of the reaction mixture changes from brown to blue-purple.

For *trans*-[$RuCl_2$(trpy)(PMe_3)], **2**, the warm reaction solution is vacuum filtered to remove insoluble ruthenium species using a Buchner funnel. The solids are washed with CH_2Cl_2 (3 × 5 mL). The original filtrate and washings are combined and eluted through a deactivated alumina column (1.0 mL of H_2O/10 mL of basic adsorption alumina, 80–200 mesh; 1 mL of alumina/0.02 g of metal complex) using CH_2Cl_2 as the eluent. The initial blue-purple band is collected. The volume of the eluent is reduced to approximately 5 mL with a rotary evaporator, then the eluent is slowly dripped into a 500-mL Erlenmeyer flask equipped a teflon-coated magnetic stir bar containing rapidly stirring hexanes (200 mL). The blue-purple solid is collected by vacuum filtration, washed with hexanes (3 × 5 mL), and allowed to air dry. The complex can be further dried in a vacuum desiccator. Yield: 0.076 g (0.158 mmol, 35%) *trans*-[$RuCl_2$(trpy)(PMe_3)], **2**.‡

* Aldrich Chemical Company, Inc., Milwaukee, WI 53233 (1.0 M PMe_3 in toluene).
† Strem Chemicals Inc., 7 Mulliken Way, Newburyport, MA 01950.
‡ The checkers obtained a 46% yield of *trans*-[$RuCl_2$(trpy)(PMe_3)], **2**.

For *trans*-[RuCl$_2$(trpy)(PBz$_3$)], **3**, the warm reaction mixture is cooled to room temperature and is vacuum filtered using a Buchner funnel to yield a purple solid. The solid is washed with CH$_2$Cl$_2$ (3 × 5 mL). This product is only slightly soluble in CH$_2$Cl$_2$ and thus is used without further purification in the synthesis of *cis*-[RuCl$_2$(trpy)(PBz$_3$)], **5**. Yield 0.193 g (0.272 mmol, 60%) *trans*-[RuCl$_2$(trpy)(PBz$_3$)], **3**.[†]

Anal. Calcd. for **2**: C$_{18}$H$_{20}$Cl$_2$N$_3$PRu: C, 44.92; H, 4.19. Found: C, 44.87; H, 4.20. Elemental analysis of **3** is not available.

Properties

Complexes **2** and **3** are air stable. Characterization is accomplished by cyclic voltammetry and UV-vis spectroscopy. For *trans*-[RuCl$_2$(trpy)(PMe$_3$)], **2**: E$_{1/2}$ [V vs. SSCE (ΔEp), CH$_2$Cl$_2$][8] = + 0.41(180). UV-Vis spectroscopy in CH$_2$Cl$_2$ gives [λ_{max}, nm (10^{-3} ε, M^{-1} cm^{-1})): 636[shoulder (sh)], 562(4.0), 409(4.5), 330(7.6), 317(15), 284(14).

Complex **3**, *trans*-[RuCl$_2$(trpy)(PBz$_3$)], is only slightly soluble in CH$_2$Cl$_2$, so characterization was completed in dilute solutions. E$_{1/2}$ [V vs. SSCE (ΔEp), CH$_2$Cl$_2$] = + 0.48(80). UV–Vis spectroscopy in CH$_2$Cl$_2$ gives (λ_{max}, nm (10^{-3} ε, M^{-1} cm^{-1})]: 596(sh), 557(4.9), 404(5.2), 330(18), 318(16), 284(sh), 274(19).

Generally, *trans*-[RuCl$_2$(trpy)(PR$_3$)] complexes are inert in the presence of a tenfold excess of a second tertiary phosphine at room temperature while the *cis*-[RuCl$_2$(trpy)(PR$_3$)] isomers undergo rapid, quantitative reaction under the same conditions to provide *trans*-[Ru(Cl)(trpy)(PR$_3$)(PR$_3'$)]$^+$ complexes. This difference in the substitutional reactivity between *trans*- and *cis*-[RuCl$_2$(trpy)(PR$_3$)] complexes is attributed to the trans effect of phosphorus ligands bound to ruthenium(II).[1] A similar *trans*-labilizing effect is observed by Franco and Taube in phosphite–amine complexes of ruthenium(II).[9] It is the ability to stop after the addition of one tertiary phosphine ligand that allows the ultimate incorporation of two different tertiary phosphine ligands within the same ruthenium coordination sphere.

C. *cis*-DICHLORO[2,2':6',2''-TERPYRIDINE](TRIMETHYL OR TRIBENZYLPHOSPHINE)RUTHENIUM(II)[1–3]

$$\textit{trans-}[RuCl_2(trpy)(PR_3)] \rightarrow \textit{cis-}[RuCl_2(trpy)(PR_3)]$$

$$[\text{where } PR_3 = PMe_3 \ (\mathbf{4}) \text{ or } PBz_3 \ (\mathbf{5})]$$

[†] The checkers obtained a 58% yield of *trans*-[RuCl$_2$(trpy)(PBz$_3$)], **3**.

Procedure

A 100-mL, round-bottomed flask containing a teflon-coated magnetic stir bar and equipped with a reflux condenser and gas inlet tube is charged with *trans*-[RuCl$_2$(trpy)(PR$_3$)] [0.070 g, 0.15 mmol where PR$_3$ = PMe$_3$ (**2**) or 0.070 g, 0.099 mmol where PR$_3$ = PBz$_3$ (**3**)] and CH$_2$Cl$_2$ (50 mL for **2**, 150 mL for **3**). The stirring solution is irradiated under N$_2$(g), with a 120-W tungsten lamp for 4 h, causing a color change from blue-purple to red-violet. The irradiation is conducted by clamping the tungsten lamp approximately 2.5 cm from the outside of the round-bottomed flask; the round-bottomed flask is wrapped in aluminum foil to redirect scattered light back into the solution. Irradiation causes the reaction mixture to reflux.

For *cis*-[RuCl$_2$(trpy)(PMe$_3$)], **4**, the solvent is reduced to a volume of 3–5 mL with a rotary evaporator and slowly dripped into a 500-mL Erlenmeyer flask equipped a teflon-coated magnetic stir bar containing rapidly stirring hexanes (200 mL). A red-violet solid is vacuum filtered, washed with hexanes (3 × 5 mL), and air dried. Yield 0.070 g (0.15 mmol, 100%) of *cis*-[RuCl$_2$(trpy)(PMe$_3$)], **4**.

For *cis*-[RuCl$_2$(trpy)(PBz$_3$)], **5**, the reaction mixture is filtered to remove insoluble impurities.* The solvent is reduced to a volume of 3–5 mL with a rotary evaporator and slowly dripped into a 500-mL Erlenmeyer flask equipped a teflon-coated magnetic stir bar containing rapidly stirring hexanes (200 mL). A red-violet solid is vacuum filtered, washed with hexanes (3 × 5 mL), and air dried. Yield 0.053 g (0.074 mmol, 75%) of *cis*-[RuCl$_2$(trpy)(PBz$_3$)], **5**.[†]

Anal. Calcd. for **4**: C$_{18}$H$_{20}$Cl$_2$N$_3$PRu: C, 44.92; H, 4.19. Found: C, 44.84; H, 4.28. Elemental analysis of **5** is not available.

Properties

The red-violet complexes **4** and **5** are air stable and are used without purification in the synthesis of **6–8**. Prolonged storage, even in the solid state, causes increasing amounts of the *trans*-isomer to be observed as an impurity; thus the *cis*-complexes are used within 1–2 days of synthesis. These complexes are readily characterized by cyclic voltammetry and UV-vis spectroscopy. For *cis*-[RuCl$_2$(trpy)(PMe$_3$)], **4**: E$_{1/2}$ [V vs. SSCE (ΔEp), CH$_2$Cl$_2$) = + 0.57(100). UV-vis spectroscopy in CH$_2$Cl$_2$ gives [λ_{max}, nm (10^{-3} ε, M^{-1} cm^{-1})]: 554(4.0), 498(sh), 370(4.2), 319(21), 273(17). The shift to higher

---

* The checkers isolated the *cis*-[RuCl$_2$(trpy)(PBz$_3$)], **5** product by filtering the reaction mixture to remove insoluble impurities and then evaporating the CH$_2$Cl$_2$ solvent.

[†] The checkers obtained a 50% yield of *cis*-[RuCl$_2$(trpy)(PBz$_3$)], **5**.

energy for the λ_{max} values (of the lower-energy adsorption bands) and the more positive $Ru^{III/II}$ potentials are consistent with the stabilization of the $d\pi$ levels in the *cis*-[RuCl$_2$(trpy)(PR$_3$)] isomer when compared to the *trans*-[RuCl$_2$(trpy)(PR$_3$)] isomer.[1]

For *cis*-[RuCl$_2$(trpy)(PBz$_3$)], **5**: $E_{1/2}$ [V vs. SSCE (ΔEp), CH$_2$Cl$_2$] = + 0.62(80). UV-Vis spectroscopy in CH$_2$Cl$_2$ gives [λ_{max}, nm ($10^{-3}\,\varepsilon$, M^{-1} cm^{-1})]: 545(5.4), 491(sh), 373(4.6), 322(26), 276(21).

D. *trans*-CHLORO[2,2':6',2''-TERPYRIDINE]BIS(TRIMETHYL AND/OR TRIBENZYLPHOSPHINE)RUTHENIUM(II) PERCHLORATE[2-4]

$$cis\text{-}[RuCl_2(trpy)(PR_3)] + PR_3' + NaClO_4 \rightarrow$$

$$trans\text{-}[Ru(Cl)(trpy)(PR_3)(PR_3')][ClO_4] + NaCl$$

[where PR$_3$ = PR$_3'$ = PMe$_3$ (**6**), PR$_3$ = PR$_3'$ = PBz$_3$ (**7**),

or PR$_3$ = PMe$_3$ and PR$_3'$ = PBz$_3$ (**8**)]

Procedure

■ **Caution.** *While perchlorate is used as the counterion with a number of ruthenium complexes without incident, perchlorate salts of metal complexes containing organic ligands are potentially explosive. Care should be exercised when using a spatula or stirring rod to mechanically agitate solid perchlorates. These complexes, as well as other perchlorate salts, should be handled in small quantities, using the appropriate safety procedures.*[10]

Trans-[Ru(Cl)(trpy)(PMe$_3$)$_2$](ClO$_4$) (6) or trans-[Ru(Cl)(trpy)(PBz$_3$)$_2$](ClO$_4$) (7). A 100-mL, round-bottomed flask equipped with a teflon-coated magnetic stir bar and a gas inlet tube is charged with **4** or **5** (0.124 mmol) and a 2:1 (v/v) solution of acetone/EtOH (30 mL). The solution is outgassed with N$_2$(g) for 5 min and then PMe$_3$ or PBz$_3$ (0.19 mmol, 1.5 equiv.) is added. The gas inlet tube is reconnected to the round-bottomed flask and the solution is allowed to stir under N$_2$(g) at room temperature for 3 h. The color of the solution changes from red-violet to yellow-brown. After this time, the solvent is completely removed by rotary evaporation. The residue is redissolved in H$_2$O (15 mL), and then a solution of NaClO$_4$ (1 g in 2 mL H$_2$O) is added to the ruthenium solution to precipitate a yellow-brown solid. The

---

* The checkers preferred not to prepare the ClO$_4^-$ salts and thus substituted KPF$_6$ for the NaClO$_4$ to successfully prepare *trans*-[Ru(Cl)(trpy)(PMe$_3$)$_2$][PF$_6$] in 39% yield.

product is collected by vacuum filtration and washed with cold H_2O (3×5 mL) and Et_2O (3×5 mL) before air drying. The yield of *trans*-[Ru(Cl)(trpy)(PMe$_3$)$_2$][ClO$_4$], **6**, is 0.061 g (0.098 mmol, 79%) and the yield of *trans*-[Ru(Cl)(trpy)(PBz$_3$)$_2$)][ClO$_4$], **7**, is 0.128 g (0.119 mmol, 96%).

Trans-[Ru(Cl)(trpy)(PMe$_3$)(PBz$_3$)][ClO$_4$] **(8)**. A 100-mL, round-bottomed flask equipped with a teflon-coated magnetic stir bar and a gas inlet tube is charged with *cis*-[RuCl$_2$(trpy)(PMe$_3$)](ClO$_4$) (0.025 g, 0.053 mmol **4**), PBz$_3$ (0.223 mmol, 1.1 equiv.) and a 2:1 (v/v) solution of acetone/EtOH (30 mL). The solution is outgassed with $N_2(g)$ for 5 min and then the gas inlet tube is reconnected to the round-bottomed flask. The solution is stirred under $N_2(g)$, at room temperature for 3 h. After this time, the solvent is completely removed by rotary evaporation. The resulting residue is redissolved in a minimal amount of 3:1(v/v) 95% EtOH/H$_2$O and then solid NaClO$_4$ (ca. 1 g) is added until precipitation is initiated. The volume of the solution is concentrated using a rotary evaporator. The yellow-brown solid is collected by vacuum filtration, washed with cold H_2O (1–2 mL), and air dried. The crude solid is dissolved in a minimal amount of CH$_3$CN and purified by column chromatography on a deactivated alumina column (1.0 mL of H$_2$O/10 mL of basic adsorption alumina, 80–200 mesh; 1 mL of alumina/0.02 g of metal complex) using 1:1 toluene/CH$_3$CN as the eluent. The initial yellow-brown band is collected and concentrated using rotary evaporator until precipitation is complete (the solution becomes colorless). The solid is collected by vacuum filtration, washed with toluene (2×5 mL) and Et$_2$O (2×5 mL) and air dried. A yield of 0.039 g (0.046 mmol, 87% yield) of *trans*-[Ru(Cl)(trpy)(PMe$_3$)(PBz$_3$)][ClO$_4$], **8**, is obtained.

Formation of (Chloro)Di(phosphine)Ruthenium Complexes via a One-Pot Synthesis—trans-[Ru(Cl)(trpy)(PMe$_3$)$_2$][ClO$_4$] **(6)** *or* *trans-[Ru(Cl)(trpy)(PBz$_3$)$_2$][ClO$_4$]* **(7)**.* A 100-mL round-bottomed flask containing a 2.5-cm teflon-coated stir bar and fitted with a reflux condenser and gas inlet tube is charged with RuCl$_3$(trpy) (0.204 g, 0.462 mmol **1**) and CH$_2$Cl$_2$ (60 mL). This stirring mixture is outgassed with $N_2(g)$ for 5 min. A sample of PMe$_3$ or PBz$_3$ (4–5 equiv.) and Zn(Hg)[11] (5.0 g) is added to the suspension, taking care not to allow any of the Zn(Hg) to adhere to the inner walls of the round-bottomed flask. This stirring mixture is again outgassed with $N_2(g)$ before reattaching the condenser to the round-bottomed flask. The brown suspension is heated at reflux under $N_2(g)$ for 24 h; during this time the color of the mixture becomes dark blue-purple. The solution is then irradiated with

---

* The checkers preferred not to prepare the ClO$_4^-$ salts and thus substituted KPF$_6$ for the NaClO$_4$ to sucessfully prepare *trans*-[Ru(Cl)(trpy)(PMe$_3$)$_2$][PF$_6$] in 39% yield.

a 120-W tungsten light for 12 h; during this time the color changes to yellow-brown. The solution is cooled and then filtered through a 3.5-cm Buchner funnel to remove the Zn(Hg) and all insoluble impurities. The solids are washed with CH_2Cl_2 (3×5 mL). The original filtrate and washings are combined and the solvent is completely removed using a rotary evaporator. The residue is dissolved in a minimal amount of 3:1 $H_2O/EtOH$ (v/v), and excess aqueous $NaClO_4$ (2.0 g in 2 mL of H_2O) is added to the ruthenium solution. The solution is concentrated through the use of a rotary evaporator until a microcrystalline, yellow-brown solid precipitates out of solution and the solution becomes colorless. The solid is collected by vacuum filtration and washed with cold H_2O (2×5 mL) and Et_2O (3×5 mL) before air drying. A yield of 0.257 g (0.414 mmol, 90%) of *trans*-[Ru(Cl)(trpy)(PMe$_3$)$_2$](ClO$_4$), **6**, or 0.319 g (0.296 mmol, 64%) of *trans*-[Ru(Cl)(trpy)(PBz$_3$)$_2$](ClO$_4$), **7**, was obtained.

Anal. Calcd. for **6**, $C_{21}H_{29}Cl_2N_3O_4P_2Ru \cdot 2H_2O$: C, 38.37; H, 5.06. Found: C, 38.01; H, 4.66.

Anal. Calcd. for **7**, $C_{57}H_{53}Cl_2N_3O_4P_2Ru$: C, 63.51; H, 4.96. Found: C, 63.34; H, 5.01.

Anal. Calcd. for **8**, $C_{39}H_{41}Cl_2N_3O_4P_2Ru$: C, 55.13; H, 4.85. Found: C, 54.90; H, 4.90.

Properties

The complexes **6–8** are air stable. Notably, two different tertiary phosphine ligands can be coordinated within the same ruthenium complex, thus providing a means of studying the additive ligand effects of dissimilar tertiary phosphine ligands.[4] The one-pot synthesis of *trans*-[Ru(Cl)(trpy)(PMe$_3$)$_2$](ClO$_4$) and *trans*-[Ru(Cl)(trpy)(PBz$_3$)$_2$](ClO$_4$) gives higher yields than the corresponding multistep syntheses.[2,3] Percentage yields of 90% of **6** and 64% of **7** [based on starting RuCl$_3$(trpy)] are obtained from the one-pot synthesis versus 34% and 43% respectively when each ruthenium product was individually isolated [a three-step synthesis from RuCl$_3$(trpy)].

For *trans*-[Ru(Cl)(trpy)(PMe$_3$)$_2$](ClO$_4$), **6**: $E_{1/2}$ [V vs. SSCE (ΔEp), CH_3CN] $= +0.73$ (70). UV-Vis spectroscopy in CH_3CN gives [λ_{max}, nm ($10^{-3} \varepsilon$, M^{-1} cm^{-1})]: 494(4.5), 451(sh), 346(sh), 308(32), 271(20). ^{1}H NMR (300 MHz, CDCl$_3$ solvent, internal reference SiMe$_4$; δ (multiplicity, integration) [J, Hz] {assignment}): 0.7 (second-order virtually coupled, 1:2:1 triplet, 18H) [3.4] {CH$_3$ of PMe$_3$}; 7.6 (t, 2 H) [6.5] {H(5,5″) of trpy)}; 8.1 (t, 3H) [7.8] {H(4,4′,4″) of trpy}; 8.6 (d, 4H) [7.8] {H(3,3′,3″,5′) of trpy}; 9.0 (d, 2H)

[5.8] {H(6,6″) of trpy}. ^{13}C NMR (300 MHz, CDCl$_3$ solvent, internal reference CDCl$_3$; δ (multiplicity) [J, Hz] {assignment}): 9.7 (t) [13.9] {CH$_3$ of PMe$_3$}; 123.2, 123.9, 127.2, 132.1, 136.8, 152.3, 157.2, 157.7 {trpy}. The presence of a second-order virtually coupled 1:2:1 triplet (PMe$_3$) is characteristic of *trans*-positioned tertiary phosphine ligands. Conductivity: $\Lambda_o = 137.8$, B = 253.6, 1:1 electrolyte.

7: $E_{1/2}$ [V vs. SSCE (ΔEp), CH$_3$CN] = + 0.86(70). UV-Vis spectroscopy in CH$_3$CN gives [λ_{max}, nm ($10^{-3}\varepsilon$, M^{-1} cm^{-1})]: 487(4.5), 436(sh), 332(sh), 313(26), 273(sh).

8: $E_{1/2}$ [V vs. SSCE (ΔEp), CH$_3$CN] = + 0.85 (75). UV-Vis spectroscopy in CH$_3$CN gives [λ_{max}, nm ($10^{-3}\varepsilon$, M^{-1} cm^{-1})]: 491(4.5), 457(sh), 350(sh), 311(30), 274(18).

E. *trans*-NITRO[2,2′:6′,2″-TERPYRIDINE]BIS(TRIMETHYL or TRIBENZYLPHOSPHINE)RUTHENIUM(II) PERCHLORATE[2–4]

trans-[Ru(Cl)(trpy)(PR$_3$)(PR$_3'$)][ClO$_4$] + NaNO$_2$ →

$$trans\text{-}[Ru(NO_2)(trpy)(PR_3)(PR_3')][ClO_4] + NaCl$$

[where PR$_3$ = PR$_3'$ = PMe$_3$ (**9**), PR$_3$ = PR$_3'$ = PBz$_3$ (**10**),

or PR$_3$ = PMe$_3$ and PR$_3'$ = PBz$_3$ (**11**)]

Procedure

■ **Caution.** *Sodium nitrite is an oxidizer and is toxic. Use with appropriate safety precautions.*

A 100-mL round-bottomed flask containing a stir bar and equipped with a reflux condenser and gas inlet tube is charged with **6**, **7**, or **8** (0.283 mmol); 1:1 (v/v) H$_2$O/EtOH solution (50 mL); and NaNO$_2$ (0.391 g, 5.67 mmol, 20 equiv.). After 5 min of outgassing with N$_2$(g), the reflux condenser is reattached to the round-bottomed flask and the mixture is heated at reflux, under N$_2$(g), for 3 h. After this time, the dark-red reaction mixture is cooled to room temperature and an excess of NaClO$_4$ (2.0 g in 20 mL of H$_2$O) is added. The volume of this solution is slowly reduced with a rotary evaporator. Two microcrystalline fractions of complexes are obtained. The first fraction crystallizes out of solution after the volume is reduced and this solid is collected by vacuum filtration, washed with cold H$_2$O (2 × 3 mL), and air dried. The second fraction is obtained by vacuum filtration after adding solid NaClO$_4$ (1 g) to the filtrate and chilling to 0°C. Again, this second fraction is washed with cold H$_2$O (2 × 3 mL), and air dried. If necessary, the product may be purified by dissolving in a minimal amount of CH_3CN and eluting through a deactivated alumina column (1.0 mL of H_2O/10 mL of basic adsorption alumina, 80–200 mesh; 1 mL of alumina/0.02 g of metal complex) using 1:1 (v/v) toluene/CH_3CN as an eluent. The red band is collected. A microcrystalline, dark red solid is obtained by concentrating the ruthenium solution with a rotary evaporator until precipitation is complete (the solvent is nearly colorless) then vacuum filtering. The solid is washed with toluene (2×5 mL) and Et_2O (3×5 mL) before air drying. Yields: 0.154 g (0.243 mmol, 86% yield) of *trans*-[$Ru(NO_2)(trpy)(PMe_3)_2$](ClO_4), **9**; 0.148 g (0.136 mmol, 48% yield) of *trans*-[$Ru(NO_2)(trpy)(PBz_3)_2$](ClO_4), **10**; or 0.0670 g (0.0822 mmol, 92% yield) of *trans*-[$Ru(NO_2)(trpy)(PMe_3)(PBz)_3$]($ClO_4$), **11**.

Anal. Calcd. for **9**, $C_{21}H_{29}ClN_4O_6P_2Ru$: C, 39.91; H, 4.63. Found: C, 39.90; H, 4.62.

Anal. Calcd. for **10**, $C_{57}H_{53}ClN_4O_6P_2Ru \cdot 0.5\ C_6H_5CH_3$: C, 64.05; H, 5.06. Found: C, 63.82; H, 5.10.

Anal. Calcd. for **11**, $C_{39}H_{41}ClN_4O_6P_2Ru$: C, 54.45; H, 4.80. Found: C, 54.65; H, 4.80.

Properties

This reaction demonstrates that once coordinated in *trans*-positions, neither of the two tertiary phosphine ligands are easily removed by typical ligand-substitution schemes (20 equiv. of incoming ligand at reflux temperatures). Thus, the steric and electronic ligand effects of the spectator tertiary phosphine ligands within the ruthenium(II) coordination sphere can be studied without any complications of spectator ligand rearrangement or substitution.

Complexes **9–11** are air stable and give satisfactory elemental analyses. These complexes are readily characterized by multiple analytical techniques. For *trans*-[$Ru(NO_2)(trpy)(PMe_3)_2$](ClO_4), **9**; $E_{1/2}$ [V vs. SSCE (ΔEp), CH_3CN] $= +1.04$ (70) ($i_{p,c}/i_{p,a} = 1.00$). Peak current ratios are used to estimate the amount of decomposition which results from the formation of the (nitro)ruthenium(III) moiety; in-depth analysis of these decomposition pathways are offered.[4] The UV-vis spectroscopy in CH_3CN gives [λ_{max}, nm ($10^{-3}\ \varepsilon$, M^{-1} cm^{-1})]: 450(sh), 430(4.9), 308(30), 272(18). Infrared spectroscopy (nujol mull on NaCl plates; ν_{as}, ν_s): 1324, 1297 cm^{-1}. ^1H NMR (300 MHz, $CDCl_3$ solvent, internal reference $SiMe_4$; δ (multiplicity, integration) [*J*, Hz] {assignment}): 0.7 (second-order virtually coupled, 1:2:1 triplet, 18 H) [3.5] {CH_3 of PMe_3}; 7.7 (t, 2H) [6.6]{H(5,5") trpy}; 8.2 (t, 3H) [7.8] {H(4,4',4") trpy}; 8.6 (d, 4H) [7.9] {H(3,3',3",5') trpy}; 9.7 (d, 2H)

[5.6]{H(6,6″)-trpy}. ^{13}C NMR (300 MHz, CDCl$_3$ solvent, reference CDCl$_3$; δ (multiplicity) [J, Hz] {assignment}): 10.0 (t) [13.4] {CH$_3$ of PMe$_3$}; 122.8, 124.0, 126.9, 135.8, 137.4, 153.4, 155.4, 156.6 {trpy}. Conductivity: Λ_o = 146.9, B = 266.2, 1:1 electrolyte. This complex is also characterized by X-ray crystal structure analysis.[4]

For *trans*-[Ru(NO$_2$)(trpy)(PBz$_3$)$_2$](ClO$_4$), **10**: E$_{1/2}$ [V vs. SSCE (ΔEp), CH$_3$CN] = + 1.17(80) (i$_{p,c}$/i$_{p,a}$ = 0.67). UV-Vis spectroscopy in CH$_3$CN gives [λ_{max}, nm (10^{-3} ε, M^{-1} cm^{-1})]: 443(4.9), 312(30), 273(19). Infrared spectroscopy (nujol mull on NaCl plates; v_{as}, v_s) = 1325, 1285 cm^{-1}.

trans-[Ru(NO$_2$)(trpy)(PMe$_3$)(PBz$_3$)](ClO$_4$), **11**: E$_{1/2}$ [V vs. SSCE (ΔEp), CH$_3$CN] = + 1.13(75) (i$_{p,c}$/i$_{p,a}$ = 0.88). UV-Vis spectroscopy in CH$_3$CN gives [λ_{max}, nm (10^{-3} ε, M^{-1} cm^{-1})]: 450(sh), 430(3.9), 330(sh), 310(22), 275(14). ^{1}H NMR spectrum (300 MHz, CDCl$_3$ solvent, internal reference SiMe$_4$; δ (multiplicity, integration) [J, Hz] {assignment}): 0.49 (d, 9H, J_{P-H} = 7.6 Hz) {PMe$_3$}; 2.8 [d, 6H, J_{P-H} = 5.5] {benzylic protons, PBz$_3$}, 6.4 (d, 6H), 7.1 (m, 9H), 7.6 (t, 2H), 7.5 (t, 1H), 8.1 (t, 2H), 8.3 (d, 2H), 8.4 (d, 2H), 9.6 (d, 2H) {trpy and PBz$_3$ protons}. ^{31}P NMR (400 MHz, CDCl$_3$ solvent, external reference H$_3$PO$_4$; δ (multiplicity) [J, Hz] {assignment}): − 1.54 (d) [$^2J_{P-P}$ = 276.4 Hz] {PMe$_3$}, 1.66 (d) [$^2J_{P-P}$ = 276.4 Hz] {PBz$_3$}.

F. *trans*-NITROSYL[2,2′:6′,2″-TERPYRIDINE]BIS(TRIMETHYL OR TRIBENZYLPHOSPHINE)RUTHENIUM(II) PERCHLORATE[3]

trans-[Ru(NO$_2$)(trpy)(PR$_3$)$_2$](ClO$_4$) + 2HClO$_4$ →

trans-[Ru(NO)(trpy)(PR$_3$)](ClO$_4$)$_3$ + H$_2$O

[where PR$_3$ = PMe$_3$ (**12**) or PBz$_3$ (**13**)]

Procedure

■ **Caution.** *Perchloric acid is an oxidizer and is corrosive. Appropriate safety precautions should be taken.*[10]

A 50-mL Erlenmeyer flask containing a teflon-coated magnetic stir bar is charged with **9** or **10** (0.236 mmol) and distilled CH$_3$CN (10 mL). While vigorously stirring, an excess of HClO$_4$ (0.3 mL of acid, 70% w/w) is added dropwise until the solution changes color from dark red to light yellow. In one portion, Et$_2$O (50 mL) is added to the rapidly stirring solution to precipitate a yellow solid which is then collected by vacuum filtration. The yellow solid is washed with Et$_2$O (3 × 5 mL) before air drying. The yield of *trans*-[Ru(NO)(trpy)(PMe$_3$)$_2$](ClO$_4$)$_3$, **12**, is 0.180 g (0.221 mmol, 94%) and the yield of *trans*-[Ru(NO)(trpy)(PBz$_3$)$_2$](ClO$_4$)$_3$, **13**, is 0.249 g (0.196 mmol, 83%).

Anal. Calcd. for **12**, C$_{21}$H$_{29}$Cl$_3$N$_4$O$_{13}$P$_2$Ru: C, 30.95; H, 3.59. Found: C, 31.02; H, 3.64.

Anal. Calcd. for **13**, C$_{57}$H$_{53}$Cl$_3$N$_4$O$_{13}$P$_2$Ru · 2H$_2$O: C, 52.36; H, 4.39. Found: C, 52.32; H, 4.27.

Properties

This reaction demonstrates the modification of a targeted ligand (NO$_2^-$ to NO$^+$) without changes in the geometry of the complex or coordination of the spectator ligands.

Complexes **12** and **13** are air stable. Characterization may be accomplished by cyclic voltammetry and UV-vis and infrared spectroscopy. For *trans*-[Ru(NO)(trpy)(PMe$_3$)$_2$](ClO$_4$)$_3$, **12**: E$_{1/2}$ [V vs. SSCE (ΔEp), CH$_3$CN] = + 0.26(70). UV-Vis spectroscopy in CH$_3$CN gives [λ$_{max}$, nm (10^{-3} ε, M^{-1} cm^{-1})]: 368(9.2), 302(sh), 284(22). Infrared spectroscopy (nujol mull on NaCl plates; v_{N-O}) = 1910 cm^{-1}.

For *trans*-[Ru(NO)(trpy)(PBz$_3$)$_2$](ClO$_4$)$_3$, **13**: E$_{1/2}$ [V vs. SSCE (ΔEp), CH$_3$CN] = + 0.42(60). UV-Vis spectroscopy in CH$_3$CN gives [λ$_{max}$, nm (10^{-3} ε, M^{-1} cm^{-1})]: 373(7.0), 332(sh), 310(25), 275(29). Infrared spectroscopy (nujol mull on NaCl plates; v_{N-O}) = 1895 cm^{-1}.

References

1. B. P. Sullivan, J. M. Calvert, and T. J. Meyer, *Inorg. Chem.*, **19**, 1404 (1980).
2. R. A. Leising, S. A. Kubow, M. R. Churchill, L. A. Buttrey, J. W. Ziller, and K. J. Takeuchi, *Inorg. Chem.*, **29**, 1306 (1990).
3. R. A. Leising, S. A. Kubow, and K. J. Takeuchi, *Inorg. Chem.*, **29**, 4569 (1990).
4. L. F. Szczepura, S. A. Kubow, R. A. Leising, W. J. Perez, M. H. V. Huynh, C. H. Lake, D. G. Churchill, and M. R. Churchill, K. J. Takeuchi, *J. Chem. Soc., Dalton Trans.*, 1463 (1996).
5. C. A. Tolman, *Chem. Rev.*, **77**, 313 (1977).
6. The products arising from the oxidation of triethylamine are not established, see ref. 1.
7. M. L. Leutkens, Jr., A. P. Sattelberger, H. H. Murray, J. D. Basil, and J. P. Fackler, Jr. *Inorg. Synth.*, **26**, 7 (1989).
8. For the electrochemical measurements reported herein, all cyclic voltammetry measurements are performed in CH$_2$Cl$_2$ with 0.1 M tetra-*n*-butylammonium tetrafluoroborate (Bu$_4$NBF$_4$) as supporting electrolyte, while measurements in CH$_3$CN use 0.1 M tetra-ethylammonium perchlorate. Cyclic voltammetry measurements are performed in a three-electrode, one-compartment cell equipped with a Pt working electrode, a Pt auxiliary electrode, and a saturated sodium chloride calomel (SSCE) reference electrode. E$_{1/2}$ = (E$_{p,a}$ + E$_{p,c}$)/2; ΔE$_p$ = E$_{p,c}$ − E$_{p,a}$. E$_{1/2}$ and ΔE$_p$ values are measured at 100 mV/sec. Ferrocene is used as a reference in the measurement of the electrochemical potentials.

9. D. W. Franco and H. Taube, *Inorg. Chem.*, **17**, 571 (1978).

10. (a) W. C. Wolsey, *J. Chem. Ed.*, **50**, A335 (1973). (b) K. Raymond, *Chem. Eng. News*, **61**, 4 (1983). (c) A. A. Schilt, *Perchloric Acid and Perchlorates*, The G. Frederick Smith Chemical Company, Columbus, OH, 1979.

11. Zinc amalgam [Zn(Hg)] is prepared as follows: A 5.00-g sample of Zn metal is added to 100-mL beaker containing 50 mL of a 6 N HCl solution which was chilled in an ice bath. Solid $HgCl_2$ (ca. 0.5 g) is quickly added to the solution until the evolution of $H_2(g)$ stops. The acid solution is decanted from the solid Zn(Hg) and the solids are rinsed with distilled H_2O (3 × 50 mL). The solid is thoroughly air dried; yields are variable. Waste should be appropriately disposed.

33. MONOMERIC TETRAHYDROFURAN-STABILIZED MOLYBDENUM(III) HALIDES

Submitted by RINALDO POLI,*[†] STEVEN T. KRUEGER,* and
SUNDEEP P. MATTAMANA*
Checked by KIM R. DUNBAR[‡] and ZHAO HANHUA[‡]

Monomeric, soluble Mo(III) precursors are extremely useful chemical intermediates, their uses ranging from transformation to other Mo(III) coordination compounds,[1] reduction to dinitrogen complexes,[2] conversion to metal–metal bonded compounds,[3] and formation of new materials.[4]

There have been several reports of the preparation of $MoCl_3(THF)_3$, all based on reductive procedures starting from $MoCl_5$.[1,2,5–7] These procedures cannot be extended to the corresponding bromide[8] and iodide[9] compounds for lack of suitable higher-valent precursors. The alternative approach developed for these two compounds[9,10] and reported here is based on the oxidative decarbonylation of lower-valent carbonyl precursors. For the preparation of the bromide compound, the commercially available $Mo(CO)_6$ is a suitable starting compound. For the preparation of the iodide compound, on the other hand, $Mo(CO)_6$ is not sufficiently reactive to be oxidized by I_2 and the more reactive $(\eta^6\text{-}C_6H_5CH_3)Mo(CO)_3$ is used instead. Because the latter compound is not commercially available and its preparation[11] has not been previously described in *Inorganic Synthesis*, we also report here a detailed description of its synthesis.

* Department of Chemistry and Biochemistry, University of Maryland, College Park, MD 20742.

† Current address: Laboratoire de Synthèse et d'Electrosynthèse Organometallique, Faculté de Sciences "Gabriel," 6 Boulevard Gabriel, 21100 Dijon, France.

‡ Department of Chemistry, Michigan State University, East Lansing, MI 48824.

A. TRIBROMOTRIS(TETRAHYDROFURAN)MOLYBDENUM(III), [MoBr₃(THF)₃][10]

$$2Mo(CO)_6 + 2Br_2 \rightarrow [Mo(CO)_4Br_2]_2 + 4CO$$

$$[Mo(CO)_4Br_2]_2 + 4THF \rightarrow 2MoBr_2(CO)_3(THF)_2 + 2CO$$

$$2MoBr_2(CO)_3(THF)_2 + Br_2 + 2THF \rightarrow 2MoBr_3(THF)_3 + 6CO$$

■ **Caution.** *Br₂ is a corrosive liquid with toxic vapors; it should be handled under a fume hood by wearing protective gloves.*

■ **Caution.** *CO is a toxic gas and all operations should be carried out in a well-ventilated fume hood.*

Hexacarbonylmolybdenum(0) (2.22 g, 8.42 mmol) is placed in a 250-mL, round-bottomed flask fitted with a ground glass neck, a side arm with ground-glass stopcock, and a magnetic stir bar. The flask is connected through the side arm to a gas inlet tube of a Schlenk line. The flask is evacuated and the atmosphere is replaced with purified nitrogen. Freshly distilled (under dinitrogen over P_4O_{10}) dichloromethane (25 mL) is introduced into the flask and the apparatus is then cooled to $-78°C$ with a dry ice–acetone bath. Then Br_2 (0.40 mL, 7.8 mmolF) is added to the mixture, which is stirred at $-80°C$ for 1 h and then allowed to warm to room temperature. The flask stopcock is left open to allow CO to escape through the Schlenk line during this procedure. Once the solution has reached room temperature, the solvent is completely removed by evaporation under reduced pressure. The solid residue of $[Mo(CO)_4Br_2]_2$ is dissolved into THF (50 mL) with the evolution of more CO gas and formation of a red solution. This solution is now cooled to 0°C in an ice water bath, followed by the addition of more Br_2 (0.20 mL, 3.9 mmol). Evolution of gas is once again observed. The mixture is stirred at 0°C for 2 h and then overnight at room temperature. This treatment results in the precipitation of the product as a salmon-pink powder. Stirring should not be continued for a longer period of time, because this causes the product to redissolve in the reaction mixture with formation of unknown, presumably oxygen-containing decomposition materials, derived by oxygen abstraction from THF. A gas-chromatographic analysis of the solution indicates the presence of 1,4-dibromobutane. The product is filtered off through a glass frit, thoroughly washed with THF, and dried under vacuum. The yield is 3.72 g (80%).

Anal. Calcd. for $C_{12}H_{24}Br_3MoO_3$: C, 26.11; H, 4.38; Br, 43.43. Found: C, 26.15; H, 4.62; Br, 43.84.

Properties

Tribromotris(tetrahydrofuran)molybdenum(III) is a salmon-pink powder. It deteriorates upon exposure to the laboratory atmosphere, but no significant change is observable upon rapid (<5 min) handling in air. It is soluble in dichloromethane and chloroform, where it rapidly loses THF to afford dinuclear edge-sharing bioctahedral and face-sharing bioctahedral products.[12] Its molecular stereochemistry is meridional both in solution[12] and in the solid state.[13] Its IR spectrum shows a strong vibration for coordinated THF at 840 cm^{-1}. The IR fingerprint region overlaps almost exactly with that of the corresponding $MoCl_3(THF)_3$. The compound is only sparingly soluble in THF, and a prolonged standing in a THF solution result in its decomposition as outlined above. This decomposition is accelerated by heating. It can, however, be conveniently used as a starting material in THF in combination with reagents that consume it rapidly at room temperature (e.g., within a few hours). Its reactions with phosphine ligands have been described.[10]

B. TRICARBONYLTOLUENEMOLYBDENUM(0), $[(\eta^6\text{-}C_6H_5CH_3)Mo(CO)_3]$[11]

$$Mo(CO)_6 + C_6H_5CH_3 \rightarrow Mo(CO)_3(\eta^6\text{-}C_6H_5CH_3)$$

■ **Caution.** *CO is a toxic gas and all operations should be carried out in a well-ventilated fume hood.*

Hexacarbonylmolybdenum(0) (9.00 g, 33.1 mmol) is placed in a 250-mL, round-bottomed flask fitted with a ground glass neck, and a side-arm with ground glass stopcock. To the ground glass neck was attached a water-cooled reflux condenser, which was connected at the top with a mineral oil bubbler. The material was suspended in 100 mL of toluene. The flask was protected from light by wrapping aluminum foil around it and the mixture was brought to a gentle reflux via an oil bath. All the starting compound dissolves before reaching the reflux temperature. Sublimation of small amounts of $Mo(CO)_6$ to the bottom of the condenser is observed at the initial stages of the reaction, but this subsequently disappears as the starting material is converted to the product, while the solution turns canary yellow. Reflux was continued until termination of the CO evolution (48 h), as is evident from the mineral oil bubbler. After cooling to room temperature, the clear yellow solution is reduced in volume by evaporation under reduced pressure. The evaporation is stopped when copious amounts of the product start to crystallize from the solution (ca. 10 mL). At this point, 50 mL of *n*-heptane are rapidly added with stirring. More product crystallizes in the form of yellow microcrystals that rapidly settle at the bottom of the flask. The precipitation is further aided by cooling to $-20°C$ overnight. The solid is now recovered by filtration through a coarse glass frit, washed with *n*-heptane (2×5 mL), and dried under vacuum. The yield is 8.82 g (95%). The material as obtained by this procedure is spectroscopically pure and can be used directly for the preparation of $MoI_3(THF)_3$ (see Section C).

A common problem of this synthesis is the decomposition of the product to a black insoluble and pyrophoric powder (presumably metallic Mo), which is possibly caused by the use of an impure starting material, to the adventitious introduction of air, to excessive exposure to light, or to a combination of those factors. If a black powder is formed, this can be removed at the end of the reaction by filtration through a glass frit; the subsequent workup is as described above. Such decomposition reduces the yield but does not lead to a less pure product. If larger amounts of $Mo(CO)_6$ are used, the sublimation of larger quantities of $Mo(CO)_6$ to the condenser occurs, which may require interruption of the procedure and manual scraping of the sublimed solid back into the flask with a spatula under a flow of N_2.

A variety of cosolvents can be used as catalysts for this reaction (e.g., diglyme, MeCN, or 1,4-dioxane). For instance, the use of 1,4-dioxane catalyst (0.5 mL, 5.8 mmol) in a synthetic procedure starting from $Mo(CO)_6$ (4.88 g, 18.48 mmol) in 50 mL of toluene has reduced the reflux time to 16 h and has led to the formation of 4.00 g of spectroscopically pure product (80% yield). In our hands, the use of MeCN leads to impurities of $Mo(CO)_3(MeCN)_3$ in the final product.

Properties

Tricarbonyltoluenemolybdenum(0) is a yellow crystalline solid, stable in air for brief periods of time as a solid but quite sensitive to air in solution. Prolonged exposure to air leads to the formation of a black insoluble material. The compound is also sensitive to light, especially in solution at high temperatures. It is sparingly soluble in saturated hydrocarbons and very soluble in arenes. The compound reacts with donor solvents with exchange of the toluene ligand. For instance, dissolution in MeCN leads to $Mo(CO)_3(MeCN)_3$. Exposure to CO at 1 atm slowly leads to $Mo(CO)_6$, but this reaction is catalyzed by a number of compounds, including I_2.[14] The compound shows two strong IR absorptions in the CO stretching region at 1969 (A_1) and 1889 (E) cm^{-1} in toluene solution, or at 1983 and 1911 cm^{-1} in *n*-heptane solution.

C. TRIIODOTRIS(TETRAHYDROFURAN)MOLYBDENUM(III), $[MoI_3(THF)_3]$ [9]

$$Mo(CO)_3(\eta^6\text{-}C_6H_5CH_3) + 3THF \rightarrow Mo(CO)_3(THF)_3$$

$$Mo(CO)_3(THF)_3 + 3/2\ I_2 \rightarrow MoI_3(THF)_3 + 3CO$$

■ **Caution.** *CO is a toxic gas and all operations should be carried out in a well-ventilated fume hood.*

A 250-mL round-bottomed flask fitted with a ground glass neck, a side arm with ground-glass stopcock, and a magnetic stir bar is connected through the side arm to a gas inlet tube of a Schlenk line and filled with dinitrogen. Under a gentle stream of nitrogen, tricarbonyl(η^6-toluene)molybdenum(0) (7.6 g, 27.4 mmol) is introduced, followed by tetrahydrofuran (60 mL). The solid dissolves to yield a pale yellow solution, which is then stirred for ca. 30 min at room temperature to complete the transformation to the tricarbonyl-tris(tetrahydrofuran)molybdenum(0) intermediate (IR bands at 1917 s and 1775 vs cm^{-1}). At this point, the solution is cooled to 0°C in an ice water bath and I_2 (10.5 g, 41.5 mmol) is added, resulting in the formation of a deep red colored solution. By keeping the stopcock open to the Schlenk line (CO is evolved during the process), the mixture is stirred at 0°C for 2 h, and then overnight at room temperature. This treatment results in the precipitation of the product as a pale brown powder. Stirring should not be continued for a longer period of time, as this causes the product to redissolve in the reaction mixture with formation of unknown, presumably oxygen-containing decomposition materials, derived by oxygen abstraction from THF. A gas-chromatographic analysis of the solution indicates the presence of 1,4-diiodobutane. The product is filtered off through a glass frit, thoroughly washed with THF, and dried under vacuum. The yield is 14.2 g (73.5%).

Anal. Calcd. for $C_{12}H_{24}I_3MoO_3$: C, 20.8; H, 3.5. Found: C, 21.1; H, 3.8.

Properties

Tribromotris(tetrahydrofuran)molybdenum(III) is a pale brown powder. It deteriorates upon exposure to the laboratory atmosphere, but no significant change is observable upon rapid (<5 min) handling in air. Its IR spectrum shows a strong vibration for coordinated THF at 835 cm^{-1}. The IR finger-print region overlaps almost exactly with that of the corresponding $MoX_3(THF)_3$ (X = Cl, Br). The compound is very sparingly soluble in any common organic solvent, including chloroform, dichloromethane, and THF.

Prolonged standing in a THF solution results in its decomposition as outlined above; however, by brief warming in THF followed by slow cooling, single crystals of the compound have been obtained. Its molecular stereochemistry is meridional in the solid state.[9] Substitution reactions of the THF ligand by phosphine and halide ligands have been described.[9,15]

References

1. M. W. Anker, J. Chatt, G. J. Leigh, and A. G. Wedd, *J. Chem. Soc., Dalton Trans.*, 2639 (1975).
2. J. R. Dilworth and R. L. Richards, *Inorg. Synth.*, **20**, 121 (1980); J. R. Dilworth and J. Zubieta, *Inorg. Synth.*, **24**, 193 (1986); *Inorg. Synth.*, **28**, 36 (1990).
3. J. E. Armstrong, D. A. Edwards, J. J. Maguire, and R. A. Walton, *Inorg. Chem.*, **18**, 1172 (1979).
4. D. Zeng and M. J. Hampden-Smith, *Chem. Mater.*, **4**, 968 (1992).
5. S.-Y. Roh and J. W. Bruno, *Inorg. Chem.*, **25**, 3105 (1986).
6. J. R. Dilworth and J. A. Zubieta, *J. Chem. Soc., Dalton Trans.*, 397 (1983).
7. D. Zeng and M. J. Hampden-Smith, *Polyhedron*, **11**, 2585 (1992).
8. E. A. Allen, K. Feenan, and G. W. A. Fowles, *J. Chem. Soc.*, 1636 (1965).
9. F. A. Cotton and R. Poli, *Inorg. Chem.*, **26**, 1514 (1987).
10. B. E. Owens, R. Poli, and A. L. Rheingold, *Inorg. Chem.*, **28**, 1456 (1989).
11. W. Strohmeier, *Chem. Ber.*, **94**, 3337 (1961).
12. R. Poli and H. D. Mui, *J. Am. Chem. Soc.*, **112**, 2446 (1990).
13. F. Calderazzo, C. Maichle-Mössmer, G. Pampaloni, and J. Strähle, *J. Chem. Soc., Dalton Trans.*, 655 (1993).
14. A Barbati, F. Calderazzo, and R. Poli, *Gazz. Chim. Ital.*, **118**, 589 (1988).
15. (a) J. C. Fettinger, S. P. Mattamana, C. J. O'Connor, R. Poli, and G. Salem, *J. Chem. Soc., Chem. Commun*, 1265 (1995). (b) J. C. Fettinger, J. C. Gordon, S. P. Mattamana, C. J. O'Connor, R. Poli, and G. Salem, *Inorg. Chem.*, **35**, 7404 (1996).

34. FACIAL MOLYBDENUM(III) TRIAMMINE COMPLEXES

Submitted by CLAUS J. H. JACOBSEN,* KARINA K. KLINKE,*
JENS HYLDTOFT,* and JØRGEN VILLADSEN*
Checked by RINALDO POLI[†]

Until recently, molybdenum(III) ammine complexes have been conspicuous, owing to their scarcity. Previously, the preparation of molybdenum(III)

* Haldor Topsøe Research Laboratories, Nymøllevej 55, DK-2800 Lyngby, Denmark.
† Laboratoire de Synthèse et d'Electrosynthèse Organometallique, Faculté des Sciences "Gabriel", 6 Boulevard Gabriel, 21100 Dijon, France.

ammine complexes has been postulated by reaction of certain molyb-
denum(III) compounds with liquid ammonia at high temperature and
pressure.[1] However, the products obtained were in all cases only poorly
characterized and it remains doubtful whether any of the reported methods
actually yielded well-defined molybdenum(III) ammines. This behavior is
contrary to that of the congeneric chromium(III), for which the rich ammine
chemistry is well known to all coordination chemists, and this difference
prompted us to attempt the preparation of new classical molybdenum(III)
amine complexes. An easily reproducible synthetic route to *fac*-
$Mo(NH_3)_3(CF_3SO_3)_3$ has been found and this compound has proved to be
a versatile precursor in the preparation of new classical molybdenum(III)
amines. The compound *fac*-$Mo(NH_3)_3(CF_3SO_3)_3$ is prepared from *mer*-
$Mo(THF)_3Cl_3$,[2] and other facial molybdenum(III) complexes are easily
obtained by the appropriate ligand substitutions of the $CF_3SO_3^-$ anion.[3]
Contrary to most of the well-known molybdenum(III) precursors frequently
used in synthesis [e.g., K_3MoCl_6,[4] *mer*-$Mo(THF)_3Cl_3$,[5] and *mer*-$Mo(py)_3$-
Cl_3,[6]], *fac*-$Mo(NH_3)_3(CF_3SO_3)_3$ does not contain coordinated halide and this
might be of importance in the synthesis of new low-valent molybdenum
compounds. It is noteworthy that syntheses completely analogous to those
described above can be carried out similarly with the congeneric
chromium(III) complexes,[2,3] thus providing easy synthetic routes to *fac*-
$Cr(NH_3)_3(CF_3SO_3)_3$, *fac*-$Cr(NH_3)_3Cl_3$, and *fac*-$Cr(NH_3)_3Br_3$. The preparation of *fac*-$Cr(NH_3)_3(CF_3SO_3)_3$ and *fac*-$Cr(NH_3)_3Cl_3$ has been previously
reported by Andersen et al.[7] by alternative routes.

General Procedures

All operations are carried out in a protective atmosphere of high-purity
dinitrogen using standard Schlenk line techniques to prevent oxidation of the
molybdenum(III) compounds *and* to avoid hydrolysis of the compounds by
reaction with water vapor. Tetrahydrofuran (THF) (HPLC grade) is dried
with 3-Å molecular sieves 48 h prior to use.

■ **Caution.** *Trifluoromethanesulfonic acid is a strong acid and extreme
care must be taken both in handling and in disposal of waste material. The
reaction between neat trifluoromethanesulfonic acid and diethyl ether is very
exothermic, causing the ether to boil and subsequently the pressure in the
Schlenk equipment to increase.*

A. *fac*-TRIAMMINETRIS(TRIFLUOROMETHANESULFONATE) MOLYBDENUM(III)

$$mer\text{-}Mo(THF)_3Cl_3 \xrightarrow[\text{2. } CF_3SO_3H]{\text{1. } NH_3(g)/THF} fac\text{-}Mo(NH_3)_3(CF_3SO_3)_3$$

fac-Mo(NH₃)₃(CF₃SO₃)₃. Three grams (7.17 mmol) of *mer*-Mo(THF)₃Cl₃,[5] is placed in a 250-mL Schlenk flask equipped with a 200-mL Schlenk dropping funnel. Then 150 mL of degassed THF is added from the dropping funnel and gaseous ammonia (99.995%) is passed through a glass tube into the orange suspension while it is heated from room temperature to 66°C and kept refluxing for 15 min. The dropping funnel is replaced with a Schlenk glass frit and the brownish powder is separated by filtration, washed twice with 25 mL of degassed THF, and dried in vacuum (ca. 1 kPa) for 2 h. Yield: 1.8 g (analyses: Mo, 35.3%; N, 16.57%, H, 4.42%; Cl, 39.08%, and C, 0.1%). For comparison the calculated analysis for Mo(NH₃)₃Cl₃ is: Mo, 37.87%; N, 16.58%; H, 3.58%, and Cl, 41.97%. The brownish product is transferred to a 25-mL Schlenk flask and 20 mL of degassed trifluoromethanesulfonic acid is added from a dropping funnel with stirring. Dinitrogen is bubbled through the solution for 12 h and finally the flask is evacuated to ca. 1 kPa for 2 h. This yields a brownish solution with a yellow tinge which is filtered through a glass frit into a 500-mL Schlenk flask (in none of our experiments was any material left on the filter). Then 300 mL of degassed diethyl ether is cautiously added with stirring and the flask is left overnight to complete the crystallization. In this way a fine, bright yellow precipitate is obtained, isolated by filtration through a glass frit, washed twice with 20 mL of degassed diethyl ether, and dried in vacuum (ca. 1 kPa) for 30 min. Yield: 1.0 g [23% from *mer*-Mo(THF)₃Cl₃]. Occasionally, the product can be contaminated with small amounts of an unidentified green impurity.* This can be eliminated by dissolving the product in 5 mL of degassed trifluoromethanesulfonic acid and reprecipitating it with 100 mL of degassed diethyl ether.

B. *fac*-TRIAMMINETRIHALOMOLYBDENUM(III)

$$fac\text{-Mo(NH}_3)_3(\text{CF}_3\text{SO}_3)_3 \xrightarrow[\text{THF/EtOH}]{\text{LiX}} fac\text{-Mo(NH}_3)_3\text{X}_3 \ (\text{X}=\text{Cl, Br, I})$$

fac-Mo(NH₃)₃Cl₃. To begin, 0.64 g (1.08 mmol) of *fac*-Mo(NH₃)₃(CF₃SO₃)₃ is placed in a 25-mL Schlenk flask equipped with a 100-mL dropping funnel. Then 0.14 g (3.3 mmol) of lithium chloride is dissolved in 15 mL of THF and degassed in the dropping funnel before the solution is added to the Schlenk flask with stirring. A clear solution quickly forms and within ca. 10 min a bright yellow precipitate forms. This is separated after 0.5 h by filtration under N₂ through a glass frit, washed twice with 5 mL of degassed THF and

---

* Checker's comment: Instead of bubbling N₂ through the solution for 12 h, we found that evacuating the Schlenk flask to ca. 1 mm Hg for 12 h gave comparable yields and seemed to reduce the amount of the unidentified green contaminant. A base trap should be used to prevent damage to the vacuum pump.

dried in vacuum (ca. 1 hP) for 1 h. Isolated yield of *fac*-Mo(NH$_3$)$_3$Cl$_3$: 0.25 g (93%).

fac-Mo(NH$_3$)$_3$Br$_3$. In a similar manner, 0.61 g (1.03 mmol) of *fac*-Mo(NH$_3$)$_3$(CF$_3$SO$_3$)$_3$ is placed in a Schlenk flask and 0.27 g (3.11 mmol) of lithium bromide dissolved in 15 mL of THF is added with stirring. After a few seconds a clear solution is formed and within ca. 15 min a bright yellow precipitate forms. This is separated after 0.5 h by filtration and washed with 5 mL of THF. The compound is dried in vacuum (ca. 1 kPa) for 1 h. Isolated yield of *fac*-Mo(NH$_3$)$_3$Br$_3$: 0.34 g (85%).

fac-Mo(NH$_3$)$_3$I$_3$. In a similar manner, 0.64 g (1.08 mmol) of *fac*-Mo(NH$_3$)$_3$-(CF$_3$SO$_3$)$_3$ is placed in a Schlenk flask and 0.44 g (3.29 mmol) of lithium iodide dissolved in 15 mL of THF is added with stirring. In this way a clear bright yellow solution is obtained and within 0.5 h a yellow precipitate forms. After 2 h the solution is filtered and the precipitate is washed with 5 mL of THF. The compound is dried in vacuum (ca. 1 kPa) for 1 h. Isolated yield of *fac*-Mo(NH$_3$)$_3$I$_3$: 0.48 g (84%). If the reaction is carried out in 99.9% ethanol instead, a yield of 21% is obtained.

Properties

The syntheses discussed above can all be scaled up by at least a factor of 3 without significantly lowering the yield. The *fac*-Mo(NH$_3$)$_3$X$_3$ (X = CF$_3$SO$_3$, Cl, Br, I) compounds are very susceptible to oxidation by the atmospheric dioxygen and should only be handled in a protective atmosphere. The halogenide complexes can, however, be handled in air for shorter (<15 min) periods of time. The compounds have all been characterized by UV-vis diffuse reflectance spectroscopy. *fac*-Mo(NH$_3$)$_3$(CF$_3$SO$_3$)$_3$: λ_{max} = 369 nm; λ_{min} = 335 nm; λ_{max} = 310 nm. *fac*-Mo(NH$_3$)$_3$Cl$_3$: λ_{max} = 406 nm; λ_{min} = 368 nm; λ_{max} = 342 nm. *fac*-Mo(NH$_3$)$_3$Br$_3$: λ_{max} = 420 nm; λ_{min} = 388 nm; λ_{max} = 356 nm. *fac*-Mo(NH$_3$)$_3$I$_3$: λ_{max} = 453 nm; λ_{min} = 425 nm. *fac*-Mo(NH$_3$)$_3$(CF$_3$SO$_3$)$_3$ was characterized by X-ray powder diffraction and was found to be hexagonal with a = 12.62(1)G Å and c = 6.86(1)G Å.

References

1. A. Rosenheim, G. Abel, and R. Lewy, *Z. Anorg. Allg. Chem.*, **197**, 189 (1931). A. Rosenheim and H. J, Braun, *Z. Anorg. Allg. Chem.*, **46**, 311 (1905). W. R. Bucknall, S. Carter, and W. Wardlaw, *J. Chem. Soc.*, 512 (1927). D. A. Edwards and G. W. A. Fowley, *J. Less-Common Met.*, **4**, 512 (1962).
2. C. H. J. Jacobsen, J. Villadsen, and H. Weihe, *Inorg. Chem.*, **32**, 5396 (1993).

3. M. Brorson, C. H. J. Jacobsen, C. M. Jensen, I. Schmidt, and J. Villadsen, *Inorg. Chim. Acta*, **247**, 189 (1995).
4. K. H. Lohmann and R. C. Young, *Inorg. Synth.*, **4**, 97 (1953).
5. J. R. Dilworth and J. Zubieta, *Inorg. Synth.*, **28**, 36 (1990).
6. H. B. Jonassen and L. J. Bailin, *Inorg. Synth.*, **7**, 140 (1963).
7. P. Andersen, A. Døssing, J. Glerup, and M. Rude, *Acta Chem. Scand.*, **44**, 346 (1990).

35. HIGH-VALENT MONO-(η^5-PENTAMETHYLCYCLO-PENTADIENYL)VANADIUM AND MOLYBDENUM COMPLEXES

Submitted by COLIN D. ABERNETHY,* FRANK BOTTOMLEY,*
JINHUA CHEN,* MICHAEL F. KEMP,* TAMMY C. MALLAIS,*
and OLUSOLA O. WOMILOJU*
Checked by SUNITA THYAGARAJAR† and KLAUS H. THEOPOLD†

High-valent mono-(cyclopentadienyl) derivatives of the early transition metals have a long history: (η^5-C_5H_5)V(O)Cl_2 was first reported in 1958[1] and (η^5-C_5H_5)Mo(O)$_2$Cl in 1963.[2] However, the lability of the η^5-C_5H_5 ligand, and the reactivity of the C–H bond, prevented the exploration of the chemistry of these compounds. The introduction of the less-labile η^5-C_5Me_5, with its inert C–H bonds, has proven of great benefit to this area. However, convenient preparations of high-valent mono-(η^5-pentamethylcyclopentadienyl)-metal derivatives are not available. The key compound [(η^5-C_5Me_5)VCl_2]$_3$ has been reported in a review,[3] but with no details. There are several routes to (η^5-C_5Me_5)V(O)Cl_2,[4] but all involve either starting materials which are difficult to prepare, or involve costly loss of C_5Me_5 on oxidation. Many reactions lead to (η^5-C_5Me_5)Mo(O)$_2$Cl, but only two are usable syntheses.[5,6] We present here the syntheses of [(η^5-C_5Me_5)-V(μ-Cl)$_2$]$_3$, (η^5-C_5Me_5)V(O)Cl_2, (η^5-C_5Me_5)Mo(O)$_2$Cl, [(η^5-C_5Me_5)Mo-(O)Cl]$_2$(μ-O), and (η^5-C_5Me_5)Mo(O)Cl_2.

All operations except the preparation of (η^5-C_5Me_5)Mo(O)$_2$Cl should be carried out under an inert atmosphere or under vacuum employing standard Schlenk techniques. The solvents toluene, hexane, tetrahydrofuran, and dichloromethane must be completely dried and freed of oxygen by standard methods before use.

* Department of Chemistry, University of New Brunswick, Fredericton, New Brunswick, E3B 5A3, Canada.
† Department of Chemistry and Biochemistry, University of Delaware, Newark, DE, 19716.

1,2,3,4,5-Pentamethylcyclopentadiene may be purchased* or synthesized by one of the procedures previously given in *Inorganic* or *Organic Syntheses*.[7-9] The synthesis of $VCl_3(THF)_3$ has been given previously.[10] The starting material $Mo(CO)_6$ and the reagent Bu_3^nSnCl are commercial materials. The latter is converted into $Bu_3^nSnC_5Me_5$ by the procedure given below.[11]

A. TRIS-(*n*-BUTYL)(PENTAMETHYLCYCLOPENTADIENYL)TIN

$$C_5Me_5H + BuLi \rightarrow C_5Me_5Li + C_4H_{10}$$

$$C_5Me_5Li + Bu_3^nSnCl \rightarrow Bu_3^nSnC_5Me_5 + LiCl$$

Procedure

■ **Caution.** *Organotin compounds are toxic. A well-ventilated hood, and a vacuum line which vents to the hood, should be used.*

To a well-stirred solution of C_5Me_5H (13.0 g, 95.6 mmol) in tetrahydrofuran (150 mL) (contained in a 500-mL flask, equipped with a side arm, a dropping funnel which is attached to the vacuum line, and a stirring bar) is added, dropwise over a period of 0.5 h, BuLi (38.5 mL of a 2.5 M hexane solution, 96.3 mmol). Butane is evolved and a white suspension of LiC_5Me_5 forms. The mixture is stirred for 2 h after complete addition of BuLi, then Bu_3^nSnCl (26 mL, 31.2 g, 95.8 mmol) is syringed onto the suspension. The white suspension dissolves, forming a yellow solution. After stirring for 1 h, the tetrahydrofuran solvent is removed under vacuum, leaving a mixture of white LiCl and yellow, oily $Bu_3^nSnC_5Me_5$. To this mixture is added hexane (100 mL). The mixture is stirred for 0.5 h, and is then filtered through a Schlenk fritted tube under argon. The hexane solvent is removed from the filtrate under vacuum and the yellow oil which remains is distilled twice under vacuum. In the first distillation, the fraction distilling between 100 and 130°C (10^{-3} torr) is collected. In the second distillation, the fraction distilling between 115 and 125°C (10^{-3} torr) is collected. Yield: 28.4 g, 70%.

Properties

The compound $Bu_3^nSnC_5Me_5$ is a viscous, yellow liquid. It may be stored in the absence of air for long periods. It is characterized by the ^{1}H NMR spectrum (400 MH₂, $CDCl_3$ solution): 1.79 ppm, singlet, 15H, with two

---

* Aldrich Chemical Co., 1001 West Saint Paul Ave., Milwaukee, WI 53233.

satellite signals of $J = 8.4\,Hz$ due to coupling to ^{117}Sn and $^{119}Sn[Sn(C_5(CH_3)_5]$; 0.87 ppm, triplet, 9H, $J = 7.2\,Hz$, $[Sn(CH_2CH_2CH_2CH_3)]$; 1.27 ppm, sextet, 6H, $J = 7.2\,Hz$ $[Sn(CH_2CH_2CH_2CH_3)]$; 1.32–1.42 ppm, multiplet, 6H, J not measured $[Sn(CH_2CH_2CH_2CH_3)]$; 0.71 ppm, triplet, 6H, $J = 8.3\,Hz$, $[Sn(CH_2CH_2CH_2CH_3)]$.

B. TRIS-[(η^5-PENTAMETHYLCYCLOPENTADIENYL)-VANADIUMDICHLORIDE)]

$3VCl_3(THF)_3 + 3Bu_3^nSnC_5Me_5 \rightarrow$

$$[(\eta^5\text{-}C_5Me_5)V(\mu\text{-}Cl)_2]_3 + 3Bu_3^nSnCl + 9THF$$

Procedure

A solution of $VCl_3(THF)_3$ (6.0 g, 16.1 mmol) in toluene (100 mL) is warmed to 60°C, and $Bu_3^nSnC_5Me_5$ (6.7 mL, 7.5 g, 17.7 mmol) is syringed into it, while stirring. The mixture is stirred for 12 h at 60°C. The toluene is removed under vacuum and (the removal is expedited if the solution is maintained at approximately 35°C in a water bath), leaving a red-brown, oily residue. This is washed twice with hexane (80-mL portions), dried in vacuum, and extracted into hot (80°C) toluene (100 mL). The toluene solution is filtered while still warm. The filtrate is reduced in volume to 20 mL by removal of toluene under vacuum at 35°C and then frozen in a liquid nitrogen bath. Hexane (80 mL) is condensed onto the frozen solution. On warming to -25°C and setting aside for 12 h, purple-black microcrystals of $[(\eta^5\text{-}C_5Me_5)V(\mu\text{-}Cl)_2]_3$ are formed. These are collected by filtration and dried in vacuum. Yield: 2.9 g, 70%. Crystals suitable for X-ray diffraction are obtained by setting aside saturated ether solutions for a few days at 4°C.

Properties

Dark purple $[(\eta^5\text{-}C_5Me_5)V(\mu\text{-}Cl)_2]_3$ is paramagnetic and air sensitive. The ^1H NMR spectrum (200 MHz, C_6D_6 solution) shows a very broad signal at -7.9 ppm. The mass spectrum (EI, 70 eV) shows as its most intense peak, m/e 257: $(\{(C_5Me_5)V(^{35}Cl)_2H\})^+$. Crystals have the diffraction data: $a = 11.80(1)$, $b = 11.785(3)$, $c = 15.96(1)$, $\alpha = 87.63(3)$, $\beta = 68.17(4)$, $\gamma = 60.13(3)$, space group $P\bar{1}$.

The conversion of $[(\eta^5\text{-}C_5Me_5)V(\mu\text{-}Cl)_2]_3$ into $(\eta^5\text{-}C_5Me_5)V(O)Cl_2$ is given below; it may also be readily converted into $[(\eta^5\text{-}C_5Me_5)VCl(\mu\text{-}N)]_2$.[12]

C. (η^5-PENTAMETHYLCYCLOPENTADIENYL)VANADIUM-(OXO)DICHLORIDE

$$2[(\eta^5\text{-}C_5Me_5)V(\mu\text{-}Cl)_2]_3 + 3O_2 \rightarrow 6(\eta^5\text{-}C_5Me_5)V(O)Cl_2$$

Procedure

A solution of $[(\eta^5\text{-}C_5Me_5)V(\mu\text{-}Cl)_2]_3$ (2.1 g, 2.7 mmol) in tetrahydrofuran (100 mL) is stirred, at room temperature, under a mixture of argon and dioxygen (Ar: O_2 ratio of 4:1, total volume 1500 mL, total pressure one atmosphere) for 1.5 h. The pressure is maintained at one atmosphere by additions of pure dioxygen at intervals. The color of the solution changes from dark red to blue-green as the reaction proceeds. The solvent is removed under vacuum leaving an oily, green residue, which is extracted with toluene (120 mL). The green extract is filtered (leaving a purple residue) and reduced in volume to 30 mL by removal of the toluene under vacuum at 35°C. The solution is rapidly frozen in a liquid nitrogen bath and hexane (100 mL) is condensed onto it. On warming to -25°C and setting aside for 12 h, bright green crystals of $(\eta^5\text{-}C_5Me_5)V(O)Cl_2$ form. These are collected by filtration and dried under vacuum. Yield: 1.45 g (65%).

Properties

Bright green $(\eta^5\text{-}C_5Me_5)V(O)Cl_2$ is moderately sensitive to water, but not to dioxygen. Dry samples may be stored indefinitely. The ^1H NMR spectrum (200 MHz, CDCl$_3$ solution) shows a singlet at 2.31 ppm, the ^{13}C spectrum shows two singlets at 13.3 ppm and 133.5 pm. The infrared spectrum (Nujol mull) shows an intense absorption band at 967 cm^{-1} [v(V=O)]. The mass spectrum (EI, 70 eV) shows the parent ion at m/e 272 ($\{(C_5Me_5)VO^{35}(Cl_2)\}^+$).

Anal. Found, C, 43.5, H, 5.6%. Calcd. for $C_{10}H_{15}Cl_2OV$; C, 43.9, H, 5.6%.

D. BIS-[(η^5-PENTAMETHYLCYCLOPENTADIENYL)-MOLYBDENUM(CARBONYL)]

$$2Mo(CO)_6 + 2C_5Me_5H \rightarrow [(\eta^5\text{-}C_5Me_5)Mo(CO)_n]_2 + (12\text{-}2n)\ CO + H_2$$

This synthesis is a modified version of one given in the literature.[13] To $Mo(CO)_6$ (8.0 g, 30.3 mmol) (contained in a 500-mL flask, with side arm, equipped with a stirring bar) is added under argon, 200 mL of *n*-decane which has been previously dried over sodium and distilled under partial vacuum. Then C_5Me_5H (2.8 ml, 30.3 mmol) is syringed into the flask under argon. A reflux condenser, on the top of which is a suba-seal rubber septa through which a 22G needle and a copper wire are inserted, is then attached to the flask. The system is purged with argon for 2 min, then heated to 150°C while stirring. The solution is swirled to bring unreacted $Mo(CO)_6$, which condenses on the sides of the flask, back into solution. This is repeated as necessary during the reaction. After 2 h, the temperature is increased to obtain reflux. If $Mo(CO_6)$ sublimes into the condenser, the copper wire is used to push it back down into the flask. After a total reaction time of 7 h, the stirring is stopped and the flask is removed from the condenser and stoppered under argon. The solution is allowed to cool to 0°C over 3 h under argon. The red crystals of $[\eta^5$-$C_5Me_5)Mo(CO)_n]_2$ which precipitate are removed by filtration (Schlenk fritted tube under argon) and washed twice with cold hexane (20-mL portions). Unreacted $Mo(CO)_6$ (2.54 g) is removed from the product by sublimation at 70°C under vacuum (0.1 torr). The yield is 3.92 g, 45% (based on the initial amount of $Mo(CO)_6$.

Properties

The product $[(\eta^5$-$C_5Me_5)Mo(CO)_n]_2$ is a mixture of the compounds with $n = 2$ and 3.[13] It is not necessary to separate the mixture, since the oxidation described below proceeds with loss of CO. The mixture is moderately air sensitive, but may be stored in an inert atmosphere. The ^1H NMR spectrum of the mixture (200 MHz, $CDCl_3$ solution) shows singlets at 1.92 and 1.89 ppm. The infrared spectrum (Nujol mull) shows $\nu(CO)$ absorption bands at 1925, 1895, 1870, and 1827 cm^{-1}.

E. (η^5-PENTAMETHYLCYCLOPENTADIENYL)MOLYBDENUM-(DIOXO)CHLORIDE

$$[(\eta^5\text{-}C_5Me_5)Mo(CO)_n]_2 + 5H_2O_2 + 2HCl \rightarrow$$

$$2(\eta^5\text{-}C_5Me_5)Mo(O)_2Cl + 2nCO + 6H_2O$$

Procedure

■ **Caution.** *30% H_2O_2 is corrosive.*

This procedure was conducted in air unless stated otherwise. To a vigorously stirring solution of $[(\eta^5$-$C_5Me_5)Mo(CO)_n]_2$ (1.21 g) in chloroform (100 mL) is added a 30% aqueous solution of hydrogen peroxide (8.7 mL), followed immediately by concentrated hydrochloric acid (3.8 mL). Within 1 h, the solution changes color from red to orange. If this color change does not occur, further amounts of hydrogen peroxide and hydrochloric acid are added. Stirring is continued until the color of the solution becomes bright yellow (approximately 12 h).

The aqueous layer is separated and extracted three times with chloroform (20-mL portions). The combined chloroform solutions are washed twice with water (150-mL portions). The first aqueous washing is extracted twice with chloroform (20-mL portions). All of the chloroform extracts are combined and treated with aqueous ferrous sulphate (5%, 150 mL). The chloroform layer is separated and treated with saturated aqueous sodium chloride (150 mL). The chloroform layer is again separated and dried over sodium sulfate for 1 h. The sodium sulfate is removed by filtration and washed with chloroform. At this stage, the filtrate is transferred to the vacuum line. The chloroform is removed under vacuum, giving a yellow solid. Yield of crude product: 1.12 g, 89%. This is dissolved in anhydrous tetrahydrofuran (15 mL) by warming to 50°C. The solution is rapidly frozen in a liquid nitrogen bath and hexane (100 mL) condensed onto the frozen solution. On warming to room temperature and setting aside for 48 h, crystals of $(\eta^5\text{-}C_5Me_5)$-$Mo(O)_2Cl$ form at the hexane/tetrahydrofuran interface. These are collected by filtration and dried in vacuum. Yield: 0.77 g, 61%.

Properties

The compound $(\eta^5\text{-}C_5Me_5)Mo(O)_2Cl$ is a yellow solid. Well-dried samples may be stored at $-40°C$ in the absence of air and light for a few days, but decomposition to an insoluble molybdenum "blue" is accelerated by traces of moisture and light. Therefore, it is preferable to use the compound as soon as possible. The ^1H NMR spectrum (CDCl$_3$ solution) shows a singlet at 2.09 ppm; the ^{13}C NMR shows two singlets at 30.2 and 122.3 ppm. The infrared spectrum (KBr pellet) shows intense absorption bands at 927 and 879 cm^{-1} $[\nu(Mo{=}O)]$ and at 568 cm^{-1} $[\vartheta(O{=}Mo{=}O)]$. The mass spectrum (EI, 70 eV) shows the parent ion at m/e 300 $(\{(C_5Me_5)(^{98}Mo)O_2(^{35}Cl)\}^+)$.

F. ANTI-(μ-OXO)-BIS[(η^5-PENTAMETHYLCYCLOPENTADIENYL)-MOLYBDENUM(OXO)CHLORIDE]

$2\ (\eta^5\text{-}C_5Me_5)Mo(O)_2Cl + P(OMe)_3 \rightarrow$

$$[(\eta^5\text{-}C_5Me_5)Mo(O)Cl]_2(\mu\text{-}O) + OP(OMe)_3$$

Procedure

■ **Caution.** *P(OMe)$_3$ is toxic and malodorous. A good fume hood is essential.*

To an ice-cold solution of (η^5-C$_5$Me$_5$)Mo(O)$_2$Cl (1.88 g, 6.3 mmol) in tetrahydrofuran/hexane (1 : 1, v : v; 60 mL) contained in a 250-mL flask (with side arm), closed with a Rotaflo joint or stopper, is added, by a syringe, P(OMe)$_3$ (3.2 mL, 3.3 g 26.2 mmol). The flask is swirled to mix the constituents, which causes a color change from bright yellow to wine-red, then set aside, without stirring, at $-25°C$ for 12 h. It is important that the temperature be maintained at $-25°C$. At higher temperatures, side reactions take place. Red-brown [(η^5-C$_5$Me$_5$)Mo(O)Cl]$_2$(μ-O) precipitates, and is removed by filtration on the vacuum line, washed twice with cold (0°C) hexane (30-mL portions), and dried under vacuum. The product is recrystallized by dissolving in tetrahydrofuran (20 mL) on warming to 50 °C, cooling rapidly in a liquid nitrogen bath, adding hexane (40 mL), and setting the solution aside at $-25°C$ for 48 h. The wine-red crystals of [(η^5-C$_5$Me$_5$)Mo(O)Cl]$_2$(μ-O) were collected by filtration and dried in vacuum. Yield: 1.47 g, 80%.

Properties

Wine-red [(η^5-C$_5$Me$_5$)Mo(O)Cl]$_2$(μ-O) is moderately air sensitive. It may be stored in the solid state, in the absence of air, at room temperature, for several days, and it is not light sensitive. In solution it undergoes disproportionation to (η^5-C$_5$Me$_5$)Mo(O)Cl$_2$ and [(η^5-C$_5$Me$_5$)Mo(O)(μ-O)]$_2$. The ^{1}H NMR spectrum (200 MHz, C$_6$D$_6$ solution) shows a singlet at 1.85 ppm. The infrared spectrum shows an intense absorption band at 930 cm^{-1} [ν(Mo=O)] and at 780 cm^{-1} [ϑ(Mo–O–Mo)]. The mass spectrum (EI, 70 eV) shows the parent ion at m/e 584 ({(C$_5$Me$_5$)$_2$(^{98}Mo)$_2$(^{35}Cl)$_2$O$_3$}$^+$).

G. (η^5-PENTAMETHYLCYCLOPENTADIENYL)-MOLYBDENUM(OXO)DICHLORIDE

[(η^5-C$_5$Me$_5$)Mo(O)Cl]$_2$(μ-O) + 2Me$_3$SiCl →

$$2(\eta^5\text{-C}_5\text{Me}_5)\text{Mo(O)Cl}_2 + [\text{Me}_3\text{Si}]_2\text{O}$$

Procedure

To an ice-cold solution of [(η^5-C$_5$Me$_5$)Mo(O)Cl]$_2$(μ-O) (0.50 g, 0.84 mmol) in tetrahydrofuran (100 mL) is added, by a syringe, Me$_3$SiCl (0.25 mL, 0.22 g,

1.9 mmol). The solution is stirred while being allowed to warm to room temperature. The color of the solution slowly changes from wine-red to brick-red. When the brick-red color is established (approximately 12 h), the solution is concentrated to 15 mL under vacuum. The resultant solution is rapidly frozen in a liquid nitrogen bath, and hexane (60 mL) is condensed onto it. On warming to room temperature and setting aside for 24 h, brick-red $(\eta^5\text{-}C_5Me_5)Mo(O)Cl_2$ is deposited. The solid is collected by filtration (Schlenk fritted tube under argon) and washed with cold (0°C) hexane (20 mL). It is recrystallized by dissolving in toluene (65 mL), filtering to remove a purple residue, and reducing the volume to 20 mL under vacuum. The solution is then frozen in a liquid nitrogen bath and hexane (60 mL) is condensed onto it. On warming to room temperature and setting aside for 72 h, brick-red crystals of $(\eta^5\text{-}C_5Me_5)Mo(O)Cl_2$ are obtained. Yield: 0.40 g, 75%.

Properties

Brick-red $(\eta^5\text{-}C_5Me_5)Mo(O)Cl_2$ is paramagnetic and sensitive to dioxygen and water. Dry samples may be stored in the absence of air indefinitely. The ^1H NMR (400 MHz, CDCl$_3$ solution) shows a very broad signal at 36.8 ppm. The infrared spectrum shows an intense absorption band at 932 cm^{-1} [$\nu(Mo=O)$]. The mass spectrum (EI, 70 eV shows the parent ion at m/e 319 ($\{(C_5Me_5)(^{98}Mo)O(^{35}Cl)_2\}^+$).

References

1. E. O. Fischer and S. Vigoureux, *Chem. Ber.*, **91**, 1342 (1958).
2. M. Cousins and M. L. H. Green, *J. Chem. Soc.*, 889 (1963).
3. L. Messerle, ref. 85 of R. Poli, *Chem. Rev.*, **91**, 509 (1991).
4. F. Bottomley, C. P. Magill, and B. Zhao, *Organometallics*, **10**, 1946 (1991) and references therein.
5. J. W. Faller and Y. Ma, *J. Organomet. Chem.*, **340**, 59 (1988).
6. F. Bottomley, P. D. Boyle, and J. Chen, *Organometallics*, **13** (1994).
7. J. M. Manriquez, P. J. Fagan, L. D. Schertz, and T. J. Marks, *Inorg. Synth.*, **21**, 181 (1982).
8. R. S. Threlkel, J. E. Bercaw, P. F. Seidler, and R. G. Bergman, *Org. Synth.*, **65**, 42 (1987).
9. C. M. Fendrick, L. D. Schertz, E. A. Mintz, and T. J. Marks, *Inorg. Synth.*, **26**, 193 (1992).
10. L. E. Manzer, *Inorg. Synth.*, **21**, 138 (1982).
11. W. A. Hermann, W. K. Alcher, H. Biersack, I. Bernal, and M. Creswick, *Chem. Ber.*, **114**, 3558 (1981).
12. T. S. Haddad, A. Aistars, J. W. Ziller, and N. M. Doherty, *Organometallics*, **12**, 2420 (1993).
13. R. B. King, M. Z. Iqbal, and A. D. King, *J. Organomet. Chem.*, **171**, 53 (1979).

36. MONOINDENYLTRICHLORIDE COMPLEXES OF TITANIUM(IV), ZIRCONIUM(IV), AND HAFNIUM(IV)

Submitted by ROBERT J. MORRIS,* SCOTT L. SHAW,* JESSE M. JEFFERIS,* JAMES J. STORHOFF,* and DEAN M. GOEDDE*
Checked by STEPHEN L. BUCHWALD[†] and BAIN CHIN[†]

The monoindenyltrichloride complexes of titanium(IV), zirconium(IV), and hafnium(IV) have recently been reported and the use of the titanium analog as an olefin polymerization cocatalyst has been demonstrated.[1,2] The preparation of these Group 4 indenyl trichloride complexes is accomplished through the use of trialkylsilicon and trialkyltin reagents, which is analogous to the preparation of cyclopentadienyl and pentamethylcyclopentadienyl trichloride complexes of Group 4 and Group 5.[3] Using 1-(trimethylsilyl)indene and 1-(tributylstannyl)indene as indenyl transfer reagents is convenient since Me_3SiCl and Bu_3SnCl are easily washed from the metal-containing products with most organic solvents.

While the preparations of 1-(trimethylsilyl)indene and 1-(trimethylstannyl)-indene have been reported,[4] there are no available preparations of 1-(tributylstannyl)indene. We chose the tributyl analog because of the decreased toxicity of tributyltin reagents vs. trimethyltin reagents.[5] A preparation for the synthesis of $Mg(C_9H_7)_2$, another indenyl transfer agent, has appeared in *Inorganic Syntheses*.[6]

General Remarks

■ **Caution.** *Chlorotributylstannane and 1-(tributylstannyl)indene should be considered toxic and thus should be handled with gloves in an adequate fume hood. All glassware and equipment should be treated with a saturated KOH/isopropanol base bath before being taken out of the fume hood. The base bath can later be disposed with the waste tin compounds.*

Solvents are distilled from sodium/benzophenone (pentane), sodium (toluene), or calcium hydride (dichloromethane) and degassed immediately prior to use. Chlorotrimethylsilane[‡] and chlorotributylstannane* are distilled under argon before use. Indene ($>90.0\%$ pure)[‡] is purified by distillation

---

* Department of Chemistry, Ball State University, Muncie, IN 47306.
† Department of Chemistry, Massachusetts Institute of Technology, Cambridge, MA 02139.
‡ Purchased from Aldrich Chemical Company, 1001 W. St. Paul Ave., Milwaukee, WI 53233.

from sodium metal (0.05 g Na/g of indene) under vacuum (0.10 mmHg) using a vacuum-jacketed distillation head with a 150-mm Vigreux column. The first 10% v/v of collected distillate is discarded and the next 70–80% v/v is collected ($>99.0\%$ pure indene, by GC). This procedure is repeated once with the collected indene. Butyllithium* is used as received; $TiCl_4$[†] is distilled before use, and $ZrCl_4$[†] and $HfCl_4$[†] are sublimed under full vacuum (<0.10 mmHg) at $270°C$.

A. 1-(TRIMETHYLSILYL)INDENE

■ **Caution.** *Butane is vigorously evolved during the reaction and it should be vented into a fume hood.*

Procedure

All reactions and manipulations should be performed under an atmosphere of nitrogen (or argon), using either a dry box or standard Schlenk techniques.[7] Degassed indene (47 mL/0.40 moles) is added to an N_2-flushed, 1-L, 3-necked, round-bottomed flask containing pentane (480 mL) and a large magnetic stir bar. Using a pressure-equalizing dropping funnel, *n*-butyllithium (250 mL of 1.6 M solution, 0.40 moles) is slowly added at ca. 1 drop sec^{-1} while stirring at room temperature with a mineral oil bubbler to vent the butane that evolves.

The solution is stirred for 12 h and, as the reaction occurs, a white precipitate of lithium indenide forms. The supernatant is removed via filtration through a filter cannula using No. 1 Whatman™ filter paper and the precipitate is washed with pentane (2×300 mL); the washes are neutralized with isopropanol and discarded. Pentane (400 mL) is added to the reaction flask containing the clean lithium indenide and, using a pressure-equalizing dropping funnel, chlorotrimethylsilane (50.8 mL, 0.40 moles) is slowly added at ca. 1 drop sec^{-1}. The yellow-white mixture is stirred for 5 h after addition is complete and the reaction mixture is then filtered through a filter cannula using No. 1 Whatman™ filter paper. The volatile material is removed from the supernatant under vacuum ($25°C$, 0.10 mmHg). The remaining yellow

---

* Purchased from Aldrich Chemical Company, 1001 W. St. Paul Ave., Milwaukee, WI 53233.
† Purchased from Cerac, Box 1178, Milwaukee, WI 53201.

liquid is distilled using vacuum-jacketed distillation head (0.10 mmHg, 150 mm Vigreux column). The first fraction (unreacted trimethylchlorosilane) is collected at a head temperature range of 29–31°C. Trimethylsilylindene, the second fraction, is collected at a head temperature range of 39–41°C. Yield: 43.5 g (61% based on indene). As with the preparation of trimethylsilyl-cyclopentadiene, more than one isomer is prepared (two with trimethyl-silylindene). Both isomers react with metal halides sufficiently well to transfer the indenyl ligand.

Anal. Calcd. for $C_{12}H_{16}Si$: C, 76.6; H, 8.56. Found: C, 76.4; H, 8.74.

Properties

Trimethylsilylindene is a clear, colorless liquid with density of 0.922 g mL^{-1} that is slightly sensitive to polymerization when exposed to light and O_2. Trimethysilylindene is best stored in the dark under an atmosphere of argon or nitrogen. ^{1}H NMR (Varian Gemini 200 NMR spectrometer at 199.975 MHz and referenced to tetramethylsilane, $CDCl_3$, 20°C): δ − 0.05 (s, 9H), δ 3.51 (m, 1H), δ 6.68–6.71 (m, 1H), δ 6.95–6.98 (m, 1H), δ 7.18–7.33 (m, 2H), δ 7.47–7.54 (m, 2H). The ^{1}H resonance for the trimethylsilyl group for the small amount of 3-(trimethylsilyl)indene isomer occurs at δ 0.31.

B. 1-(TRIBUTYLSTANNYL)INDENE

■ **Caution.** *Butane is vigorously evolved during the reaction and it should be vented into a fume hood.*

Procedure

All reactions and manipulations are performed under an atmosphere of nitrogen (or argon), using either a dry box or standard Schlenk techniques.[7] Degassed indene (47.0 mL/0.40 moles) is added to a 1-L, 3-necked, round-bottomed flask containing pentane (480 mL) and equipped with a N_2 inlet, a gas outlet connected to an oil bubbler, a pressure-equalizing dropping funnel, and a large magnetic stir bar. Through the pressure-equalizing funnel, *n*-butyllithium (250 mL of 1.6 M solution, 0.40 moles) is slowly added at roughly 1 drop sec^{-1} while stirring at room temperature, with a mineral oil bubbler to vent the butane that evolves. The solution is stirred for 12 h and, as the reaction occurs, a white precipitate of lithium indenide forms. The supernatant is removed via filtration through a filter cannula using No. 1 Whatman™ filter paper and the precipitate is washed with pentane (2×300 mL); the washes are neutralized with isopropanol and discarded. Pentane (400 mL) is added to the reaction flask containing the clean lithium indenide and, using a pressure-equalizing dropping funnel, chlorotributylstannane (110 mL, 0.41 moles) is added dropwise at 1 drop sec^{-1}. The yellow-white mixture is stirred for 5 h after the addition is complete and the reaction mixture is filtered via a filter cannula using No. 1 Whatman™ filter paper. The volatile material is removed from the supernatant under vacuum (25°C, 0.10 mmHg). The remaining yellow liquid is distilled using vacuum-jacketed distillation head (0.10 mmHg, 150 mm Vigreux column). The first fraction (unreacted tributylstannylchloride) is collected at a head temperature range of 92–95°C. The second fraction (tributylstannylindene) is collected at a head temperature range of 140–142°C. Yield: 107 g, 65%, $d = 1.070\,\mathrm{g\,mL}^{-1}$.

Anal. Calcd. for $C_{21}H_{34}Sn$: C, 62.3; H, 8.46. Found: C, 62.1; H, 8.37. ^1H NMR (Varian Gemini 200 NMR spectrometer at 199.975 MHz and referenced to tetramethylsilane, toluene-d_8, 20°C): δ 0.97 (m, 6H), δ 1.08 (t, J = 6.9 Hz, 9H), δ 1.34–1.63 (m, 12H), δ 6.92 (t, J = 3.4 Hz, 1H), δ 7.37–7.41 (m, 2H), δ 7.62–7.78 (br, m, 2H). Tributylstannyl-indene is fluxional; the tributylstannyl group migrates rapidly between the 1- and 3-positions at 20°C. While the 3-proton is unobservable at this temperature, the 1-proton is a very broad peak centered at δ 4.24.

Properties

Tributylstannylindene is a clear, pale yellow liquid with a density of 1.070 g/mL. As with 1-(trimethylsilyl)indene, 1-(tributylstannyl)indene is best stored in the dark under an atmosphere of argon (or nitrogen).

C. (η^5-INDENYL)TRICHLOROTITANIUM(IV)

$$TiCl_4 + \text{(indene-SiMe}_3\text{)} \xrightarrow{-ClSiMe_3} (\eta^5\text{-}C_9H_7)TiCl_3$$

Procedure

All reactions and manipulations should be performed under an atmosphere of nitrogen (or argon), using either a dry box or standard Schlenk techniques.[7] A solution of 1-(trimethylsilyl)indene (25.0 g, 0.133 mol) in dichloromethane (50 mL) is added to a solution of $TiCl_4$ (17.5 mL, 0.16 mol) in dichloromethane (150 mL). The reaction solution immediately turns dark red. The reaction mixture is then stirred at room temperature for 12 h and filtered via a filter cannula using No. 1 Whatman™ filter paper. The filtered solution is cooled overnight at $-20°C$ to give dark red crystals of $(\eta^5\text{-}C_9H_7)TiCl_3$; the supernatant is reduced and a second crop of crystals is collected. Yield: 28.6 g (79.9%).

Anal. Calcd. for $C_9H_7Cl_3Ti$: C, 40.1; H, 2.62; Cl, 39.5. Found: C, 40.3; H, 2.70; Cl, 39.4.

Properties

The deep red crystals of $(\eta^5\text{-}C_9H_7)TiCl_3$ are air- and moisture-stable for short periods of time, but the complex is more sensitive in solution. The complex displays the following physical properties. Mp (under argon, in sealed capillaries, using a Gallenkamp melting point apparatus): 160°C. ^{1}H NMR (Varian Gemini 200 NMR spectrometer at 199.975 MHz and referenced to tetramethylsilane, $CDCl_3$, 20°C): δ 7.82 (m, 2H), δ 7.53 (m, 2H), δ 7.21 (d, 2H), δ 7.12 (t, 1H). IR (Nicolet 5ZDX FT-IR spectrometer, cm^{-1}, Nujol, KBr): 435 m, 467 s, 540 w, 590 w, 628 vw, 672 vw, 742 w, 754 s, 799 w sh, 834 s, 843 m, 854 m, 868 w, 940 w, 965 w, 1005 w, 1030 w sh, 1045 m, 1052 m, 1088 w sh, 1210 w, 1340 m, 1399 m, 1445 m, 1490 w, 1530 w, 1545 w, 1570 vw, 1580 vw, 1612 w, 1620 w, 1645 vw, 1655 w, 1665 w, 1697 w, 1770 vw, 1800 vw, 1825 vw, 1870 w, 1920 w, 1953 w, 1983 w, 3060 w, 3091 m, 3103 m, 3111 m.

Following the procedure outlined above, there is a 3–5% impurity (by ^{1}H NMR) of an η^5-indenyl product; the triplet of the five-membered ring of the impurity is centered at δ 6.87. The initial crude product is suitable for further use. However, sublimation at 0.10 torr and 85°C lowers the amount of impurity to <1 with $>85\%$ compound recovery.

D. (η^5-INDENYL)TRICHLOROZIRCONIUM(IV)

Procedure

All reactions and manipulations should be performed under an atmosphere of nitrogen (or argon), using either a dry box or standard Schlenk techniques.[7] Neat 1-(tributylstannyl)indene (5.6 mL, 14.7 mmol) is added to a room-temperature suspension of $ZrCl_4$ (3.43 g, 14.7 mmol) in toluene (160 mL). The reaction mixture slowly turns clear orange. After stirring overnight at room temperature, a fine yellow precipitate of $[(\eta^5\text{-}C_9H_7)ZrCl_3]_x$ forms. The supernatant is removed via a filter cannula using No. 1 Whatman™ filter paper and it is discarded. The remaining yellow precipitate is washed with pentane (4 × 50 mL) and dried overnight under vacuum at room temperature. Yield: 3.54 g (77.0%).

Anal. Calcd. for $C_9H_7Cl_3Zr$: C, 34.6; H, 2.26; Cl, 34.0. Found: C, 34.6; H, 2.31; Cl, 33.9 (34.01). Mp: 194°C. The reaction between zirconium tetrachloride and 1-(trimethylsilyl)indene leads to multiple products and is not suitable for the preparation of $[(\eta^5\text{-}C_9H_7)ZrCl_3]_x$.

Properties

The canary-yellow powder of $(\eta^5\text{-}C_9H_7)ZrCl_3$ is air- and moisture-stable for short periods of time. Accounting for its low solubility in common organic solvents, the complex is likely a polymer. It displays the following physical properties: mp (under argon, in sealed capillaries, using a Gallenkamp melting point apparatus): 194°C, decomposition. ^{1}H NMR (Varian Gemini 200 NMR spectrometer at 199.975 MHz and referenced to tetramethylsilane, $CDCl_3$, 20°C): δ 7.70 (m, 2H), δ 7.38 (m, 2H), δ 7.02 (t, 1H), δ 6.80 (d, 2H). IR (Nicolet 5ZDX FT-IR spectrometer, cm^{-1}, Nujol, KBr): 450 w, 460 w, 594 vw, 620 vw, 730 w sh, 742 m, 750 w sh, 830 m, 847 m, 860 w, 872 w, 982 vw, 999 vw, 1050 w, 1125 vw, 1218 w, 1344 w, 1364 w, 1398 w, 1471 m, 3060 w, 3092 w, 3110 vw, 3129 vw.

E. (η^5-INDENYL)TRICHLOROHAFNIUM(IV)

Procedure

All reactions and manipulations should be performed under an atmosphere of nitrogen (or argon), using either a dry box or standard Schlenk techniques.[7] Neat 1-(trimethylsilyl)indene (4.7 mL, 22.8 mmol) is added to a room-temperature suspension of $HfCl_4$ (7.29 g, 22.8 mmol) in toluene

(160 mL) contained in a 250-mL round-bottomed flask equipped with a reflux condensor. The cloudy orange reaction mixture is stirred at reflux overnight to give a cloudy, yellow-brown mixture. The hot supernatant is filtered via a filter cannula using No. 1 Whatman™ filter paper and cooled at − 20°C to give bright yellow crystals of the product. Yield: 7.33g (80.4%).

Anal. Calcd. for $C_9H_7Cl_3Hf$: C, 27.0; H, 1.76; Cl, 26.6. Found: C, 27.3; H, 1.80; Cl, 26.3.

Properties

The yellow crystals of $[(\eta^5\text{-}C_9H_7)HfCl_2(\mu\text{-}Cl)]_2$ are air- and moisture-sensitive. The complex is a chloride-bridged dimer in the solid state, but very small amounts of monomeric $(\eta^5\text{-}C_9H_7)HfCl_3$ can be detected in the ^1H NMR spectrum of $[(\eta^5\text{-}C_9H_7)HfCl_2 (\mu\text{-}Cl)]_2$. Physical properties for $[(\eta^5\text{-}C_9H_7)\text{-}HfCl_2(\mu\text{-}Cl)]_2$ are: mp (under argon, in sealed capillaries, using a Gallenkamp melting point apparatus): 160°C. ^1H NMR (Varian Gemini 200 NMR spectrometer at 199.975 MHz and referenced to tetramethylsilane, $CDCl_3$, 20°C): δ 7.79(m, 2H), δ 7.38(m, 2H), δ 7.05(t, 1H), δ 6.85(d, 2H). IR (Nicolet 5ZDX FT-IR spectrometer, cm^{-1}, Nujol, KBr): 456 s, 505 w, 550 vw, 585 w, 715 w, 730 w, 754 s, 822 s, 832 s, 845 m sh, 875 w, 892 vw, 915 w, 965 w, 994 w, 1000 vw, 1051 m, 1087 vw, 1140 w, 1199 w, 1215 m, 1297 vw, 1341 m, 1400 m, 1405 w sh, 1447 m, 1490 w sh, 1508 vw, 1537 w, 1546 w, 1560 vw, 1587 vw sh, 1600 w, 1620 w, 1650 w, 1657 w, 1707 w, 1741 w, 1802 w, 1820 w, 1903 w, 1942 w, 1967 w, 3062 m, 3093 m, 3101 m, 3112 w sh.

References

1. T. E. Ready, R. O. Day, J. C. W. Chien, and M. D. Rausch, *Macromolecules*, **26**, 5822 (1993).
2. S. L. Shaw, R. J. Morris, and J. C. Huffman, *J. Organomet. Chem.*, **489**, C4–C7 (1995).
3. A. M. Cardoso, R. J. H. Clark, and S. Moorehouse, *J. Chem. Soc., Dalton Trans.*, 1156 (1980).
4. P. E. Rakita and A. Davison *Inorg. Chem.*, **8**, 1164 (1969).
5. J. M. Barnes and L. Magos, *Organomet. Chem. Rev.*, **3**, 137 (1968).
6. K. D. Smith and J. L. Atwood, *Inorg. Synth.*, **16**, 137 (1976).
7. D. F. Shriver and M. A. Drezdzon, *The Manipulation of Air-Sensitive Compounds*, 2nd ed., Wiley, New York, 1986.

37. LABILE COPPER(I) CHLORIDE COMPLEXES: PREPARATION AND HANDLING

Submitted by MIKAEL HAKANSSON*
Checked by SHARON NIEZGODA[†] and DONALD J. DARENSBOURG[†]

Previous literature methods for the preparation of CuCl complexes with neutral ligands involve either direct reaction between the copper(I) halide and the ligand, or reduction of a cupric salt—in the presence of the ligand—either chemically[1] or electrochemically.[2] Methods based on reduction of $CuCl_2$ benefit from the fact that cupric salts are more soluble than cuprous salts in most common solvents. However, reduction methods are less advantageous for syntheses using reactive and loosely bound ligands, because the ligand may be reduced and complicated workup procedures may be necessary. The heterogeneous direct-reaction method also has drawbacks, such as prolonged reaction time, product contamination with unreacted CuCl, and frequent failure to produce high-quality single crystals for X-ray diffraction analysis. These difficulties can be overcome by the use of a soluble and labile precursor complex. Nitriles, such as aceto- or benzonitrile form complexes that will not react with olefins, for example, because of the strong Cu–N interactions; the situation is even worse with sulfur–donor ligands such as dimethylsulfide. Most cyclic mono- or diolefins form insoluble complexes with CuCl or are too tightly bound, as in the tetrahapto $[CuCl(COD)]_2$ dimer,[3] to effectively function in precursors for labile CuCl complexes. The reaction between CuCl and ethylene at 60 bar produces $CuCl(C_2H_4)$,[4] but this compound decomposes far too rapidly to be of any preparative use. While neutral ligands with only oxygen donor atoms (nitromethane, acetone, THF, dioxane, DME) do not dissolve CuCl, the combination of an alkene moiety with an oxygen donor atom, as in ethyl vinyl ketone (EVK), results in a ligand that forms a soluble and reactive adduct with copper(I) chloride. Unfortunately, the resulting solid CuCl(EVK) is so labile that it only exists at temperatures below 0°C.[5]

In the preparative method described, the precursor complex is formed in situ, using the precursor ligand as solvent, which minimizes ligand dissociation. The [CuCl(EVK)] precursor solution in EVK (designated as $[CuCl(EVK)]_{EVK}$) is highly reactive but stable at −25°C under inert

* Department of Inorganic Chemistry, Chalmers University of Technology, S-41296 Gothenburg, Sweden.
† Department of Chemistry, Texas A&M University, College Station, TX 77843.

atmosphere for extended periods of time. A wide range of labile CuCl complexes can be prepared from the $[CuCl(EVK)]_{EVK}$ reagent, with neutral ligands such as α,β-unsaturated carbonyl compounds,[6] conjugated dienes,[7] nonconjugated cyclic and acyclic dienes,[8] alkynes,[9] and carbon monoxide.[10] The syntheses of two complexes using the $[CuCl(EVK)]_{EVK}$ reagent are presented below: the CuCl adducts with carbon monoxide (**A**) and isoprene (**B**).

Carbonylchlorocopper(I) (**A**) is a valuable precursor for the synthesis of organometallic and coordination compounds of copper(I). Previous methods[11] utilizing water, tetrahydrofuran, methanol, or benzene as solvent may yield lower-purity products because copper(I) chloride is not soluble in these solvents without excess halide ligand. The synthetic method described here also provides a convenient way of growing large crystals of X-ray quality of carbonylchlorocopper(I).

Dichloro[*catena*-2-methyl-1,4-butadiene]dicopper(I) (**B**) can be prepared from direct reaction between isoprene and copper(I) chloride, analogously to the synthesis of the complex between butadiene and CuCl.[12] The advantage of the synthesis described here is that the product is not contaminated with unreacted CuCl, and can be obtained as high-quality crystals instead of a microcrystalline powder.

General Procedure—Preparation

The precursor solution $[CuCl(EVK)]_{EVK}$ is prepared by dissolving approximately 1.0 g CuCl in 50 mL ethyl vinyl ketone* at ambient temperature. Centrifugation may sometimes be necessary in order to obtain a perfectly clear solution. Reaction can then be performed with solid, liquid, or gaseous ligands, simply by adding the new neutral ligand slowly to the $[CuCl(EVK)]_{EVK}$ precursor solution. However, when preparing very labile complexes, the reverse order of addition is used: the precursor solution is added dropwise to the new neutral ligand, at -70 to $0°C$, thus ensuring a high concentration of the product ligand in the reacting solution. Choice of reaction temperature can influence the crystallization rate. The isolation and handling of labile complexes are done at low temperature. Complexes prepared in this way are high-purity and frequently high-quality crystals, suitable for X-ray analysis or further reactions.

General Procedure—Handling

This technique for handling labile complexes depends on retarding decomposition rates to a minimum by keeping the compounds close to a liquid

---

* Checkers' note: the scale of the reaction was reduced to one-fifth with only a minor reduction of yield.

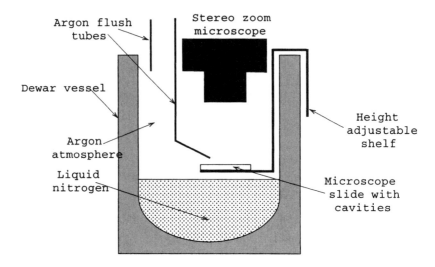

Figure 1. The experimental set-up for handling sensitive solids.

nitrogen surface. A wide Dewar vessel (approximate dimensions: height 200 mm; diameter 170 mm) is adapted according to Figure 1.

Before filling the Dewar vessel with liquid nitrogen, it should be flushed with argon to minimize ice formation; thereafter a slow argon stream is maintained directly over the shelf, as shown in Figure 1. The temperature experienced by the labile complexes is regulated by altering the distance between the shelf and the liquid nitrogen surface, and/or varying the flow rate in the argon flush tubes. The actual temperature at the microscope slide can be monitored by the use of a thermocouple. The shelf, which can be constructed from a piece of aluminum or steel (75×300 mm), is best inserted into the Dewar after filling it with liquid nitrogen.

The labile complex, together with some solvent/mother liquor, are transferred from a Schlenk tube to the metal shelf by the use of a carefully argon-flushed Pasteur pipette or syringe. The complex is then washed with several portions of an inert solvent with a low freezing point. The large majority of sensitive compounds can be handled at $-120\,°C$ for a short period of time, so that using common solvents like THF, toluene, or hexane is often fully possible. The operations are performed on microscope slides (with two or three cavities) placed on the shelf, using needles, spatulas, and glass rods. When handling larger amounts of solids or solutions, medicinal black embryo bowls for microscopy are useful. The slides and the Dewar vessel should be flushed with argon to exclude oxygen and to prevent ice formation on the shelf.* It is essential to mount a standard stereo-zoom microscope (10–60 × magnification, pillar stand) above the Dewar vessel, since the effect of all operations should be controlled in the search for signs of decomposition.

For obtaining IR spectra, for example, solid samples are transferred, washed, and selected at low temperature using the described technique. The windows of an IR mull cell are placed on the metal shelf and the sensitive compound is ground to a mull using pentane—or, at lower temperatures, a condensed hydrocarbon gas—as mulling agent. When possible, the use of CaF_2 mull cell windows is recommended because of their general insensitivity. The mull is applied to the windows and the cell is assembled at low temperature, whereafter the entire Dewar vessel is transferred to an inert atmosphere-flushed FTIR spectrometer, and the mull cell is quickly inserted. Spectra should be recorded continuously as the mull cell is warming up, in order to monitor how decomposition affects the spectral features. Using similar techniques, samples for studies NMR or x-ray diffraction can be prepared.

A. CARBONYL CHLOROCOPPER(I), Cu(CO)Cl

$$CuCl + EVK \Rightarrow CuCl(EVK)$$

$$CuCl(EVK) + CO \Rightarrow Cu(CO)Cl + EVK$$

Procedure

■ **Caution.** *Carbon monoxide is an odorless, extremely toxic gas and must be handled in an efficient hood. Solid Cu(CO)Cl exhibits a very high CO dissociation pressure and should be handled with the same care as carbon monoxide. Ethyl vinyl ketone (EVK) is a strong lachrymator and must be handled in a well-ventilated hood. EVK may contain peroxides and caution must be exercised when distilling.*

All operations are performed under argon or carbon monoxide atmosphere using Schlenk or special low-temperature,[13] techniques. Ethyl vinyl ketone

---

* Keeping the Dewar in a well-ventilated hood is *not* recommended on acccount of extensive ice formation and decreased visibility, but due to the low temperature and the small amount of chemicals used, the release of harmful vapors is minimal. The set-up described can be built inside a glove box if the problem of handling liquid nitrogen without contaminating the glove-box atmosphere can be solved. The main advantage would be that ice formation is eliminated, while the disadvantages of not having an open system—in the ease of handling and consequently time consumption—are obvious.

(1-penten-3one) is degassed, dried with molecular sieves (4 Å), and distilled shortly prior to use. Copper(I) chloride is purified according to a literature method.[11a] Commercial carbon monoxide (99.998%) is used without further purification.

Copper(I) chloride (1.00 g, 10.1 mmol) is dissolved in 50 mL EVK in a 250-mL Schlenk flask using a magnetic stir bar at ambient temperature for 15 min. If necessary, the solution is centrifuged to remove the last traces of undissolved CuCl. Carbon monoxide is introduced into the flask via a large needle through a rubber septum. A slow stream of carbon monoxide is bubbled into the solution for approximately 30 min to ensure complete reaction; precipitation of Cu(CO)Cl starts quickly. Unreacted CO is vented through the pressure-release valve which is connected to the manifold used. The needle is removed and the white precipitate is allowed to settle, whereafter the solvent is withdrawn by the use of a syringe and Cu(CO)Cl is washed with three 10-mL portions of dry, carbon monoxide-saturated diethyl ether. The remaining ether suspension can be used directly for preparing spectroscopy samples at low temperature. Alternatively, the last ether washing is withdrawn with a syringe and the white solid is dried under CO atmosphere. To this effect, the argon connected to the manifold is replaced with CO and a very slow stream of carbon monoxide is passed into the manifold and out through the pressure release valve. Yield: 1.17 g (91%). Single crystals for X-ray diffraction can be grown by introducing carbon monoxide into the Schlenk flask *above* the liquid surface. The gaseous ligand will slowly diffuse into the solution and crystals will start growing at the solution/CO interface. It is of paramount importance to keep the flask completely still. Larger amounts of Cu(CO)Cl may be prepared by adding more copper(I) chloride before withdrawing the solvent and repeating the subsequent operations, but care must be taken to ensure complete dissolution of copper(I) chloride in order to obtain a pure product.

Properties

Carbonylchlorocopper(I) is a colorless crystalline substance that decomposes rapidly in the absence of a carbon monoxide atmosphere to give copper(I) chloride and carbon monoxide. The compound is, however, stable for long periods of time if stored under carbon monoxide. Cu(CO)Cl has a polymeric structure,[10] which may be described as layers of fused, six-membered, copper–chloride rings in the chair conformation, with terminally bonded carbonyl ligands. The infrared spectrum of Cu(CO)Cl (Nujol mull at 0°C) displays a characteristic large peak at 2127 cm^{-1} and a vibrational analysis has been reported.[13]

B. DICHLORO [*catena*-2-METHYL-1,4-BUTADIENE]DICOPPER(I), $Cu_2Cl_2(C_5H_8)$

$$CuCl + EVK \Rightarrow CuCl(EVK)$$

$$2CuCl(EVK) + C_5H_8 \Rightarrow Cu_2Cl_2(C_5H_8) + 2EVK$$

Procedure

▪ **Caution.** *Isoprene (2-methyl-1,4-butadiene) is a low-boiling (bp 34°C), flammable liquid which is suspected to be carcinogenic. Isoprene must therefore be handled in a well-ventilated hood and care must be taken to avoid overpressurizing glass containers.*

Isoprene is degassed, dried with molecular sieves (4 Å), and distilled shortly prior to use. Following the precautions and procedures in Part A, copper(I) chloride (1.00 g, 10.1 mmol) is dissolved in 50 mL EVK in a 250-mL Schlenk flask. Cold (0°C) isoprene (10 mL) is added dropwise over the course of 10 min to the solution via a syringe; yellow plates of $Cu_2Cl_2(C_5H_8)$ are deposited after a few hours. If a larger volume of isoprene is added, instantaneous precipitation of microcrystalline $Cu_2Cl_2(C_5H_8)$ results. The solvent is withdrawn with a syringe and the solid is quickly washed with three 10-mL portions of cold isoprene, taking care that the solid never completely dries. Crystals of $Cu_2Cl_2(C_5H_8)$ can be isolated, dried, and manipulated without decompositon at temperatures below $-50°C$, and if kept under isoprene they can be allowed to reach ambient temperature for limited periods of time. Yield: 1.28 g (96%).

Properties

Dichloro[*catena*-2-methyl-1,4-butadiene]dicopper(I) is a bright yellow crystalline substance that decomposes rapidly at ambient temperature in absence of isoprene to give copper(I) chloride and isoprene. The compound is, however, stable for long periods of time if stored at $-80°C$ under isoprene. Dichloro[*catena*-2-methyl-1,4-butadiene]dicopper(I) is insoluble, owing to its polymeric structure,[7] which may be described in terms of copper–chloride layers with peripherally bound isoprene ligands, exhibiting the s-*trans* conformation and bridging two copper(I) atoms. Rows of isoprene ligands are positioned by the Cu–Cl framework in such a way that there are several short van der Waal contacts (3.4–3.5 Å) between ligands in adjacent layers. IR: $v(C=C-C=C)$ 1569, 1520 cm^{-1}. Corresponding frequencies for uncomplexed isoprene: 1640, 1598 cm^{-1}.

References

1. (a) E. A. Abel, M. A. Bennet, and G. Wilkinson, *J. Chem. Soc.*, 3178 (1959); (b) B. W. Cook, R. G. J. Miller, and P. F. Todd, *J. Organomet. Chem.*, **19**, 421 (1969).
2. (a) S. E. Manahan, *Inorg. Chem.*, **5**, 2063 (1966); (b) S. E. Manahan, *Nucl. Chem. Lett.*, **3**, 383 (1967).
3. J. H. van der Hende and W. C. Baird, Jr., *J. Am. Chem. Soc.*, **35**, 1009 (1963).
4. H. Tropsch and W. J. Mattox, *J. Am. Chem. Soc.*, **57**, 1102 (1935).
5. M. Håkansson, S. Jagner, and M. Nilsson, *J. Organomet. Chem.*, **336**, 279 (1987).
6. M. Håkansson and S. Jagner, *J. Organomet. Chem.*, **361**, 269 (1989).
7. M. Håkansson, S. Jagner, and D. Walther, *Organometallics*, **10**, 1319 (1990).
8. M. Håkansson and S. Jagner, *J. Organomet. Chem.*, **397**, 383 (1990).
9. M. Håkansson, K. Wettström, and S. Jagner, *J. Organomet. Chem.*, **421**, 347 (1991).
10. M. Håkansson and S. Jagner, *Inorg. Chem.*, **29**, 5241 (1990).
11. (a) R. N. Keller and H. D. Wycoff, *Inorg. Synth.*, **2**, 1 (1946); (b) P. Fiaschi, C. Floriani, M. Pasquali, A. Chiesi-Villa, and C. Guastini, *Inorg. Chem.*, **25**, 462 (1986).
12. J. R. Doyle, P. E. Slade, and H. B. Jonassen, *Inorg. Synth.*, **6**, 216 (1960).
13. M. Håkansson, S. Jagner, and S. F. A. Kettle, *Spectr. Acta*, **48A**, 1149 (1992).

Chapter Four

MAIN GROUP AND TRANSITION METAL CLUSTER COMPOUNDS

38. 7,8-DICARBAUNDECABORANE(13)

Submitted by GREGORY G. HLATKY* and DONNA J. CROWTHER[†]
Checked by RICHARD F. JORDAN[‡] and DANIEL E. BOWEN[‡]

$$1,2\text{-}C_2B_{10}H_{12} + KOH \xrightarrow{MeOH} K[C_2B_9H_{12}] + B(OMe)_3 + H_2O + H_2$$

$$K[C_2B_9H_{12}] + H_3PO_4 \xrightarrow{Toluene} C_2B_9H_{13} + KH_2PO_4$$

The $nido$-$[C_2B_9H_{12}]^-$ anion was first prepared in 1964 by the alkaline degradation of 1,2-dicarbaundecaborane $(1,2\text{-}C_2B_{10}H_{12})$.[1] Further deprotonation yields the $[C_2B_9H_{11}]^{2-}$ anion, which is isolobal to $[C_5H_5]^-$ and widely used in the synthesis of metallacarborane complexes.[2] The diprotic acid 7,8-dicarbaundecaborane(13), $C_2B_9H_{13}$, is formed when $[C_2B_9H_{12}]^-$ is protonated. This compound has been found to be useful in the preparation of metallacarborane compounds of the early in the preparation of metallacarborane compounds of the early transition metals[3-4] and was first prepared using HCl in diethyl ether.[1] An improved preparation used phosphoric acid for the protonation.[5] The preparation described here is a modification of the latter procedure and affords high yields of the diacid.

* Lyondell Petrochemical Company, Lyondell Technology Center, P.O. Box 2917, Alvin, TX 77512.
[†] Exxon Chemical Company, Baytown Polymers Center, P.O. Box 5200, Baytown, TX 77522.
[‡] Department of Chemistry, University of Iowa, Iowa City, IA 52242.

229

Reagents and Apparatus

Methanol and benzene are deaerates by purging with nitrogen. The synthetic procedure was performed in an inert nitrogen atmosphere.

■ **Caution.** *Benzene is a suspected human carcinogen and should be used only in a well-ventilated fume hood; protective gloves should be worn. Toluene can be substituted for benzene in this procedure, but yields are frequently lower.*

Procedure

A 250-mL, three-necked, round-bottomed flask is equipped with a nitrogen inlet and a condenser attached to a pressure-relief bubbler. The flask is flushed with nitrogen and potassium hydroxide (3.2 g, 57 moles) and anhydrous methanol (80 mL) are added. 1,2-Dicarbadodecaborane (1,2-$C_2B_{10}H_{12}$)* (4.7 g, 32.6 moles) is added and the mixture is stirred at room temperature until all the carborane dissolves (about 30 min).

The mixture is refluxed for 15 h under a nitrogen atmosphere. After cooling to room temperature, the methanol is removed by rotary evaporation to yield a white semisolid containing residual methanol. Benzene (150 mL) is added and the mixture is distilled to remove residual methanol until an overhead temperature of 78°C (1 atm) is maintained.

The solution, containing about 60 mL of benzene, is cooled to room temperature and 85% phosphoric acid (20 mL) is added. The two-phase mixture is stirred for 15 h under nitrogen. The benzene layer is decanted from the phosphoric acid, which is extracted with an additional 50 mL of benzene. At this point, a nitrogen atmosphere no longer need be maintained, but exposure to air should be minimized.

The combined benzene extracts are dried with anhydrous $MgSO_4$ (about 5 g) and filtered. The benzene in the filtrate is removed by rotary evaporation under reduced pressure at 25°C to yield an off-white solid. The nearly pure $C_2B_9H_{13}$ is sublimed at 40–60°C at 10^{-3} mm onto a probe cooled to −78°C. The yield of pure product is 3.66 g (83.6%).[†]

---

* Available from Dexsil Corp., 1 Hamden Park Dr., Hamden, CT 06517.

[†] Checkers' comment: Using the submitted procedure, we obtained a comparable yield (76% vs. the reported 84%) but our product was contaminated with 6% starting material (1,2-$C_2B_{10}H_{12}$). When the amount of KOH was increased by 20%, we obtained a lower yield (51%), but the product was pure.

Using 20% more KOH, increasing the reaction time, stripping the benzene solution to dryness after the azeotropic distillation, and removing any unreacted 1,2-$C_2B_{10}H_{12}$ by sublimation at 40°C prior to H_3PO_4 treatment affords high yields (80–90%) of pure product.

Properties

7,8-Dicarbaundecaborane(13) is a white hygroscopic solid which is soluble in water, alcohols, and aromatic solvents. It is best stored cold in a dry, inert atmosphere. The proton-decoupled ^{11}B NMR spectrum (115 MHz, C_6D_6, referenced to external $BF_3 \cdot OEt_2$) shows peaks at d 4.0 (2 B), -4.4 (2 B), -16.4 (1 B), -17.0 (1 B), and -27.4 (3 B). The complex is a diprotic acid with pK_{a1} of 2.98 and pK_{a2} of 14.25.[6] Deprotonation of $C_2B_9H_{13}$ with butyl lithium in benzene affords anhydrous $Li_2C_2B_9H_{11}$.[7]

7,8-Dicarbaundecaborane(13) reacts with the metal–methyl bonds of $[Sc(C_5Me_5)Me_2]_x$ and $Zr(C_5Me_5)Me_3$ to form the metallacarborane complexes $Sc(C_5Me_5)(C_2B_9H_{11})$[4] and $Zr(C_5Me_5)(C_2B_9H_{11})Me$,[3] respectively. The reaction with $Zr(C_5Me_5)_2Me_2$ affords $Zr(C_5Me_5)_2Me(C_2B_9H_{12})$, in which the carborane functions as a "non-interfering" anion to a $Zr(C_5Me_5)_2Me^+$ cation.[8] The two zirconium complexes are active catalysts for ethylene polymerization.

When pyrolyzed at 100°C, 7,8-dicarbaundecaborane(13) loses hydrogen to form *closo*-1,2-dicarbaundecaborane(11), $C_2B_9H_{11}$.[9]

References

1. R. A. Wiesboeck and M. F. Hawthorne, *J. Am. Chem. Soc.*, **86**, 1642 (1964).
2. M. F. Hawthorne, *Acc. Chem. Res.*, **1**, 281 (1968).
3. D. J. Crowther, N. C. Baenziger, and R. F. Jordan, *J. Am. Chem. Soc.*, **113**, 1455 (1991).
4. G. C. Bazan, W. P. Schaefer, and J. E. Bercaw, *Organometallics*, **12**, 2126 (1993).
5. (a) D. A. T. Young, R. J. Wiersma, and M. F. Hawthorne, *J. Am. Chem. Soc.*, **93**, 5687 (1971). (b) J. Plesek, S. Hermanek, and B. Stibr, *Inorg. Synth.*, **22**, 231 (1983).
6. N. A. Truba and B. I. Nabivanets, *Zh. Obshch. Khim.*, **49**, 1333 (1979).
7. R. Uhrhammer, D. J. Crowther, J. D. Olson, D. C. Swenson, and R. F. Jordan, *Organometallics*, **11**, 3098 (1992).
8. G. G. Hlatky, H. W. Turner, and R. R. Eckman, *J. Am. Chem. Soc.*, **111**, 2728 (1989).
9. F. N. Tebbe, P. M. Garrett, and M. F. Hawthorne, *J. Am. Chem. Soc.*, **90**, 869 (1968).

39. BORAZINE, POLYBORAZYLENE, B-VINYLBORAZINE, AND POLY(B-VINYLBORAZINE)*

Submitted by THOMAS WIDEMAN,[†] PAUL J. FAZEN,[†]
ANNE T. LYNCH,[†] KAI SU,[†] EDWARD E. REMSEN,[‡]
and LARRY G. SNEDDON[‡]
Checked by TUQIANG CHEN[§] and ROBERT T. PAINE[§]

Borazine, $B_3N_3H_6$, originally discovered by Stock in 1926,[1] is an inorganic analog of benzene.[2] The recent design and synthesis of a range of new borazine-based polymers that serve as "single-source" precursors to BN ceramics[3] have sparked new interest in the chemistry and properties of this ring compound. However, chemical studies, as well as practical applications, of borazine compounds and polymers have been held back by the absence of efficient and economical synthetic routes. We report below convenient procedures for the multigram syntheses of borazine and B-vinylborazine (an inorganic analog of styrene), as well as their homopolymers, polyborazylene and poly(B-vinylborazine).

A. BORAZINE, $B_3N_3H_6$

$$3\,(NH_4)_2SO_4 + 6\,NaBH_4 \xrightarrow[\text{tetraglyme}]{120-140^\circ C,\,3h} 2\,B_3N_3H_6 + 3Na_2SO_4 + 18\,H_2 \quad (1)$$

Previous preparations of borazine in this series[4] and elsewhere[5,6] have suffered from low yields, difficult purification, and the handling of air-sensitive compounds such as 2,4,6-trichloroborazine and diborane. Described below[7] is a convenient, one-step synthetic procedure for the laboratory production of multigram quantities of borazine, $B_3N_3H_6$, in 55–65% yields. The method is based on the reaction of $NaBH_4$ and $(NH_4)_2SO_4$ in tetraglyme solution at 120–140°C with continuous removal of the product in vacuo. Borazine is the only volatile product and is obtained in excellent purity by a single vacuum fractionation.

* Work supported by the U.S. Department of Energy, Division of Chemical Sciences, Office of Basic Energy Sciences, Office of Energy Research, and the National Science Foundation, Washington, DC.
† Department of Chemistry, University of Pennsylvania, Philadelphia, PA 19104.
‡ Analytical Sciences Center, Monsanto Corporate Research, St. Louis, MO 63167.
§ Department of Chemistry, University of New Mexico, Albuquerque, NM 87131.

Procedure

▪ **Caution.** *Relatively large amounts of hydrogen are produced from the reaction. The size of the liquid nitrogen traps should be adequate to retain the borazine without clogging.*

All synthetic manipulations are carried out using standard vacuum or inert atmosphere techniques as described by Shriver.[8]

Sodium borohydride* (99%, powder) is used as received. Ammonium sulfate* (anhydrous, 99 + %) is used without further purification;[†] however, *it is very important that it is ground to a very fine powder.* Tetraethylene glycol dimethyl ether* (tetraglyme, 2,5,8,11,14-pentaoxapentadecane)(99%) *must* be vacuum distilled (110–120°C, ≤ 100 μTorr) from molten sodium before use.

A simplified sketch of the apparatus used in the preparation is shown in Figure 1. An intimate mixture of 30.7 g (0.81 mol) of $NaBH_4$ and 82.3 g (0.62 mol) of $(NH_4)_2SO_4$ is prepared by shaking the fine powders together in a 500-mL screw-topped plastic bottle. The mixture is added, under a nitrogen flow from **A**, to 350 mL of tetraglyme in a 2-L, three-necked, round-bottomed flask (**D**), which is fitted with a thermometer (**E**) and distilling column (**F**) (such as Chemglass Part No. CG-1231-18[‡]). The exit of the reflux condenser is connected via vacuum tubing (**G**) to a mercury manometer (**H**), followed by four traps (**I**) (~ 200 mL each, with the distance between the trap wall and the inner tube ≥ 2 cm, such as Chemglass Part No. CG-4512-04[‡] or CG-4516-07[‡]).[§]

Hydrogen evolution begins immediately at room temperature and is released through the oil bubbler (**B**) (such as Chemglass Part No. AF-0513-20[‡]). Foaming is observed as the reaction mixture is gradually warmed under nitrogen, using an electric heating mantle, to 70°C over the course of 1 h. When the hydrogen evolution and foaming subsides, the valve (**C**) leading to the oil bubbler is closed. Liquid nitrogen dewars are placed around the four traps (**I**). A mechanical pump is then used to apply vacuum through valve **J**. The pressure, as measured by the manometer (**H**), is maintained at 2–5 Torr by regulating the opening in valve **J**, as the evolved hydrogen and borazine product is continuously removed through the vacuum line. The reaction is further heated to 135°C, where it is held for an additional 1–2 h under a dynamic vacuum of 2–5 Torr. When hydrogen evolution ceases, borazine is further stripped from the heated reaction mixture at ~ 0.1 Torr for 20 min.

---

* Aldrich Chemical Company, Milwaukee, WI 53233.
[†] The checkers suggested vacuum drying the ammonium sulfate before use.
[‡] Chemglass, Inc., 3861 N. Mill Rd., Vineland, NJ 08360.
[§] The reviewers used traps similar to AF-0203-03 from Chemglass, Inc.

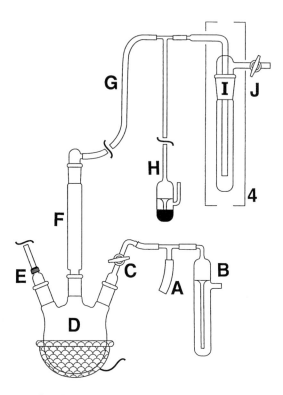

Figure 1. Apparatus for the synthesis of borazine. (**A**) nitrogen supply from tank; (**B**) oil bubbler; (**C**) valve; (**D**) 2-L, three-necked, round-bottomed flask; (**E**) thermometer; (**F**) distilling column; (**G**) vacuum tubing; (**H**) mercury manometer; (**I**) four traps for the collection of the borazine product; (**J**) valve to mechanical vacuum pump.

Following the reaction, the borazine collected in the liquid-nitrogen traps (**I**) is further purified by a single vacuum fractionation,[9] as shown in Figure 2, through a U-trap series cooled at $-45°C$ (**A**), $-78°C$ (**B**), and $-196°C$ (**C**). In the $-78°C$ trap (**B**), 13.1 g (0.16 mol, 60% yield based on starting BH_4^-) of borazine is condensed. No other products are detected in the IR, ^{11}B NMR, ^{1}H NMR (see below), or GC/MS spectra of the product, and its vapor pressure (85 Torr at 0°C) matches the literature value,[10] indicating that the borazine is obtained in excellent purity.

The solvent is reused without purification in future preparations, with slightly improved yields, until the accumulation of solids in the solvent prevents the wetting of the reactants.

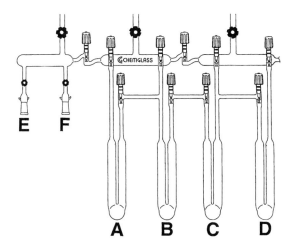

Figure 2. Vacuum line used for the transfer of reactants and the purification of borazine or B-vinylborazine. (**A, B, C,** and **D**) U-traps; (**E** and **F**) ports for attaching reaction flasks to the vacuum line.

Properties

Borazine is a clear, colorless liquid (mp $-58°C$; bp $55°C$)[11] that is readily hydrolyzed on exposure to atmospheric moisture or alcohols. It is miscible with polar and nonpolar organic solvents. Significant decomposition with evolution of hydrogen gas is observed when borazine is stored at room temperature over a period of months. It is recommended[12] that borazine be stored at or below 4°C.

Characteristic Spectroscopic Data. IR (gas phase, cm^{-1}): 3450 (s, m), 2500 (m), 1420 (br, m), 910 (m), 710 (m). ^{11}B (64.1 MHz, C_6D_6, ppm): 30.5 (d, J_{HB} = 135 Hz). ^1H NMR (200.1 MHz, C_6D_6, ppm): 5.10 (t, J_{NH} = 55 Hz), 4.48 (q, J_{BH} = 135 Hz). Other physical properties are provided in reference 10.

B. POLYBORAZYLENE,[13] $[B_3N_3H_{\sim4}]_x$

$$X \quad \text{Borazine} \quad \xrightarrow{70°C} \quad \text{Polyborazylene} \quad + \quad XH_2 \quad (2)$$

Borazine　　　　　　**Polyborazylene**

Procedure

■ **Caution.** *Heating gram quantities of borazine will produce large volumes of hydrogen. If the polymerization is not carried out in a high-pressure vessel, frequent degassings of the reaction vessel are necessary.*

All manipulations are carried out using inert atmosphere techniques.[8] Solvents must be distilled from sodium–benzophenone ketyl. Borazine is prepared by the procedure described above.

On a preparative vacuum line (Figure 2), a 2.97-g (36.9 mmol) sample of borazine is condensed at − 196°C into an evacuated 88-mL Andrews Glass pressure-reaction vessel (part no. 110-207-0003)* equipped with a Whitey needle valve (part no. B-1RM4)[†] and a magnetic stir bar. The valve is closed and the flask is removed to a shielded hood, where it is lowered into a 70°C oil bath until the glass portion of the flask is covered and stirring is commenced. Periodically (every 6 h), the flask is removed to the vacuum line,[‡] where the bottom half of the flask is immersed in liquid nitrogen, freezing the contents at − 196°C. The valve is then opened to the vacuum line to release evolved hydrogen. The valve is closed again and the flask is returned to the oil bath. The reaction is continued until the liquid becomes so viscous that stirring cannot be continued (2–3 days). If the reaction is allowed to continue further, cross-linking produces an insoluble gel. The reaction flask is degassed at − 196°C, then warmed to room temperature, and the volatile components are vacuum evaporated into a − 196°C trap for 12 h. A white polymeric solid (**I**) (2.27 g, 91% yield based on reacting borazine) remains in the flask. Unreacted borazine (0.48 g, 6.0 mmol) is recovered from the materials in the − 196°C trap by vacuum fractionation through a − 78°C trap.

Anal. Found for **I**: B, 40.39%; N, 52.64%; H, 4.22%, consistent with a $B_{3.00}N_{3.02}H_{3.36}$ formula.

A 1.25-g sample of polymer **I** is dissolved in 10 mL of glyme and precipitated by slowly adding the solution to 100 mL of dry pentane. Filtration under nitrogen and vacuum drying for 12 h of the precipitate gives a fine white powder (**II**) (0.839 g, 67% yield).

Anal. Found for **II**: B, 40.74%; N, 53.80%; H, 4.10%, consistent with a $B_{3.00}N_{3.06}H_{3.25}$ formula.

---

* Andrews Glass Co., 3740 N. West Blvd., Vineland, NJ, 08360.
[†] Penn Valve and Fitting Company, Willow Grove, PA, 19090.
[‡] The checkers suggested keeping both the reaction and vacuum line under the same shielded hood.

Properties

Polyborazylene is a white, moisture-sensitive solid which decomposes before melting. It is soluble in dry glyme or THF. It is insoluble in pentane, benzene, diethyl ether, and methylene chloride. The polymer undergoes rapid hydrolysis in solution and should therefore be handled only in *anhydrous* solvents, but may be handled for short periods (min) in air as a solid without significant decomposition. Extended storage (months) at room temperature results in further cross-linking and reduced solubility. Polyborazylene has a complex structure containing linear, chain-branched, and fused cyclic segments.[13] The polymer is an excellent single-source precursor for the formation of turbostratic boron nitride in high chemical and ceramic yields ($\sim 90\%$).[14]

Characteristic Spectroscopic Data. (**I**)—IR (DRIFT, KBr powder, cm^{-1}): 3445 (m), 2505 (m), 1450 (br, s), 1200 (m), 900 (m), 750 (m), 690 (m). ^{11}B NMR (160.5 MHz, THF, ppm) 31 (br). **II**—spectroscopic data are similar to that reported for **I**.

Molecular weights for representative samples of the polymers using viscometry and size exclusion chromatography/low angle laser light scattering (SEC/LALLS) are as follows. **I**: $M_n = 506$, $M_w = 1476$, and $M_w/M_n = 2.90$ (viscometry); $M_n = 1400$, $M_w = 4000$, and $M_w/M_n = 2.86$ (SEC/LALLS). **II**: $M_n = 637$, $M_w = 1554$, and $M_w/M_n = 2.44$ (viscometry); $M_n = 3400$, $M_w = 7600$, and $M_w/M_n = 2.23$ (SEC/LALLS).

C. B-VINYLBORAZINE OR 2-ETHENYLBORAZINE, $H_3C_2B_3N_3H_5$

Although vinyl derivatives of alkylated borazines have been known for some time,[15] the methods used for their syntheses cannot be used for the parent B-vinylborazine because of extensive side reactions. The transition metal catalyzed synthetic route described below gives B-alkenylborazines in high yields.[16]

Procedure

■ **Caution.** *Acetylene is a flammable and explosive gas and should be handled with care either in vacuo or in an efficient fume hood. Pure acetylene should not be pressurized.*

Since the pressure in the reaction vessel could initially exceed one atmosphere, a suitable pressure reactor must be used.

All manipulations are carried out using inert atmosphere techniques.[8] The vacuum line used in the following preparation is similar to that shown in Figure 2.

Borazine is prepared by the procedure described above. Impurities in commercially obtained acetylene can result in the precipitation of the catalyst, and in such cases, no reaction is observed. Acetylene is therefore purified by a process similar to that described in the literature[17] in which acetone and other impurities are removed by a purification train consisting of two bubblers of concentrated sulfuric acid followed by two U-traps containing NaOH pellets and white Drierite, respectively, to remove moisture. The inlet of the train is connected to a cylinder of acetylene, and its outlet to the vacuum line.

In a typical reaction, a sample of 50.1 mg $(5.45 \times 10^{-2}$ mmol) of $RhH(CO)(PPh_3)_3$* is placed in an 88-mL Andrews Glass glass pressure-reaction vessel (part no. 110-207-0003)[†] equipped with a Whitey needle valve (part no. B-1RM4)[‡] and magnetic stir bar. The vessel is evacuated and 9.59 g (119.1 mmol) of borazine and 32.5 mmol of acetylene[§] (purified, 99.6%) (measured by expansion into a calibrated vacuum line) are vacuum transferred into the reactor which is cooled at $-196°C$. The reactor valve is closed, the vessel is removed to a shielded hood, and the reaction mixture is allowed to warm to room temperature. *Caution should be taken as the pressure in the vessel could exceed 1 atm until the acetylene dissolves.* Within minutes the yellow solution turns reddish brown and becomes completely homogeneous. After stirring for 21 h (longer reaction times result in decreased yields), the reaction flask is attached to the vacuum line, frozen at $-196°C$, and degassed to remove any small amounts of noncondensable gas. The volatile materials are then vacuum fractionated,[9] as shown in Figure 2, through U-traps cooled at $-18°C$ (**A**), $-45°C$ (**B**), $-78°C$ (**C**), and $-196°C$ (**D**). The material in the $-18°C$ trap (**A**) contains B-vinylborazine along with a small

---

* Strem Chemicals, Inc., Newburyport, MA 01950.
† Andrews Glass Co., 3740 N. West Blvd., Vineland, NJ, 08360.
‡ Penn Valve and Fitting Company, Willow Grove, PA 19090.
§ MG Industries, Malvern, PA 19355.

amount of B,B-divinylborazine, while the $-45°C$ trap (**B**), contains both B-vinylborazine and unreacted borazine. Most of the unreacted borazine is collected in the $-78°C$ trap (**C**). After fractionation, the liquid collected in the $-18°C$ (**A**) and $-45°C$ (**B**) traps is fractionated through the trap series again. It is normally necessary to repeat the process three times to obtain pure B-vinylborazine. This procedure gives 2.21 g (20.7 mmol, 70.6% yield relative to consumed borazine) of B-vinylborazine in the $-45°C$ trap (**B**), and 7.23 g (89.8 mmol) of borazine is recovered in the combined $-78°C$ (**C**) and $-196°C$ (**D**) traps. The compound is obtained in pure form according to its IR, ^{11}B NMR, and ^{1}H NMR spectra (see below).

The catalyst is still active upon removal of the products and can be used repeatedly with no signs of decreased yields or turnover rates.

Properties

B-Vinylborazine is an air-sensitive volatile liquid and is soluble in dry aprotic organic solvents. It decomposes rapidly in contact with water and alcohols.

Characteristic Spectroscopic Data. IR (gas phase, cm^{-1}): 3480 (s), 3080 (sh), 3070 (m), 2970 (m), 2520 (vs), 1620 (s), 1475 (vs), 1425 (s), 1380 (vs), 1350 (s), 955 (s), 930 (vs), 920 (vs), 735 (s), 720 (vs). ^{11}B-NMR (160.5 MHz, C$_6$D$_6$, ppm): 31.2 (s, B2) and 31.5 (d, $J_{BH} = 129$ Hz, B4,6). ^{1}H-NMR (500.1 MHz, C$_6$D$_6$, ppm): 4.4 (q, $J_{BH} = 136$ Hz, 2H), 5.26 (t, $J_{NH} = 51$ Hz, 3H), and 5.85 (m, CH$_{a,b,c}$, 3H).

Exact mass. Calcd. for ^{11}B$_3$^{14}N$_3$^{12}C$_2$^{1}H$_8$: 107.0997. Found: 107.1000.

D. POLY(B-VINYLBORAZINE) OR 2-ETHENYLBORAZINE HOMOPOLYMER

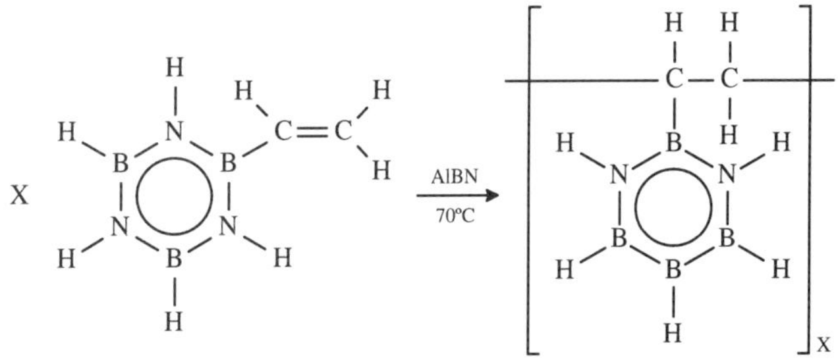

B-Vinylborazine   Poly(B-Vinylborazine)

The procedure[18] described below gives lightly cross-linked, pure poly(B-vinyl-borazine) (an inorganic analog of polystyrene) in good yields.

■ **Caution.** *Benzene is a known carcinogen and should be handled with care either in vacuo or in an efficient fume hood.*

All manipulations are carried out using inert-atmosphere techniques.[8] The vacuum line used in the following preparation is similar to that shown in Figure 2.

A 0.056-g sample of AIBN (2,2′-azobisisobutyronitrile)* is placed into a one-piece 100-mL flask (such as Chemglass AF-0522-02)[†] equipped with a Teflon vacuum stopcock and a magnetic stir bar, and the flask is evacuated at −196°C. A 1.69-g (15.9 mmol) sample of B-vinylborazine (prepared by the procedure described above) is vacuum distilled into the flask. Three freeze-pump-thaw cycles are performed in order to remove any traces of oxygen. The stopcock is closed and the reaction flask is removed to a shielded hood[‡] where it is heated in an oil bath at 70°C for about 3 h, at which point the material is sufficiently viscous that the stir bar stops. Then 5 mL of benzene is condensed into the flask and the solution is heated at 70°C for another 9 h. Slow addition of the benzene solution into 40 mL of pentane under inert atmosphere affords the precipitation of 0.73 g (43.2% yield) of poly(B-vinyl-borazine). The polymer is filtered under nitrogen and dried in vacuo for about 5 min.

Anal. Calcd. for poly(B-vinylborazine): C, 23.63%; H, 6.60%; B, 28.44%; N, 37.82%. Found: C, 22.6%; H, 4.7%; B, 30.5%; N, 26.3%.

Properties

Poly(B-vinylborazine) is an air-sensitive white solid. Lightly cross-linked, freshly prepared samples are soluble and thermally stable in most dry aprotic solvents, such as benzene and ethers. If stored as a solid at room temperature, the polymer may become insoluble in several hours, but in benzene or glyme solution it is stable and can be used as a stock solution. The polymer decomposes rapidly in contact with water and alcohols.

Characteristic Spectroscopic Data. IR spectrum (DRIFT, KBr, cm^{-1}): 3450 (s), 2950 (s), 2925 (s), 2900 (s), 2500 (s), 1460 (vs, br), 1100 (s), 1000 (s), 900 (vs), 700 (s). ^{1}H NMR (200.1 MHz, THF-d_8, ppm): broad peaks in the ranges of 4.2 to 6.6 ppm, arising from the N–H and B–H protons, and 2.1 to 0.5 ppm,

---

* Aldrich Chemical Company, Milwaukee, WI 53233.
† Chemglass, Inc., 3861 N. Mill Rd., Vineland, NJ 08360.
‡ The reviewers suggested keeping both the reaction and vacuum line under the same shielded hood.

attributed to the backbone C–H groups. The ^{11}B NMR spectrum of the polymer is too broad to allow definitive assignment.

Molecular weights, as determined by SEC/LALLS, of representative materials that are prepared as described above are $M_n = 13,900$, $M_w = 39,250$, and $M_w/M_n = 2.82$.

References

1. A. Stock and E. Pohland, *Chem. Ber.*, **59B**, 2215 (1926).
2. For a review of borazine chemistry see *Gmelin Handbush der Anorganishen Chemie, Borazine and Its Derivatives*, Springer-Verlag, New York, 1978, p. 48.
3. (a) R. T. Paine and C. K. Narula, *Chem. Rev.*, **90**, 73 (1990). (b) R. T. Paine and L. G. Sneddon, in *Inorganic and Organometallic Polymers II: Advanced Materials and Intermediates*, P. W. Neilson, H. R. Allcock, and K. J. Wynne, Eds, American Chemical Society, Washington, DC, 1994, p. 358. (c) Y. Kimura and Y. Kubo, in *Inorganic and Organometallic Polymers II: Advanced Materials and Intermediates*, P. W. Neilson, H. R. Allcock, and K. J. Wynne, Eds, American Chemical Society, Washington, DC, 1994, p. 375.
4. K. Niedenzu and J. W. Dawson, *Inorg. Synth.*, **10**, 142 (1967).
5. (a) C. A. Brown and A. W. Laubengayer, *J. Am. Chem. Soc.*, **77**, 3699 (1955). (b) R. Schaeffer, M. Steindler, L. Hohnstedt, H. S. Smith, Jr, L. B. Eddy, and H. I. Schlesinger, *J. Am. Chem. Soc.*, **76**, 3303 (1954). (b) L. F. Hohnstedt and D. T. Haworth, *J. Am. Chem. Soc.*, **82**, 89 (1960). (c) G. H. Dahl and R. Schaeffer, *J. Inorg. Nucl. Chem.*, **12**, 380 (1960). (d) L. I. Zakharkin and V. A. Ol'shevskaya, *Metallorg. Khim.*, **6**, 381 (1993).
6. (a) W. V. Hough, C. R. Guibert, and G. T. Hefferan, U.S. Patent No. 4,150,097, April 17, 1979. (b) G. W. Schaeffer, R. Schaeffer, and H. I. Schlesinger, *J. Am. Chem. Soc.*, **73**, 1612 (1951). (c) V. I. Mikheeva and V. Y. Markina, *J. Inorg. Chem., USSR*, **56** (1956). See also V. I. Mikheeva and V. Y. Markina, *Zh. Neorg. Khim.*, **1**, 2700 (1956). (d) V. V. Volkov, G. I. Bagryantsev, and K. G. Myakishev, *Russ. J. Inorg. Chem.*, **15**, 1510 (1970). (e) V. V. Volkov, A. A. Pukhov, and K. G. Myakishev, *Izv. Sib. Otd. Akad. Nauk. SSSR, Ser. Khim. Nauk.*, **3**, 116 (1983). (f) D. T. Haworth and L. F. Hohnstedt, *Chem. Ind. (London)*, 559 (1960).
7. T. Wideman and L. G. Sneddon, *Inorg. Chem.*, **32**, 1002 (1995).
8. D. F. Shriver and M. A. Drezdzon, *The Manipulation of Air Sensitive Compounds*, 2nd Ed., Wiley: New York, 1986.
9. See reference 8, pp. 105–106.
10. A. Stock, in *Hydrides of Boron and Silicon*, Cornell University Press, Ithaca, NY, 1933, p. 94.
11. K. Niedenzu and J. W. Dawson, *Boron–Nitrogen Compounds*, Academic Press, New York, 1965, chapter 3.
12. Material Safety Data Sheet, Callery Chemical Company, Division of Mine Safety Appliances Company, P.O. Box 429, Pittsburg, PA 15230.
13. An IUPAC approved name for the polymer would be poly(borazinediyl).
14. (a) P. J. Fazen, E. E. Remsen, J. S. Beck, P. J. Carroll, A. R. McGhie, and L. G. Sneddon, *Chem. Mater.*, **7**, 1942 (1995). (b) P. J. Fazen, J. S. Beck, A. T. Lynch, E. E. Remsen, and L. G. Sneddon, *Chem. Mater.*, **2**, 96 (1990). (c) P. J. Fazen, E. E. Remsen, and L. G. Sneddon, *Poly. Prepr.*, **32**, 544 (1991).
15. (a) A. J. Klancia, J. P. Faust, and C. S. King, *Inorg. Chem.*, **6**, 840 (1967). (b) A. J. Klancia and J. P. Faust, *Inorg. Chem.*, **7**, 1037 (1968). (c) L. A. Jackson and C. W. Allen, *J. Chem. Soc., Dalton Trans.*, 1989, 2423. (d) J. Pellon, W. G. Deichert, and W. M. Thomas, *J. Polym. Sci.*, **55**, 153 (1961). (e) L. A. Jackson and C. W. Allen, *Phosphorus, Sulfur, Silicon*, **41**, 341 (1989). (f) P. Fritz, K. Niedenzu, and J. W. Dawson, *Inorg. Chem.*, **3**, 626 (1964).

16. (a) A. T. Lynch and L. G. Sneddon, *J. Am. Chem. Soc.*, **109**, 5867 (1987). (b) A. T. Lynch and L. G. Sneddon, *J. Am. Chem. Soc.*, **111**, 6201 (1989).
17. James C. W. Chien, *Polyacetylene: Chemistry, Physics, and Material Science*, Academic Press, New York, 1984, pp. 29–31.
18. K. Su, E. E. Remsen, H. M. Thompson, and L. G. Sneddon, *Macromolecules*, **24**, 3760 (1991).

40. TRANSITION METAL COMPLEXES OF THE LACUNARY HETEROPOLYTUNGSTATE, $[P_2W_{17}O_{61}]^{10-}$

Submitted by WILLIAM J. RANDALL,* DAVID K. LYON,[†‡]
PETER J. DOMAILLE,[§] and RICHARD G. FINKE[†¶]
Checked by ALEXANDER M. KHENKIN[#] and CRAIG HILL[#]

The synthesis and characterization of isolated, isomerically pure α_2-$[P_2W_{17}O_{61}(M^{n+} \cdot L)]^{y-}$ ($M^{n+} = Mn^{3+}$, Fe^{3+}, Co^{2+}, Ni^{2+}, and Cu^{2+}) as aqueous-soluble potassium salts ($L = H_2O$) and as the more novel, organic-solvent-soluble tetrabutylammonium salts ($L = Br$) are described herein. These preparations yield mono-metal-substituted complexes of the Dawson α_2 isomer, α_2-$[P_2W_{17}O_{61}]^{10-}$, that are typically $>98\%$ free of the α_1 isomer.[1,2] The X-ray crystallographic structures of the α_2-$[P_2W_{17}O_{61}]^{10-}$ lacunary precursor and the α_2-$[P_2W_{17}O_{61}(Co^{II} \cdot OH_2)]^{8-}$ complex are known.[3]

One practical application of these compounds is as oxidation catalysts, where they serve as "inorganic porphyrin analogs."[2] The availability of organic-solvent-soluble Bu_4N^+ salts is important in this regard. The "inorganic porphyrin" analogy is supported not only by the finding that these α_2-$[P_2W_{17}O_{61}(M^{n+} \cdot L)]^{y-}$ complexes are highly effective oxidation catalysts in comparison to some of the best metalloporphyrin catalysts (examined as controls under identical conditions),[2] but also because these inorganic porphyrin analogs have the important and highly desirable property of being *highly oxidation resistant, if not nearly inert*.[1,2] Further information,

* Department of Chemistry, Lewis & Clark College, Portland, OR 97219.
† Department of Chemistry, University of Oregon, Eugene, OR 97403.
‡ Present address: Bend Research, Bend, OR 97701.
§ Central Research & Development, The DuPont Company, Experimental Station, Wilmington, DE 19898.
¶ Present address: Department of Chemistry, Colorado State University, Fort Collins, CO 80523.
# Department of Chemistry, Emory University, Atlanta, GA 30322.

characterization, catalytic activity in olefin epoxidations and alkane and aromatic oxygenations,[2] and important prior work are provided in references 1–9.

The isomerically pure compounds described below were obtained by adaptation of important prior work which detailed the syntheses of α_1- and α_2-$[P_2W_{17}O_{61}(M^{n+})]^{n-10}$ complexes[10] along with the following considerations: (1) the use of crystalline, isomerically pure[11] α_2-$[P_2W_{17}O_{61}]^{10-}$ as the lacunary precursor; (2) the purification by recrystallization at the *aqueous-soluble* K^+ *salt stage* for each α_2-$[P_2W_{17}O_{61}(M^{n+} \cdot L)]^{10-}$ (purification at the $[(n\text{-}C_4H_9)_4N]^+$ stage is quite difficult, if not impossible); (3) careful control of the solution pH during metathesis to the $[(n\text{-}C_4H_9)_4N]^+$ salt; (4) use of a CH_2Cl_2/CH_3CN as extracting solvents to avoid extended, slow filtrations of the $[(n\text{-}C_4H_9)_4N]^+$ salts; and (5) extensive use of the ^{31}P NMR to survey conditions for the best route to pure α_2-$P_2W_{17}M$ products. It is well known to workers in the area that ^{31}P NMR and similar techniques are more reliable than elemental analysis, for example, in quantitating the purity of polyoxoanions. The α_2-$[P_2W_{17}O_{61}$-$(M^{n+} \cdot OH_2)]^{(n-10)} \cdot xH_2O$ complexes were precipitated as their potassium salts and recrystallized from hot water (pH of 6–7) to remove minor impurities.

The IR spectrum (KBr pellet) is the simplest way to identify these complexes. It is essential that the spectrum for each complex exactly match to ensure purity. For this reason, the IR spectra for these complexes are included.[12]

A. DECAPOTASSIUM α_2-TRIPENTACONTAOXOBIS-[PHOSPHATO(3−)]HEPTADECATUNGSTATE(10−) PENTADECAHYDRATE, $K_{10}[\alpha_2\text{-}P_2W_{17}O_{61}] \cdot 15H_2O$

$$K_6\alpha_2\text{-}[P_2W_{18}O_{62}] \cdot 14H_2O + (n\text{-}14)H_2O + KHCO_3 \rightarrow$$

$$K_{10}\alpha_1,\alpha_2\text{-}[P_2W_{17}O_{61}] \cdot nH_2O$$

$$K_{10}\alpha_1,\alpha_2\text{-}[P_2W_{17}O_{61}] \cdot nH_2O \xrightarrow[\text{from boiling water}]{\text{recrystallized}}$$

$$K_{10}\alpha_2\text{-}[P_2W_{17}O_{61}] \cdot 15H_2O + (n\text{-}15)H_2O$$

This preparation requires pure α-$[P_2W_{18}O_{62}]^{6-}$ as a starting material.[1a,12] The synthesis given here for $K_{10}\alpha_2\text{-}[P_2W_{17}O_{61}]$ is on a larger scale than the previous one in this series[11] and is based on the fact that α_2-$[P_2W_{17}O_{61}]^{10-}$ is the first product formed by base degradation[11,13,14] of α-$[P_2W_{18}O_{62}]$.[6−] The monolacunary polyoxometalate is isolated as the potassium salt and is recrystallized once from boiling water to yield isomerically pure α_2-$[P_2W_{17}O_{61}]$.$^{10-}$

Procedure

In a 1000-mL Erlenmeyer flask, 135 g (0.0293 mmole) $K_6\alpha$-$[P_2W_{18}O_{62}]\cdot 15H_2O$ is dissolved in 300 mL of 40°C H_2O with stirring on a stirrer/hot plate. A water solution of $KHCO_3$ (500 mL of a 1 M solution, 0.5 mole) is added with vigorous stirring. A white precipitate begins to form after about 50 mL of the base is added. After the base addition is complete, the mixture is stirred for an additional 60 min. The suspension is cooled to $<5°C$ in an ice-water bath, the white precipitate is collected on a medium glass frit, and the product is dried by aspiration for ~ 1 h. The crude white solid is recrystallized by dissolving it in 200 mL of boiling H_2O (any insoluble material is removed from the hot solution by filtration through a celite pad, ~ 1.5 cm thick with a piece of Whatman No. 4 filter paper on top, or a funnel with a medium sintered-glass frit and moderate aspirator suction), followed by cooling in a refrigerator at 5°C overnight. The resulting white crystals are collected on a medium glass frit and washed with 150 mL of cold (2–3°C) $(3 \times 50$ mL$)$ H_2O, 150 mL $(3 \times 50$ mL$)$ anhydrous ethanol and 150 mL $(3 \times 50$ mL$)$ anhydrous diethyl ether. The solid is dried under vacuum at $22 \pm 1°C$ for 8 h, or in a 50°C drying oven overnight. Yield: 108 g (0.022 mole, 75%). This procedure yields 70–85% of the α_2 isomer, which is greater than 98% pure by ^{31}P NMR. This preparation is successfully scaled up to ~ 190 g of starting $K_6\alpha$-$[P_2W_{18}O_{62}]\cdot 15H_2O$ by proportionately increasing each of the concomitant reagent amounts and solution volumes.

Anal. Calcd. for $K_{10}\alpha_2$-$[P_2W_{17}O_{61}]\cdot 15H_2O$: H, 0.63; O, 25.20; P, 1.28; K, 8.11; W, 64.78; H_2O, 5.66. Found: H, 0.64; O, 25.0; P, 1.37; K, 8.13; W, 64.9; H_2O by TGA, 5.60.

Properties[1]

The ^{31}P NMR in 1:1 $H_2O:D_2O$ at 20°C of the Li^+ salt (in order to increase the solubility of the anion, the K^+ is replaced by Li^+ by metathesis of 1–2 g of sample with 1–2 g of $LiClO_4$ in 6–10 mL of 1:1 $H_2O:D_2O$ followed by cooling to 2–3°C and gravity filtration through Whatman or No. 5 filter paper to remove the solid $KClO_4$) indicates the product is greater than 98% pure α_2-$[P_2W_{17}O_{61}]^{10-}$ with peaks at -7.27, -14.11 ppm (relative to 85% H_3PO_4 at $22 \pm 1°C$ by the substitution method). Confirmation that the preparation yields the α_2 isomer (and not the α_1 isomer) was obtained by

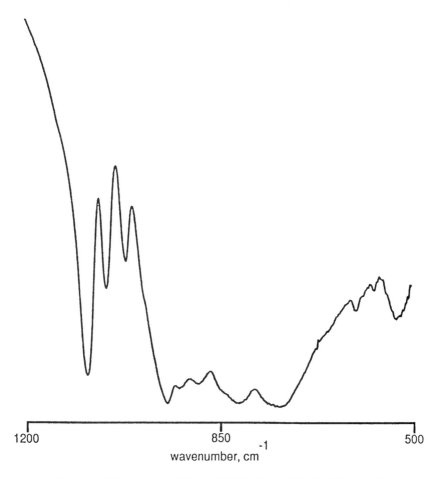

Figure 1. IR Spectrum of $K_{10}\alpha_2\text{-}[P_2W_{17}O_{61}] \cdot 15H_2O$ (KBr pellet).

^{183}W NMR which gave the expected 9-line spectrum[1a,11,15] of the Li^+ salt (the sample is prepared as described above) which gave peaks at -120, -143, -154, -183, -185, -220, -222, -226 and -245 ppm (relative to a 2M solution of $Na_2WO_4 \cdot 2H_2O$ at $22 \pm 1°C$ by the substitution method). The infrared spectrum from a KBr pellet showed peaks at: 740(s), 805(s), 880(m), 905(sh), 940(s), 985(m), 1022(m), 1084(s) cm^{-1}. (See ref. 12 if details on the preparation of the KBr pellet are desired and ref. 18 for possible complications when using KBr pellets.)

B. HEPTAPOTASSIUM α_2-[AQUAMANGANATE(III)]-TRIPENTACONTAOXOBIS[PHOSPHO(3−)]HEPTADECATUNGSTATE(7−) TRIDECAHYDRATE, $K_7\alpha_2$-[$P_2W_{17}O_{61}(Mn^{III}\cdot OH_2)]\cdot 13H_2O$

$$K_{10}\alpha_2\text{-}[P_2W_{17}O_{61}]\cdot 15H_2O + MnCl_2\cdot 4H_2O + (n\text{-}19)H_2O \rightarrow$$

$$K_8\alpha_2\text{-}[P_2W_{17}O_{61}(Mn^{II}\cdot OH_2)]\cdot nH_2O + 2KCl$$

$$2\,K_8\alpha_2\text{-}[P_2W_{17}O_{61}(Mn^{II}\cdot OH_2)]\cdot nH_2O + K_2S_2O_8 \rightarrow$$

$$2\,K_7\alpha_2\text{-}[P_2W_{17}O_{61}(Mn^{III}\cdot OH_2)]\cdot 13H_2O + (2n\text{-}26)H_2O + 2K_2SO_4$$

This compound may be prepared successfully by three procedures:[1,11] (1) direct incorporation of the Mn^{3+} into the lacunary polyoxotungstate; (2) direct incorporation of Mn^{2+} into the lacunary polyoxotungstate followed by isolation and characterization of $K_8\alpha_2$-[$P_2W_{17}O_{61}(Mn^{II}\cdot OH_2)$] by ^{31}P NMR, then oxidation of the Mn^{2+} to the Mn^{3+} with potassium persulfate; and (3) a one-pot, two-step procedure for the preparation of the Mn^{2+} complex followed by persulfate oxidation to the Mn^{3+} complex. Each synthetic strategy gives identical products and similar yields. We present the last procedure because of its directness, simplicity, and ease of completion.

Procedure

In a 500-mL flask, 52.0 g (10.8 mmole) $K_{10}\alpha_2$-[$P_2W_{17}O_{61}$]·15H_2O is dissolved in 150 mL of 90°C H_2O. A solution of 2.40 g (12.1 mmole) $MnCl_2\cdot 4H_2O$ dissolved in 40 mL of H_2O is added with vigorous stirring to give a dark brown solution. After mixing of the two reagent solutions is complete (< 1 min), 1.64 g $K_2S_2O_8$ (6.07 mmole) is dissolved in 25 mL of H_2O and added while the uncovered solution is stirred slowly at 90°C for 60 min. During this time the solution volume is reduced to ∼80 mL. The oxidation of the manganese is complete (as monitored by visible spectroscopy) after 60 min. Solid KCl (20 g, 0.268 mole) is added to the hot solution and the solution is cooled to room temperature. The solution is then placed in a refrigerator at 5°C overnight. The resultant purple crystals are collected on a medium glass frit, dried with aspiration ∼30 min, and recrystallized by dissolving them in ∼50 mL of boiling H_2O followed by cooling the solution in a refrigerator at 5°C overnight. The crystals are collected on a medium frit, washed with 50 mL of cold (2–3°C) distilled H_2O and dried with aspiration for ∼1 hr. They are then dried at 22 ± 1°C under dynamic vacuum for 6 hr. Yield: 36.5 g (7.6 mmole, 70%).

Anal. Calcd. for $K_7\alpha_2\text{-}[P_2W_{17}O_{61}(Mn^{III}\cdot OH_2)]\cdot 13H_2O$: K, 5.72; Mn, 1.15; P, 1.29; W, 65.38; H_2O, 5.32. Found: K, 5.71; Mn, 1.15; P, 1.10; W, 65.51; H_2O by TGA, 5.21.

Properties[1]

The homogeneity of $\alpha_2\text{-}[P_2W_{17}O_{61}(Mn^{III}\cdot Br^-)]^{7-}$ is demonstrated by the ^{31}P NMR spectrum of the Li^+ salt (see Properties under Section A for details of this NMR sample preparation) which gives a single P(2) phosphorus resonance at $\delta = -10.0$ ppm ($\Delta v_{1/2} = 90 \pm 1$ Hz). The ^{31}P NMR of the K^+ salt at 30°C gives a P(2) resonance at $\delta - 12.3$ ppm ($\Delta v_{1/2} = 66$ Hz), and a P(1) resonance of $+564$ ppm ($\Delta v_{1/2} = 12{,}300$) (both spectral values are relative to 85% H_3PO_4 by the substitution method). The ^{183}W NMR of the K^+ salt in 1:1 $H_2O:D_2O$ at 30°C gives six of the nine possible resonances. Those observed are [δ in ppm ($\Delta v_{1/2}$ in Hz)]: $-74(26)$, $-128(32)$, $-238(12)$, $-412(30)$, $-531(56)$, $-653(68)$. The ^{183}W NMR spectrum of the Li^+ salt using a wide-bore probe and a NT-360 spectrophotometer at 27°C in 1:1 $H_2O:D_2O$ (see Properties under Section A for details of this NMR sample preparation) also shows six resonances, but slightly upfield relative to the K^+ spectrum. The peaks are [δ in ppm ($\Delta v_{1/2}$ in Hz)]: $-77.4(10)$, $-132.2(6)$, $-243.8(9)$, $-423.1(14)$, $-524.2(27)$, $-666.3(23)$. (Both ^{183}W spectra are relative to a 2 M solution of $Na_2[WO_4]\cdot 2H_2O$ as 0.0 ppm at 22 ± 1°C by the substitution method). The infrared spectrum from a KBr pellet shows peaks or shoulders at: 780(s), 915(s), 948(s), 970(sh), 1020(m), 1085(s) cm^{-1}. The UV/visible spectrum has one peak at $\lambda_{max} = 484$ nm and $\varepsilon_{484} = 400$ cm^{-1} M^{-1}.

C. TETRABUTYLAMMONIUM HYDROGEN α_2-[BROMOMANGANATE(III)]TRIPENTACONTAOXOBIS-[PHOSPHATO(3−)] HEPTADECTUNGSTATE(8−) (7.3:0.7:1), $[(n\text{-}C_4H_9)_4N]_{7.3}H_{0.7}\alpha_2\text{-}[P_2W_{17}O_{61}(Mn^{III}\cdot Br)]$

$K_7\alpha_2\text{-}[P_2W_{17}O_{61}(Mn^{III}\cdot OH_2)]\cdot 13H_2O$

$$+ 7.3[(n\text{-}C_4H_9)_4N]Br + 0.7H_2SO_4 \rightarrow$$

$[(n\text{-}C_4H_9)_4N]_{7.3}H_{0.7}\alpha_2\text{-}[P_2W_{17}O_{61}(Mn^{III}\cdot Br)]$

$$+ 6.3KBr + 0.7KHSO_4 + 13H_2O$$

If the $\alpha_2\text{-}P_2W_{17}O_{61}(Mn^{III}\cdot OH_2)^{7-}$ is metathesized with tetrabutyl ammonium bromide at a pH < 4, the compound partially degrades (10–20%, as seen by ^{31}P NMR, showing the importance of ^{31}P NMR to follow these

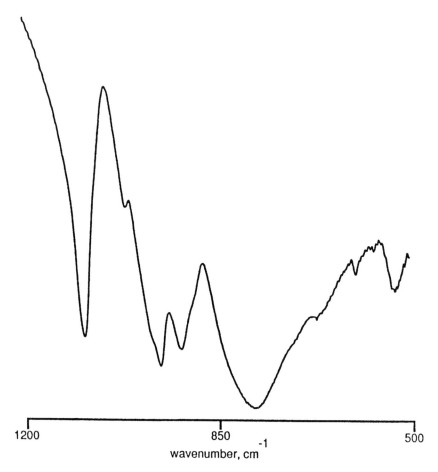

Figure 2. IR Spectrum of $K_7\alpha_2\text{-}[P_2W_{17}O_{61}(Mn^{III} \cdot OH_2)] \cdot 13H_2O$ (KBr pellet).

processes) to the starting materials $\alpha_2\text{-}[P_2W_{17}O_{61}]^{10-}$ and an unidentified manganese compound. At a pH > 8, the Mn^{3+} complex appears to hydrolyze to either an oxo- or hydroxomanganese polyoxotungstate, although the presence of a Mn=O or Mn–OH species has not been unambiguously established.[1,16]

Procedure

In a 1500-mL flask product from Section B, Procedure, $K_7\alpha_2\text{-}[P_2W_{17}O_{61}(Mn^{3+} \cdot OH_2)] \cdot 13H_2O$ (10.4 g, 2.32 mmole), is dissolved in

500 mL H_2O. Solid $[(n-C_4H_9)_4N]Br$ (4.9 g, 15.2 mmole, 7 equiv) is added with vigorous stirring of the solution in $\sim$250-mg portions concurrently with 0.18 M H_2SO_4 to keep the pH of the solution between 6 and 7 at all times (monitored by a pH meter with a standard glass/calomel combination electrode). After addition of the salt, the solution has a small amount of suspended precipitate and the pH is 6.1. The water suspension is extracted with a mixture of CH_2Cl_2 (200 mL) and CH_3CN (200 mL), which when added forms a slurry causing the dissolution of the small amount of precipitate. The mixture is stirred vigorously for 5 min and allowed to separate into a cloudy aqueous layer (pH 6.5) and a dark purple organic layer. The two layers are separated with a large separatory funnel and the organic layer is evaporated under vacuum to dryness with a rotovaporator at 60°C. The resulting solid is dissolved in a minimum amount of CH_3CN (less than 15 mL). Anhydrous diethyl ether (200 mL) is added precipitating a small amount of brown solid and a purple oil. The supernatant diethyl ether is decanted away from the oil and passed through a medium frit to remove the small amount of brown solid (less than 0.5 g). The purple oil is treated with 75 mL fresh anhydrous diethyl ether and triturated until a fine powder is obtained ($\sim$3 × 25 mL of the ether and 1–2 h is needed to complete this step). This brown solid is collected on the same frit as the original solid and washed with 75 mL (3 × 25 mL) anhydrous diethyl ether. The powder is air dried on the frit with aspiration until it is easily manipulated and then transferred to a weighing pan and dried under dynamic vacuum for 24 h at 50°C. Yield: 11.4 g (18.1 mmole, 81%).

Anal. Calcd. for $[(n-C_4H_9)_4N]_{7.3}H_{0.7}\alpha_2\text{-}[P_2W_{17}O_{61}(Mn^{III} \cdot Br)]$: C, 23.12; H, 4.38; N, 1.83; P, 1.02; W, 51.5; Mn, 0.90; Br, 1.32; O, 16.1. Found: C, 23.32, 23.12; 23.06; H, 4.37; N, 1.72; P, 0.84; W, 50.0, 51.4; Mn, 0.85; Br, 1.00; O, 15.5.

Properties[1]

The ^{31}P NMR in 1:1 $CH_3CN:CD_3CN$ at 20°C shows that the product is homogeneous by the single P(2) resonance of $\delta = -10.0$ ppm ($\Delta v_{1/2} = 90 \pm 1$ Hz) (relative to 85% H_3PO_4 by the substitution method). The P(1) resonance (the phosphorus closest to the paramagnetic center) for this salt in CH_3CN is not seen (i.e., under the NMR parameters employed[1]). The ^{183}W NMR at 20°C shows peaks at: -64, -116, -136, -202, -295 ppm (relative to a 2 M solution of $Na_2[WO_4] \cdot 2H_2O$ at 22 ± 1°C by the substitution method). The infrared spectrum has peaks at: 791(s), 888(s), 945(s), 956(sh), 1016(m), and 1089(s) cm^{-1}. The UV/visible spectrum in CH_3CN has a $\lambda_{max} = 478$ nm, $\varepsilon_{478} = 380$ cm^{-1} M^{-1}; and in CH_2Cl_2 gives a $\lambda_{max} = 453$ nm, $\varepsilon_{453} = 250$ $cm^{-1} M^{-1}$. The calculated weight loss for 3.65

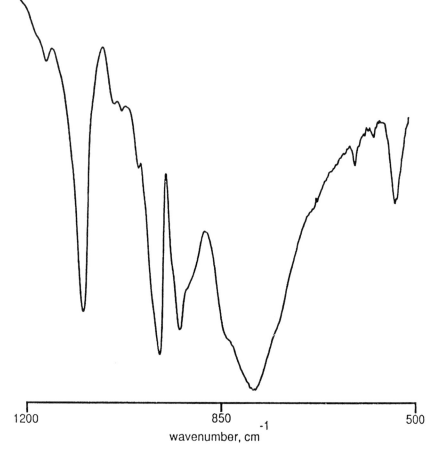

Figure 3. IR Spectrum of $[(n\text{-}C_4H_9)_4N]_{7.3}H_{0.7}\alpha_2\text{-}[P_2W_{17}O_{61}(Mn^{III}\cdot Br)]$ (KBr pellet).

$[(n\text{-}C_4H_9)_4N]_2O$ and for 1.00 P_2O_5 is 32.7%. The observed total weight loss by TGA (240–940°C) is 32.4%. The molecular weight calculated for $[(n\text{-}C_4H_9)_4N]_{7.3}H_{0.7}\alpha_2\text{-}[P_2W_{17}O_{61}(Mn\cdot Br)]$ is 6069 amu. The molecular weight determined by the sedimentation equilibrium method[19] for a CH_3CN solution which is 1×10^{-5} M in complex and 0.10 M in $[(n\text{-}C_4H_9)_4N][PF_6]$ as the bulk electrolyte was 6500 ± 600 amu.

D. HEPTAPOTASSIUM α_2-[AQUAFERRATE(III)]-TRIPENTACONTAOXOBIS[PHOSPHATO(3−)]HEPTADECATUNGSTATE(7−) OCTAHYDRATE, $K_7\alpha_2\text{-}[P_2W_{17}O_{61}(Fe^{III}\cdot OH_2)]\cdot 8H_2O$

$$K_{10}\alpha_2\text{-}[P_2W_{17}O_{61}]\cdot 15H_2O + Fe(NO_3)_3\cdot 9H_2O + (n\text{-}15)H_2O \rightarrow$$

$$K_7\alpha_2\text{-}[P_2W_{17}O_{61}(Fe^{III}\cdot OH_2)]\cdot 8H_2O + 2KNO_3 + 16H_2O$$

The monosubstituted iron polyoxotungstate is prepared by the addition of a solution of Fe^{3+} to a solution of $K_{10}\alpha_2\text{-}[P_2W_{17}O_{61}]\cdot 15H_2O$, which yields a dark yellow-orange solution.

Procedure

Pure $K_{10}\alpha_2\text{-}[P_2W_{17}O_{61}]\cdot 15H_2O$ (50 g, 10.35 mmole) is dissolved in 150 mL 90°C H_2O in a 400-mL beaker. A solution of 4.3 g (10.65 mmole) $Fe(NO_3)_3\cdot 9H_2O$ in 20 mL H_2O is added with vigorous stirring to give a dark yellow-orange solution.

■ **Caution.** *Bromine is corrosive and a health hazard. This procedure should be completed in a good fume hood and rubber gloves should be worn!*

To ensure a fully oxidized product three drops of bromine is added to the solution with a Pasteur pipet. The solvent volume is reduced to 100 mL by gentle boiling and then allowed to cool. It is important to cool the reaction solution to $22 \pm 1°C$ to permit a small amount of an unidentified dark-orange impurity to precipitate, which in turn is removed by filtration through a Celite pad (~ 1.5 cm thick with a piece of Whatman No. 4 filter paper on top of the Celite) before the addition of the KCl. Solid KCl (25 g, 335 mmole) is added to the filtrate and the solution is warmed until it becomes homogeneous ($\sim 60°C$). The solution is then cooled in a refrigerator at 5°C overnight. The resulting fine yellow crystals are collected on a medium glass frit. The product is recrystallized from a minimum of boiling H_2O (about 35 mL) and cooled again in a refrigerator at 5°C overnight. The yellow, crystalline product is collected on a medium frit, washed with 25 mL of cold (2–3°C) H_2O and dried under vacuum at room temperature for 6 h or in a 50°C oven overnight. Yield: 75.7 g (16.0 mmole, 78%).

Anal. Calcd. for $K_7\alpha_2\text{-}[P_2W_{17}O_{61}(Fe^{III}\cdot OH_2)]\cdot 8H_2O$: K 5.88, Fe 1.20; for $9H_2O$, 3.48%; Found: K, 6.10; Fe, 1.22; H_2O by TGA, 3.43%.

Properties[1]

The ^{31}P NMR of the Li^+ salt (see Properties under Section A for the description of this NMR sample preparation) at 20°C gives a P(2) resonance $\delta = -11.9$ ppm ($\Delta v_{1/2} = 65$ Hz). The ^{183}W NMR at 27°C (obtained on an NT-360 spectrometer using a wide-bore probe) in 1:1 $H_2O:D_2O$ on the Li^+ salt (see above for sample preparation) has resonances [δ in ppm, ($\Delta v_{1/2}$ in Hz)]: $-144.6(2)$, $-146.3(3)$, $-151.8(5)$, $-190.7(7)$, $-445.9(41)$ (see Properties under Section A for standard reference information). The infrared spectrum from a KBr pellet shows peaks at: 800(s), 914(s), 945(s), 965(sh), 1012(m), and 1083(s) cm^{-1}. The UV/visible spectrum shows no maxima but a shoulder of moderate intensity in the UV at ~ 250 nm. The molecular weight, as measured by the sedimentation equilibrium method[19] for a solution of 8.0×10^{-5} M complex in 0.20 M LiCl and pH $= 4$ in water at 22 ± 1°C indicates that the anion is a monomer. The calculated MW is 4673 amu and the experimentally found MW is 5453 ± 900 amu.

E. [AQUA-BROMO-FERRATE(III)] MIXED LIGAND α_2-TRIPENTACONTAOXOBIS[PHOSPHATO(3−)] HEPTADECATUNGSTATE, $[(n\text{-}C_4H_9)_4N]_{6.75}H_{0.5}\alpha_2\text{-} [P_2W_{17}O_{61}(Fe^{III} \cdot L)]$ (L=0.75H$_2$O, 0.25Br)

$$K_7\alpha_2\text{-}[P_2W_{17}O_{61}(Fe \cdot OH_2)] \cdot 8H_2O + 6.75[(n\text{-}C_4H_9)_4N]Br \rightarrow$$

$$[(n\text{-}C_4H_9)_4N]_{6.75}H_{0.5}\alpha_2\text{-}[P_2W_{17}O_{61}(Fe^{III} \cdot L)]$$

$$(L = 0.75H_2O, 0.25Br) + 6.5KBr + 0.5KOH + 7.75H_2O$$

This iron complex, which is soluble in organic solvents, can be prepared in the pH range ~ 6.5 to pH 5. No degradation to $\alpha_2\text{-}[P_2W_{17}O_{61}]^{10-}$ and free heterometal was observed as in the case of the manganese complex.

Procedure

In a 1500-mL Erlenmeyer flask, $K_7\alpha_2\text{-}[P_2W_{17}O_{61}(Fe \cdot OH_2)] \cdot 8H_2O$ (9.64 g, 2.07 mmole) is dissolved in 500 mL of H_2O. The pH of the solution is adjusted to 6.25 by the addition of 1 mL of 0.4 M aqueous $[(n\text{-}C_4H_9)_4N]OH$. The solid salt $[(n\text{-}C_4H_9)_4N]Br$ (4.68 g, 14.5 mmole, 7 equiv) is added to the solution, which causes the pH to rise to 6.5 and the solution to become cloudy. The extracting solvent, CH_2Cl_2 (200 mL), is added to the solution to form a suspension which is shaken vigorously for 5 min. Upon standing, the mixture separates into a clear-yellow organic layer and a cloudy white aqueous layer (pH 8.2 by pH meter, pH 6–7 by pH paper). The two layers are

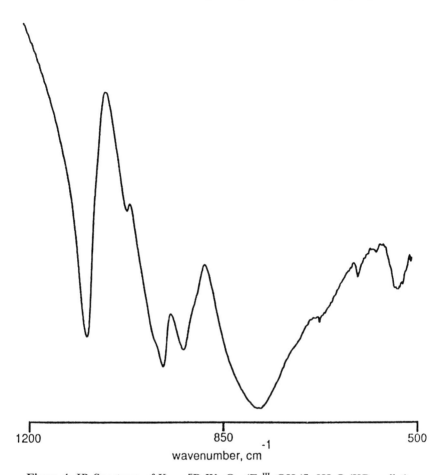

Figure 4. IR Spectrum of $K_7\alpha_2\text{-}[P_2W_{17}O_{61}(Fe^{III}\cdot OH_2)]\cdot 8H_2O$ (KBr pellet).

separated using a large separatory funnel and the organic layer is evaporated to dryness at 60°C with a rotovaporator. The resulting yellow oil is dissolved in 30 mL CH_3CN. Anhydrous diethyl ether (200 mL) is added to the solution causing immediate precipitation of a yellow oil. The liquid is decanted away from the oil which is then repeatedly triturated in diethyl ether (~ 75 mL total) until a fine yellow-green solid is obtained. The solid is collected on a medium glass frit and air dried until it is easily manipulated. The solid is then vacuum dried for 24 h at 60°C. Yield: 6.72 g (1.13 mmole, 54.6%).

Anal. Calcd. for $[(n\text{-}C_4H_9)_4N]_{6.75}H_{0.5}\alpha_2\text{-}[P_2W_{17}O_{61}]Fe^{III}\cdot L)$ (L = 0.75 H_2O and 0.25 Br): C, 22.07; H, 4.18; N, 1.61; P, 1.05; W, 53.2; Br, 0.34; Fe, 0.95;

O, 16.6. Found: C, 22.02 and 22.00; H, 4.08 and 4.29; N, 1.64 and 1.80; P, 0.95; W, 50.08 and 53.1; Br, 0.30; Fe, 0.86; O, 16.8.

Properties[1]

The ^{31}P NMR in 1:1 $CH_3CN:CD_3CN$ at 20°C, gives a single P(2) resonance $\delta = -12.7$ ppm ($\Delta\nu_{1/2} = 100 \pm 5$ Hz) which demonstrates the homogeneity of the product. The visible spectrum shows no maxima. The infrared spectrum exhibits peaks at: 791(s), 888(s), 945(s), 956(sh), 1016(m), and 1089(s) cm^{-1}. The calculated weight loss for 3.375 $[(n\text{-}C_4H_9)_4N]_2O + 1.00$ P_2O_5 is 31.1%. The total observed weight loss by TGA (240–940°C) is 30.6%. These data are consistent with 6.75 $[(n\text{-}C_4H_9)_4N]^+$/anion. The MW determined by the sedimentation equilibrium method[19] is 6200 $\pm$ 600 amu compared to the calculated MW of 5876 amu. Thus, solution molecular weight measurements demonstrate that the anion is monomeric for 1×10^{-5} M complex and 0.1 M $[(n\text{-}C_4H_9)_4N][PF_6]$ electrolyte in CH_3CN.

F. OCTAPOTASSIUM α_2-[AQUACOBALTATE(II)]-TRIPENTACONTAOXOBIS[PHOSPHATO(3−)] HEPTADECATUNGSTATE(8−) HEXADECAHYDRATE, $K_8\alpha_2\text{-}[P_2W_{17}O_{61}(Co^{II} \cdot OH_2)] \cdot 16H_2O$

$$K_{10}\alpha_2\text{-}[P_2W_{17}O_{61}] \cdot 15H_2O + Co(NO_3)_2 \cdot 6H_2O \rightarrow$$
$$K_8\alpha_2\text{-}[P_2W_{17}O_{61}(Co^{II} \cdot OH_2)] \cdot 16H_2O + 2KNO_3 + 5H_2O$$

This complex is prepared by the addition of a water solution of Co^{2+} to a solution of the monolacunary polyoxotungstate.

Procedure

In a 500-mL flask, $K_{10}\alpha_2\text{-}[P_2W_{17}O_{61}] \cdot 15H_2O$ (51.0 g, 10.6 mmole) is dissolved in 200 mL of 90°C H_2O. A solution of $Co(NO_3)_2 \cdot 6H_2O$ (3.36 g, 11.5 mmole) dissolved in 40 mL of H_2O is added with vigorous stirring, giving a dark-red solution. After 15 min, solid KCl (30 g, 0.40 mmole) is added and the solution is cooled to 22 $\pm$ 1°C. The resulting light red crystals are collected on a medium glass frit and recrystallized twice from a minimum amount ($\sim$50 mL) of boiling H_2O. The mother liquor solution in each crystallization is cooled in a refrigerator at 5°C overnight before filtration. The product is collected on a medium glass frit, washed with 50 mL of H_2O, and dried under dynamic vacuum at 22 $\pm$ 1°C for 6 h. Yield: 36.2 g (7.5 mmole, 71%). Yields for this reaction vary from 55 to 75%.

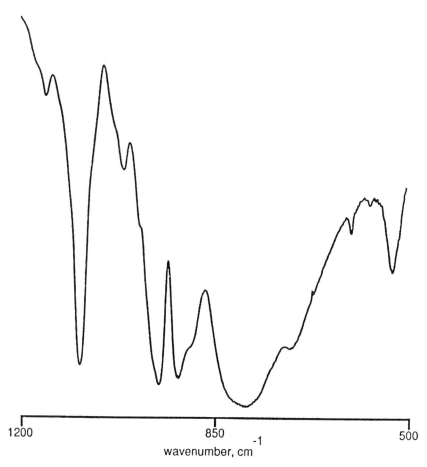

Figure 5. IR Spectrum of $[(n\text{-}C_4H_9)_4N]_{6.75}H_{0.5}\alpha_2\text{-}[P_2W_{17}O_{61}(Fe^{III}\cdot L)]$ $(L = 0.75H_2O,$ $0.25Br)$ (KBr pellet).

Anal. Calcd. for $K_8\alpha_2\text{-}[P_2W_{17}O_{61}(Co^{II}\cdot OH_2)]\cdot 16H_2O$: K, 6.46; W, 64.56; Co, 1.22; H_2O, 6.32. Found: K, 6.35; W; 64.43; Co, 1.22; H_2O by TGA, 6.75.

Properties[1]

The ^{31}P NMR of the Li^+ salt (see Properties under Section A for details of this NMR sample preparation) gives a single P(2) resonance at $\delta -22.6$ ppm for the product recrystallized twice from water. If the crude material is recrystallized only once an unassigned resonance ($< 5\%$) is observed at $\delta -34.0$ ppm. (Note the importance of the recrystallization step(s) prior to

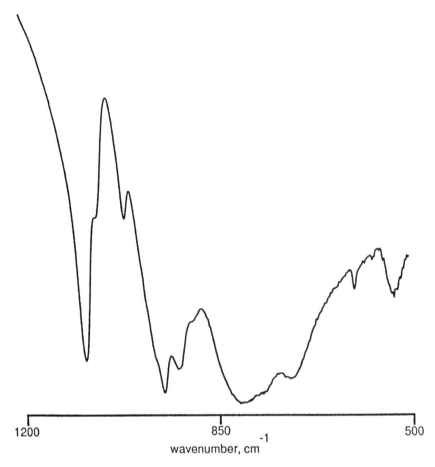

Figure 6. IR Spectrum of $K_8\alpha_2$-$[P_2W_{17}O_{61}(Co^{II}\cdot OH_2)]\cdot 16H_2O$ (KBr pellet).

making the $[(n\text{-}C_4H_9)_4N]^+$ salt.) The infrared spectrum from a KBr pellet has peaks at: 796(s), 912(s), 942(s), 965(sh), 1010(m), and 1082(s) cm^{-1}. The UV/visible spectrum shows a $\lambda_{max} = 544$ nm, $\varepsilon_{544} = 108$ cm^{-1} M^{-1}.

G. NONAKIS(TETRABUTYLAMMONIUM) α_2-[BROMOCOBALTATE(II)] TRIPENTACONTAOXOBIS-[PHOSPHATO(3−)] HEPTADECATUNGSTATE(9−), $[(n\text{-}C_4H_9)_4N]_9\,\alpha_2$-$[P_2W_{17}O_{61}(Co^{II}\cdot Br)]$

$$K_8\alpha_2\text{-}[P_2W_{17}O_{61}(Co^{II}\cdot OH_2)]\cdot 16H_2O + 9[(n\text{-}C_4H_9)_4N]Br \rightarrow$$

$$[(n\text{-}C_4H_9)_4N]_9\,\alpha_2\text{-}[P_2W_{17}O_{61}(Co^{II}\cdot Br)] + 8KBr$$

This product salt was obtained by exchange of the cation and isolation of the product after extraction into dichloromethane and acetonitrile.

Procedure

In a 1500-mL flask $K_8\alpha_2\text{-}[P_2W_{17}O_{61}(Co^{II} \cdot OH_2)] \cdot 16H_2O$ (15.2 g, 3.13 mmole) is dissolved in 150 mL of H_2O (pH 6.6). The pH is adjusted to 5.6 (monitored by a pH meter with a standard combination glass/calomel electrode) by the addition of one drop of 0.18 M H_2SO_4. The solid salt $[(n\text{-}C_4H_9)_4N]Br$ (7.91 g, 24.5 mmole, 8 equiv) is added to the solution causing the pH to rise to 5.9. The extracting solvents, CH_2Cl_2 (200 mL) and CH_3CN (100 mL), are added to the aqueous solution, the flask is stoppered, and the solution is vigorously shaken for 5 min. The mixture separates into two layers upon standing a few minutes. The aqueous layer is pink and cloudy (pH is 9.7 by pH meter) while the organic layer is clear and dark red. These two layers are separated using a large separatory funnel and the organic layer is evaporated to dryness at 60°C with a rotovaporator. The resulting light-brown solid is dissolved in a minimum amount ($\sim$20 mL) of CH_2Cl_2. This solution is then treated with 75 mL anhydrous diethyl ether which causes a deep brown oil to form. The oil is then triturated in diethyl ether until a light brown powder is obtained. This product is collected on a medium glass frit and washed with 25 mL anhydrous diethyl ether. The product is air dried with aspiration until it is easily manipulated and then dried at 50°C under dynamic vacuum for 24 h. Yield: 11.7 g (1.88 mmole, 60.1%). This cobalt complex may be prepared at a variety of pH values and concentrations. If the metathesis of $K_8\alpha_2\text{-}[P_2W_{17}O_{61}(Co^{II} \cdot OH_2)] \cdot 16H_2O$ is attempted at a pH < 5, decomposition yields $\alpha_2\text{-}[P_2W_{17}O_{61}]^{10-}$ (by ^{31}P NMR), accompanied by an undetermined cobalt product. At pH <4.5, the decomposition is observed to be as high as 50%. This compound appears relatively insensitive to base hydrolysis with no decomposition products being detected for pH < 9 and > 5.

Anal. Calcd. for $[(n\text{-}C_4H_9)_4N]_9\alpha_2\text{-}[P_2W_{17}O_{61}(Co^{II} \cdot Br)]$: C, 26.67; H, 5.03; N, 1.94; P, 0.96; W, 48.2; Co, 0.87; Br, 1.23; O, 15.0. Found: C, 26.88; H, 5.23; N, 2.01; P, 0.74; W, 45.1, 44.5, 46.2, 50.0; Co, 0.91; Br, 1.11; O, 13.9, 15.0.

Properties[1]

The isomeric purity of the product is demonstrated by a single resonance in the ^{31}P NMR spectrum at $\delta = -26.4$ (25 ± 1 Hz). The infrared spectrum from a KBr pellet shows peaks at: 816(s), 914(s), 947(s), 956(sh), 1017(m), and 1087(s) cm^{-1}. The UV/visible spectrum in CH_3CN has a $\lambda_{max} = 484.0$ nm,

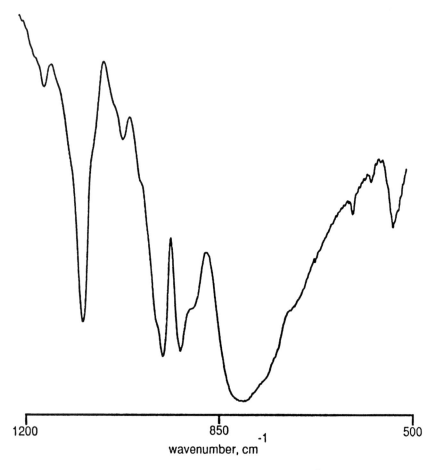

Figure 7. IR Spectrum of $[(n\text{-}C_4H_9)_4N]_9\alpha_2\text{-}[P_2W_{17}O_{61}(Co^{II}\cdot Br)]$ (KBr pellet).

$\varepsilon_{484} = 325\ \text{cm}^{-1}\text{M}^{-1}$; and in CH_2Cl_2 has a $\lambda_{max} = 574.0$ nm, $\varepsilon_{474} = 76\ \text{cm}^{-1}\text{M}^{-1}$. The calculated combined weight loss for 4.5 $[(n\text{-}C_4H_9)_4N]_2O + 1.0P_2O_5$ is 36.9%, and the observed weight loss by TGA (240–940°C) is 36.5%. The elemental analysis and TGA results are consistent with the presence of 9 $[(n\text{-}C_4H_9)_4N]^+$ ions. The molecular weight calculated for $[(n\text{-}C_4H_9)_4N]_9\alpha_2\text{-}[P_2W_{17}O_{61}(Co^{II}\cdot Br)]$: is 6479 amu and determined by the sedimentation equilibrium method[19] in CH_3CN is 6500 ± 500 amu; thus, the anion is a monomer.

H. OCTAPOTASSIUM α_2-[AQUANICKELATE(II)]-TRIPENTACONTAOXOBIS[PHOSPHATO(3−)] HEPTADECATUNGSTATE(8−) HEPTADECAHYDRATE, $K_8\alpha_2$-[$P_2W_{17}O_{61}(Ni^{II} \cdot OH_2)$]·$17H_2O$

$$K_{10}\alpha_2\text{-}[P_2W_{17}O_{61}]\cdot 15H_2O + Ni(NO_3)_2\cdot 6H_2O \rightarrow$$

$$K_8\alpha_2\text{-}[P_2W_{17}O_{61}(Ni^{II}\cdot OH_2)]\cdot 17H_2O + 2KNO_3 + 3H_2O$$

This complex is prepared by the addition of a water solution of Ni^{2+} to a solution of the lacunary polyoxometalate.

Procedure

In a 250-mL flask $K_{10}\alpha_2$-[$P_2W_{17}O_{61}$]·$15H_2O$ (50.1 g, 10.4 mmole) is dissolved in 100 mL of 90°C H_2O. A solution of $Ni(NO_3)_2\cdot 6H_2O$ (3.45 g, 11.9 mmole) dissolved in 40 mL of H_2O is added with vigorous stirring, giving a light green solution. After stirring for 15 min, the solution is cooled in a refrigerator at 5°C overnight. The light green crystals are collected on a medium frit and recrystallized by dissolving them in a minimum amount (~100 mL) of boiling H_2O and cooling again in a refrigerator at 5°C overnight. The product is collected on a medium glass frit, washed with 50 mL of cold (2–3°C) H_2O, dried with aspiration until an easily manipulated powder is formed which is dried at 22 ± 1°C under dynamic vacuum for 6 h. Yield: 37.9 g (7.8 mmole, 75%). Yields are ~75% but can be improved to as high as 95% by the addition of solid KCl to the mother liquor. However, the addition of excess KCl tends to give products 1–2% high in the potassium analysis.

Anal. Calcd. for $K_8\alpha_2$-[$P_2W_{17}O_{61}(Ni^{II}\cdot OH_2)$]·$17H_2O$: K, 6.44; Ni, 1.21; P, 1.28; H_2O, 6.32%. Found: K, 6.52; Ni, 1.23; P, 1.27; H_2O by TGA, 6.75.

Properties[1]

The product is determined to be isomerically pure with the ^{31}P NMR of the Li^+ salt (see Properties under Section A for details of this NMR sample preparation) by a single P(2) resonance at $\delta - 14.0$ ppm. The infrared spectrum from a KBr pellet has peaks at: 800(s), 910(s), 947(s), 960(sh), 1007(m), and 1080(s) cm^{-1}. The UV/visible spectrum has a maximum at $\lambda = $ ~680 nm, $\varepsilon_{680} = 10$ cm^{-1} M^{-1}. These data confirm the incorporation of the nickel into the lacunary $K_{10}\alpha_2$-[$P_2W_{17}O_{61}$]. The higher K analyses from the use of excess KCl in the precipitation step may be interpreted to mean that

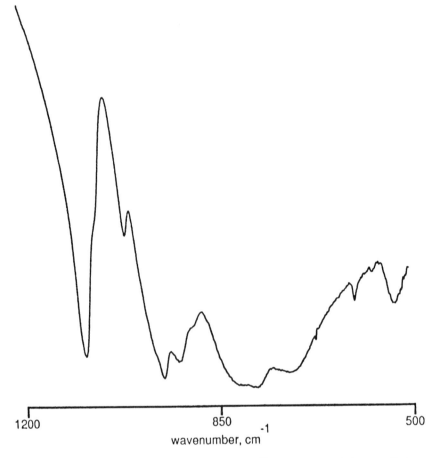

Figure 8. IR Spectrum of $K_8\alpha_2\text{-}[P_2W_{17}O_{61}(Ni^{II}\cdot OH_2)]\cdot 17H_2O$ (KBr pellet).

some KCl is occluded in the solid, or Cl is ligated to some of the Ni^{2+} sites which would require extra potassium.[1a]

I. NONAKIS(TETRABUTYLAMMONIUM) α_2-[BROMONICKELATE(II)]TRIPENTACONTAOXOBIS-[PHOSPATO(3−)] HEPTADECATUNGSTATE(9−), $[(n\text{-}C_4H_9)_4N]_9\text{-}[\alpha_2P_2W_{17}O_{61}(Ni^{II}\cdot Br)]$

$$K_8\alpha_2\text{-}[P_2W_{17}O_{61}(Ni^{II}\cdot OH_2)]\cdot 17H_2O + 9[(n\text{-}C_4H_9)_4N]Br \rightarrow$$

$$[(n\text{-}C_4H_9)_4N]_9\alpha_2\text{-}[P_2W_{17}O_{61}(Ni^{II}\cdot Br)] + 8KBr + 18H_2O$$

In contrast to Mn^{3+} and Co^{2+} complexes, the Ni^{2+} complex shows almost no tolerance to a range of pH conditions for this preparation. Thus, careful pH control in the preparation and recrystallization steps is essential.

Procedure

In a 1500-mL flask $K_8\alpha_2\text{-}[P_2W_{17}O_{61}(Ni^{II}\cdot OH_2)]\cdot 17H_2O$ (4.0 g, 0.823 mmole) is dissolved in 200 mL of H_2O (pH 6.55). Solid $[(n\text{-}C_4H_9)_4N]Br$ (2.11 g, 6.5 mmole, 8 equiv) is added in small portions concurrently with 0.18 M H_2SO_4 to maintain the pH between 6 and 7. The final pH should be around 6.0 (monitored by a pH meter and a standard glass/calomel combination electrode). The extracting solvent, CH_2Cl_2 (200 mL), is added and the solution is stirred or shaken vigorously for 5 min. Upon standing, the solution separates into a cloudy, white aqueous layer (pH 6.5) and a yellow organic layer. The layers are separated using a large separatory funnel and the organic layer is then evaporated to dryness at 60°C with a rotovaporator. The yellow solid is dissolved in a minimum amount (~ 20 mL) of CH_3CN. This solution is treated with 50 mL of anhydrous diethyl ether, which causes a small amount of yellow solid and a yellow oil to form. The supernatant diethyl ether layer is decanted and passed through a medium frit to collect the small amount of solid (less than 400 mg). The oil is triturated with fresh diethyl ether (3×50 mL) until a fine light yellow powder is obtained. The diethyl ether supernate is decanted into the filter funnel before addition to the next portion used in the trituration procedure. The powder is collected on the same frit as the initial solid and finally is washed with 75 ml (3×25 mL) anhydrous diethyl ether. The yellow solid is air dried with aspiration, until it is easily manipulated and then dried at 50°C under dynamic vacuum for 24 hr. Yield: 2.54 g (0.39 mmole, 47.4%).

Anal. Calcd. for $[(n\text{-}C_4H_9)_4N]_9\alpha_2\text{-}[P_2W_{17}O_{61}(Ni^{II}\cdot Br)]$: C, 26.67; H, 5.04; N, 1.94; P, 0.95; W, 48.2; Ni, 0.91; Br, 1.23; O, 15.0. Found: C, 26.75; H, 4.79; N, 2.09; P, 0.84; W, 47.1, 47.4; Ni, 0.78; Br, 0.88; O, 14.3, 15.0.

Properties[1]

The ^{31}P NMR spectrum in 1:1 $CH_3CN:CD_3CN$ gives a P(2) resonance of $\delta = -12.0$ ($\Delta v_{1/2} = 21 \pm 1$ Hz). The infrared spectrum from a KBr pellet shows peaks at: 812(s), 909(s), 945(s), 958(sh), 1028(m), and 1087(s) cm^{-1}. The UV/visible spectrum is sensitive to the coordinating ability of the solvent since in CH_3CH there is a $\lambda_{max} = 671$ nm, $\varepsilon_{671} = 12.4$ cm^{-1} M^{-1}; and in CH_2Cl_2 the $\lambda_{max} = 699.5$ nm, $\varepsilon_{699.5} = 10.9$ cm^{-1} M^{-1}. The weight loss calculated for 4.5 $[(n\text{-}C_4H_9)_4N]_2O + 1.0P_2O_5$ is 36.9% and the observed

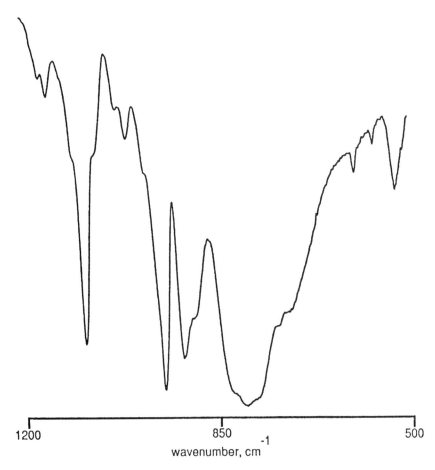

Figure 9. IR Spectrum of $[(n\text{-}C_4H_9)_4N]_9\alpha_2\text{-}[P_2W_{17}O_{61}(Ni^{II}\cdot Br)]$ (KBr pellet).

weight loss by TGA (240–940°C) is 36.1%. The MW calculated for $[(n\text{-}C_4H_9)_4N]_9\alpha_2\text{-}[P_2W_{17}O_{61}(Ni^{II}\cdot Br)]$ is 4736 amu and the MW determined by the sedimentation equilibrium method[19] is 4700 $\pm$ 400 amu, which confirms that the anion is monomeric in CH_3CN.

If the metathesis is done at 5.5 < pH > 6, decomposition is ~5%, but reaches ~50% when the pH is lowered to 5.0. At pH > 7 the compound appears to form several different hydrolysis products, depending upon the concentration and solution pH, which are not completely characterized, but show different ^{31}P NMR and UV/visible spectra.[1] The visible spectrum of the Ni^{II} complex depends upon the coordinating ability of the solvents, which suggests that the Ni^{II} ion in the polyoxometalate is six-coordinate, or pseudo-five-coordinate, with the sixth position occupied by solvent.

J. OCTAPOTASSIUM α_2-[AQUACUPRATE(II)]-TRIPENTACONTAOXOBIS[PHOSPHATO(3−)] HEPTADECATUNGSTATE(8−) HEXADECAHYDRATE, $K_8\alpha_2$-[$P_2W_{17}O_{61}(Cu^{II} \cdot OH_2)$] $\cdot 16H_2O$

$$K_{10}\alpha_2\text{-}[P_2W_{17}O_{61}] \cdot 15H_2O + CuSO_4 \cdot 5H_2O \rightarrow$$

$$K_8\alpha_2\text{-}[P_2W_{17}O_{61}(Cu^{II} \cdot OH_2)] \cdot 16H_2O + K_2SO_4 + 3H_2O$$

As in the previous monosubstituted complexes, this complex was prepared by the addition of a solution of Cu^{2+} to a solution of $K_{10}\alpha_2$-[$P_2W_{17}O_{61}$].

Procedure

In a 500-mL flask $K_{10}\alpha_2$-[$P_2W_{17}O_{61}$] $\cdot 15H_2O$ (75.0 g, 15.6 mmole) is dissolved in 150 mL of 90°C H_2O. A solution of 4.5 g (18.0 mmole) $CuSO_4 \cdot 4H_2O$ in 40 mL of H_2O is added with vigorous stirring, giving a green solution. After stirring for 15 min, the solution is cooled in a refrigerator at 5°C overnight. The light-green crystals are collected on a medium frit. This crude product is recrystallized by dissolving it in a minimum amount (~ 75 mL) of boiling H_2O and cooling the solution in a refrigerator at 5°C overnight. The product is collected on a medium frit, washed with 100 mL (2 × 50 mL) of cold (2–3°C) H_2O and then dried at 22 ± 1°C under dynamic vacuum for 6 hr. Yield: 54.7 g (11.3 mmole, 72%). This synthesis gives yields of $\sim 70\%$. Yields can be improved by addition of excess KCl, but extra KCl causes the potassium analysis to be high by 1–2%. Even so, multiple recrystallizations are required, otherwise samples tend to analyze slightly high for potassium.

Anal. Calcd. for $K_8\alpha_2$-[$P_2W_{17}O_{61}(Cu^{II} \cdot OH_2)$] $\cdot 16H_2O$: K, 6.45; W, 64.50; Cu, 1.31; P, 1.27; H_2O, 6.31. Found: K, 6.57; W, 64.93; Cu, 1.20; P, 1.38; H_2O by TGA, 6.26.

Properties[1]

The product is determined to be isomerically pure with the ^{31}P NMR of the Li^+ salt at 20°C (see Properties under Section A for details of this NMR sample preparation) by a single P(2) resonance at $\delta = -13.0$ ppm ($\Delta\nu_{1/2} = 11$), and a P(1) signal at $\delta = -35$ ppm ($\Delta\nu_{1/2} = 900$) (relative to

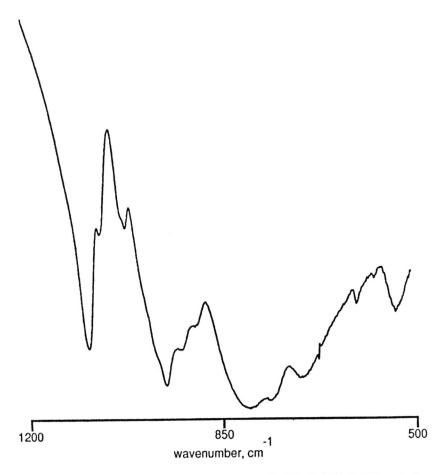

Figure 10. IR Spectrum of $K_8\alpha_2\text{-}[P_2W_{17}O_{61}(Cu^{II}\cdot OH_2)]\cdot 16H_2O$ (KBr pellet).

85% H_3PO_4 by the substitution method). The ^{183}W NMR spectrum at 27°C obtained on a NT-360 wide bore spectrophotometer in 1:1 H_2O/D_2O on the Li^+ salt gives peaks at [δ in ppm ($\Delta\nu_{1/2}$ in Hz)]: $-151.5(3)$, $-198.1(9)$, $-205.0(3)$, $-220.5(11)$, $-230.7(3)$. The infrared spectrum from a KBr pellet has peaks at: 788(s), 905(s), 942(s), 962(sh), 1005(m), and 1074(s) cm^{-1}. These data confirm the incorporation of the heterometal into the lacunary polyoxometalate. The UV/visible spectrum shows a broad absorption that extends into the near IR with a $\lambda_{max} = 885$ nm, $\varepsilon_{885} = 55$ $cm^{-1}M^{-1}$.

K. NONAKIS(TETRABUTYLAMMONIUM)
α_2-[BROMOCUPRATE(II)]TRIPENTACONTAOXOBIS-
[PHOSPHATO(3−)] HEPTADECATUNGSTATE(9−),
[(n-C$_4$H$_9$)$_4$N]$_9$ α_2-[P$_2$W$_{17}$O$_{61}$(Cu^{II}·Br)]

$$K_8\alpha_2\text{-}[P_2W_{17}O_{61}(Cu^{II}\cdot OH_2)] \cdot 16H_2O + 9[(n\text{-}C_4H_9)_4N]Br \rightarrow$$

$$[(n\text{-}C_4H_9)_4N]_9\alpha_2\text{-}[P_2W_{17}O_{61}(Cu^{II}\cdot Br)] + 8KBr + 17H_2O$$

This compound has been prepared at 4.0 < pH > 8.5. In contrast to Mn^{3+} and Co^{2+}, the potassium salt shows almost no tolerance to pH outside the range 5.5 < pH > 6.0 during metathesis.

Procedure

In a 1500-mL flask K$_8\alpha_2$-[P$_2$W$_{17}$O$_{61}$(Cu^{II}·OH$_2$)]·16H$_2$O (10.24 g, 2.11 mmole) is dissolved in 500 mL of 40°C H$_2$O (pH 6.2) with stirring on a hot-plate/stirrer. Solid [(n-C$_4$H$_9$)$_4$N]Br (5.43 g, 16.9 mmole, 8 equiv) is added with stirring, causing a slight precipitate to form. The extracting solvents, CH$_2$Cl$_2$ (200 mL) and CH$_3$CN (200 mL), are added and vigorously stirred for 5 min. The solution forms an emulsion that slowly (15–30 min) separates into a cloudy aqueous layer (pH = 9.4; monitored by a pH meter with a combination glass/calomel electrode) and a bright-green organic layer. The layers are separated using a large separatory funnel and the organic layer is evaporated to dryness at 60°C with a rotovaporator. The green solid is dissolved in a minimum of CH$_3$CN (<30 mL) and the solution is treated with 100 mL anhydrous diethyl ether. The diethyl ether treatment causes a small amount of light green precipitate and a dark green oil to form. The supernatant diethyl ether is passed through a medium-glass frit to collect the solid material. The oil is then triturated with 150 mL (3 × 50 mL) diethyl ether until a fine light-green powder is formed. The powder is then collected on the same frit as the initial solid and washed with 75 mL (3 × 25 mL) anhydrous diethyl ether. The solid is air dried with aspiration until it is easy to manipulate and then further dried at 50°C under dynamic vacuum for 24 h. Yield: 9.56 g (1.47 mmole, 69.8%).

Anal. Calcd. for [(n-C$_4$H$_9$)$_4$N]$_9\alpha_2$-[P$_2$W$_{17}$O$_{61}$(Cu^{II}·Br)]: C, 26.65; H, 5.03; N, 2.04; P, 0.95; W, 48.2; Cu, 0.98; Br, 1.23; O, 15.0. Found: C, 26.08; H, 4.88; N, 2.06; P, 0.86; W, 48.1; Cu, 0.79; Br, 1.04; O, 14.4.

Properties[1]

The homogeneity of the α_2-[P$_2$W$_{17}$O$_{61}$(Cu^{II}·Br)]$^{9-}$ compound is confirmed by a single P(2) ^{31}P NMR resonance at $\delta = -9.9$ ppm ($\Delta v_{1/2} = 53 \pm 1$). The

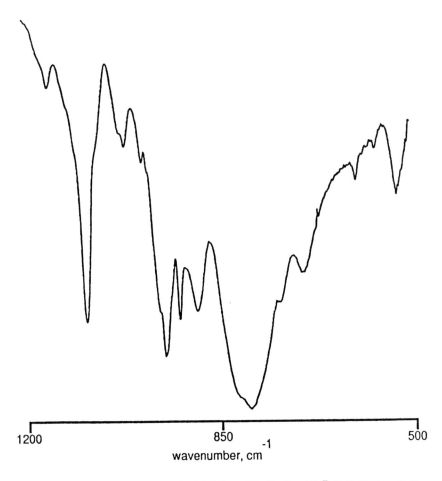

Figure 11. IR Spectrum of $[(n\text{-}C_4H_9)_4N]_9\alpha_2\text{-}[P_2W_{17}O_{61}(Cu^{II}\cdot Br)]$ (KBr pellet).

infrared spectrum from a KBr pellet has peaks at: 792(s), 820(s), 916(s), 952(s), 962(s), 1017(m), 1087(s) cm^{-1}. Unlike the case with the Ni^{2+} complex, the UV/visible spectrum for this Cu^{2+} complex is not sensitive to the co-ordinating ability of the solvent, as illustrated by the fact that in CH_3CN $\lambda_{max} = 706.0$ nm, $\varepsilon_{706} = 35.0$ $cm^{-1}\,M^{-1}$; and in CH_2Cl_2 $\lambda_{max} = 706.0$ nm, $\varepsilon_{706} = 38.0$ $cm^{-1}\,M^{-1}$. The weight loss calculated for $4.5\,[(n\text{-}C_4H_9)_4N]_2O + 1.0P_2O_5$ is 37.0% and the observed weight loss by TGA (240–940°C) is 37.2%. The MW calculated for $\{[(n\text{-}C_4H_9)_4N]_5$ $8[\alpha_2\text{-}P_2W_{17}O_{61}(Cu^{II}\cdot Br)]\}^{4-}$ is 5437 amu, and the MW determined by the sedimentation equilibrium method[19] is 5500 ± 500 amu; thus, the anion is monomeric in CH_3CN. If the metathesis is attempted at $5.5 < pH > 6$, decomposition is $\sim 5\%$, as judged by ^{31}P, but becomes as high as 50% if the pH is reduced to 5.0. At $pH > 7$, uncharacterized hydrolysis products, which show unique ^{31}P NMR and UV/visible spectra, are formed depending upon the concentration and the solution pH.[1] The visible spectrum of the α_2-$[P_2W_{17}O_{61}(Cu^{II} \cdot OH_2)]^{8-}$ ion shows almost no difference between coordinating and noncoordinating solvents. This observation is consistent with the behavior of an analogous compound, $[SiW_{11}O_{39}(Cu^{II} \cdot OH_2)]^{6-}$, extracted into toluene.[17]

References

1. (a) D. K. Lyon, W. K. Miller, T. Novet, P. J. Domaille, E. Evitt, D. C. Johnson, and R. G. Finke, *J. Am. Chem. Soc.*, **113**, 7209 (1991). (b) T. L. Jorris, M. Kozik, N. Casan-Pastor, P. J. Domaille, R. G. Finke, W. K. Miller, and L. C. W. Baker, *J. Am. Chem. Soc.*, **109**, 7402 (1987).
2. D. Mansuy, J. F. Bartoli, D. K. Lyon, and R. G. Finke, *J. Am. Chem. Soc.*, **113**, 7222 (1991).
3. T. J. R. Weakley, *Polyhedron*, **6**, 931 (1987).
4. L. C. W. Baker, *Plenary Lecture, XV Int. Conf. on Coord. Chem.*, Proceedings, Moscow, 1973.
5. (a) D. E. Katsoulis and M. T. Pope, *J. Am. Chem. Soc.*, **106**, 2737 (1984). (b) D. E. Katsoulis and M. T. Pope, *J. Chem. Soc., Dalton Trans.*, 1483 (1989). (c) D. E. Katsoulis and M. T. Pope, *J. Chem. Soc., Chem. Commun.*, 1186 (1986).
6. (a) C. L. Hill and R. B. Brown, Jr., *J. Am. Chem. Soc.*, **108**, 536 (1986); *Activation and Functionalization of Alkanes*, C. L. Hill, Ed., Wiley-Interscience, New York, 1989. (b) M. Faraj and C. L. Hill, *J. Chem. Soc., Chem. Commun.*, 1487 (1987).
7. J. E. Lyons, P. E. Ellis, Jr., H. K. Myers, Jr., G. Slud, and W. A. Langdale, US Patent, 4,803, 187, 1989.
8. (a) R. Neumann and C. Abu-Gnim, *J. Chem. Soc., Chem. Commun.*, 1324 (1989). (b) R. Neumann and C. Abu-Gnim, *J. Am. Chem. Soc.*, **112**, 6025 (1990). (c) C. Rong and M. T. Pope, *J. Am. Chem. Soc.*, **114**, 2932 (1992).
9. M. T. Pope, *Heteropoly and Isopoly Oxometalates*, Springer-Verlag, Berlin, 1983.
10. (a) T. J. R. Weakley and S. A. Malik, *J. Inorg. Nucl. Chem.*, **29**, 2935 (1967). (b) S. A. Malik and T. J. R. Weakley, *J. Chem. Soc. (A)*, 2647 (1968). (c) C. M. Tourne, S. A. Malik, and T. J. R. Weakley, *J. Inorg. Nucl. Chem.*, **32**, 3875 (1970). (d) R. Massart, R. Contant, J.-M. Fruchart, J.-P. Ciabrini, and M. Fourier, *Inorg. Chem.*, **16**, 2916 (1977).
11. R. Contant, *Inorg. Synth.*, **27**, 107 (1990).
12. W. J. Randall, T. J. R. Weakley, P. J. Domaille, and R. G. Finke, *Inorg. Synth.*, **31**, 167 (1997).
13. R. G. Finke, M. W. Droege, and P. J. Domaille, *Inorg. Chem.*, **26**, 3886 (1987).
14. H. Wu, *J. Biol. Chem.*, **43**, 189 (1920).
15. (a) R. Acerete, S. Harmalker, C. F. Hammer, M. T. Pope, and L. C. W. Baker, *J. Chem. Soc., Chem. Commun.*, 777 (1979). (b) M. Kozik, R. Acerete, C. F. Hammer, and L. C. W. Baker, *Inorg. Chem.*, **30**, 4429 (1991).
16. (a) L. C. W. Baker and T. J. R. Weakley, *J. Inorg. Nucl. Chem.*, **28**, 447 (1966). (b) B. W. Dale, J. M. Buckley, and M. T. Pope, *J. Chem. Soc. (A)*, 301 (1969). (c) H. Ichida, K. Nagai, Y. Sasaki, and M. T. Pope, *J. Am. Chem. Soc.*, **111**, 586 (1989).
17. D. E. Katsoulis, V. S. Tausch, and M. T. Pope, *Inorg. Chem.*, **26**, 215 (1987).

18. K. Nakamoto, *Infrared Spectra of Inorganic Compounds*, 2nd ed., Wiley-Interscience, New York, 1970, pp. 259–260.
19. (a) H. K. Schachman, *Methods in Enzymology*, Vol. 4, S. P. Colowick and N. O. Kaplan, Eds., Academic Press, New York, 1957, pp. 32–103. (b) D. G. Rhodes and T. M. Laue, *Methods in Enzymology, Guide to Protein Purification*, S. P. Colowick and N. O. Kaplan, Eds., Academic Press, Vol. 182, New York, 1990, pp. 564–565 and 571–574.

41. BIS(TRIPHENYLPHOSPHORANYLIDENE)AMMONIUM μ-CARBONYL-1κC:2κC-DECACARBONYL-1κ^3C,2κ^3,3κ^4C-μ-HYDRIDO-1κ:2κ-*TRIANGULO*-TRIRUTHENATE(1−)

$$6RuCl_3 \cdot 3H_2O + 33CO + 2KOMe + 2[PPN]Cl \rightarrow$$

$$2[PPN][HRu_3(CO)_{11}] + 2KCl + 11CO_2 + 2MeOH + 18HCl + 7H_2O$$

Submitted by ALAIN BÉGUIN,* JEAN-MARC SOULIÉ,*
and GEORG SÜSS-FINK*
Checked by JEROME B. KEISTER† and DAVID J. BIEREDERMAN†

There are many reports in literature on the synthesis of the trinuclear cluster anion [HRu$_3$(CO)$_{11}$]$^-$, all of which utilize reactions of Ru$_3$(CO)$_{12}$ with basic reagents.[1–8] The most efficient methods up to now are the reactions with NaBH$_4$,[2] [NEt$_4$][BH$_4$],[3] and KH.[4] One of the best yields described was obtained by the NaBH$_4$ reduction of Ru$_3$(CO)$_{12}$ [yield up to 85% with respect to Ru$_3$(CO)$_{12}$], followed by a crystallization in methanol as the tetraethyl-ammonium salt [NEt$_4$][HRu$_3$(CO)$_{11}$].[8] Nevertheless, all the known synthetic pathways that provide this cluster anion need at least two-step reactions, starting from RuCl$_3$ · n-H$_2$O, the first one of which is the carbonylation of ruthenium chloride, giving the cluster Ru$_3$(CO)$_{12}$ in a yield up to 80%.[9]

Oro and co-workers have shown that the "one-pot" reductive carbonylation of RuCl$_3$ · n-H$_2$O with zinc, in the presence of a pyrazole derivative, yields dinuclear, pyrazolato-bridged, carbonyl complexes.[10] To our knowledge, this was the first report on a facile, one-step synthesis of substituted carbonyl clusters, starting from ruthenium chloride. The "one-pot" synthetic route described here gives [PPN][HRu$_3$(CO)$_{11}$] in a yield of 83%, with respect to RuCl$_3$ · n-H$_2$O as the starting material.

* Institut de Chimie, Université de Neuchâtel, CH-2000 Neuchâtel, Switzerland.
† Department of Chemistry, State University of New York at Buffalo, Buffalo, NY 142-3000.

Procedure

■ **Caution.** *The synthesis requires autoclave equipment in a well-ventilated laboratory adequate for high-pressure reactions with carbon monoxide.*

The reaction is conducted in a 100-mL stainless-steel autoclave containing a glass insert. The autoclave is stirred magnetically and heated with an electrical mantle. The purification was carried out with standard Schlenk techniques, under purified nitrogen. The methanol was distilled over potassium methanolate and saturated with nitrogen prior to use.[11]

In a 100-mL autoclave 1.57 g (6 mmol) of $RuCl_3 \cdot n\text{-}H_2O$ $(n \approx 3)$ [Johnson-Matthey], 2.5 g of KOMe (10 mL of a 25% methanolic solution, 35.6 mmol) [Fluka],* and 1.72 g (3 mmol) of [PPN]Cl [Aldrich] are dissolved in 40 mL of methanol. The contents of the autoclave are put under CO pressure (60 bar) and the autoclave is heated at 120°C for 14 h. After cooling and venting, the deep red solution, containing already solid product, is filtered under a nitrogen atmosphere through a 2-cm layer of filter pulp using a 250-mL Schlenk frit. The residue is washed with methanol (10 mL), and the methanol extract is combined with the filtrate. The product is allowed to crystallize from the resulting solution at $-35°C$. After 14 h, the crystalline precipitate is isolated, washed three times with methanol (4 mL) at $-75°C$, and dried under vacuum (10^{-3} mbar) for 2 h. The yield is 1.7 g (74%) of $[PPN][HRu_3(CO)_{11}]$. An additional crop of 220 mg (9%) is obtained by concentrating the filtrate to about 10 mL and crystallizing at $-35°C$; the total yield is 83%.

Anal. Calcd. for $C_{47}H_{31}NO_{11}P_2Ru_3$: C, 49.05; H, 2.71; N, 1.22. Found: C, 48.93; H, 2.83; N, 1.14.

Properties

The title compound, $[PPN][HRu_3(CO)_{11}]$, is a deep red crystalline material. The crystals are slightly air sensitive and decompose over the range of 153–158°C. They dissolve in polar solvents such as THF, N,N-dimethylformamide, CH_2Cl_2, CH_3CN, or CH_3OH. The blood-red solutions that result are much more sensitive to oxygen than the solid. The IR spectrum of $[PPN][HRu_3(CO)_{11}]$ (in THF) displays characteristic absorptions at 2072 (vw), 2014 (vs), 1987 (s), 1952 (m), and 1733 (w, br) cm^{-1}.

---

* The checkers substituted KO^tBu for KOMe with the same results.

References

1. J. Knight and M. J. Mays, *J. Chem. Soc., Dalton Trans.*, **1972**, 1022.
2. B. F. G. Johnson, J. Lewis, P. R. Raithby, and G. Süss, *J. Chem. Soc., Dalton Trans.*, **1979**, 1356.
3. D. H. Gibson, F. U. Ahmed, and K. R. Phillips, *J. Organomet. Chem.*, **218**, 325 (1981).
4. J. C. Bricker, C. C. Nagel, and S. G. Shore, *J. Am. Chem. Soc.*, **104**, 1444 (1982).
5. J.B. Keister, *J. Chem. Soc., Chem. Commun.*, **1979**, 214.
6. C. Ungermann, V. Landis, S. A. Moya, H. Cohen, H. Walter, R. G. Pearson, R. G. Rinker, and P. C. Ford, *J. Am. Chem. Soc.*, **101**, 5922 (1979).
7. C. R. Eady, P. F. Jackson, B. F. G. Johnson, J. Lewis, M. C. Malatesta, M. McPartlin, and W. J. H. Nelson, *J. Chem. Soc., Dalton Trans.*, **1980**, 383.
8. G. Süss-Fink, *Inorg. Synth.*, **24**, 168 (1986).
9. M. I. Bruce, J. G. Matisons, R. C. Wallis, B. W. Skelton, and A. H. White, *J. Chem. Soc., Dalton Trans.*, **1983**, 2365.
10. J. A. Cabeza, C. Landazuri, L. A. Oro, A. Tiripicchio, and M. Tiripicchio-Camellini, *J. Organomet. Chem.*, **322**, C16 (1987).
11. D. D. Perrin and W. L. F. Armarego, *Purification of Laboratory Chemicals*, 3rd ed., Pergamon Press, Oxford, 1988.

42. DIPOTASSIUM UNDECACARBONYL TRIMETALLATE(2−) CLUSTERS, $K_2[M_3(CO)_{11}]$ (M = Ru, Os)

Submitted by TIM J. COFFY,* DEBORAH A. McCARTHY,*
EDWIN P. BOYD,* and SHELDON G. SHORE*
Checked by D. F. SHRIVER[†] and R. EVELAND[†]

Metal carbonyl cluster compounds which contain three ruthenium or three osmium atoms in the cluster core are common.[1] Potentially useful reagents for syntheses of these compounds are the triruthenium and triosmium dianions $[M_3(CO)_{11}]^{2-}$ (M = Ru, Os).[2] Therefore, it is desirable to develop good synthetic routes to obtain $[M_3(CO)_{11}]^{2-}$ (M = Ru, Os) of high purity in high yields. A method that is particularly useful for generating $[M_3(CO)_{11}]^{2-}$ (M = Ru, Os) is the designed stoichiometric reduction of $M_3(CO)_{12}$ (M = Ru, Os) using an electron carrier such as potassium-benzophenone.[3]

$$M_3(CO)_{12} + 2K[(C_6H_5)_2CO \rightarrow K_2[M_3(CO)_{11}] + 2(C_6H_5)2CO + CO$$

* Department of Chemistry, The Ohio State University, Columbus, OH 43210.
[†] Department of Chemistry, Northwestern University, Evanston, 60208-3113.

This method has the advantage of generating the desired anions quantitatively and avoids the formidable task of separating mixtures of polynuclear anions into their analytically pure components. The trinuclear anions may be isolated as alkali metal salts and used in situ or may be quantitatively metathesized to form alkyl ammonium, arsonium, or phosphonium salts. Therefore, we have found these alkali metal dianions to be a very useful, general starting material for a variety of reactions.

Syntheses of $K_2[Ru_3(CO)_{11}]$ and $K_2[Os_3(CO)_{11}]$ differ only in certain details.[3] The pure anion $[Ru_3(CO)_{11}]^{2-}$ is prepared by adding $Ru_3(CO)_{12}$ in small increments to a solution of alkali metal–benzophenone in THF until the appropriate molar ratio for the formation of $[Ru_3(CO)_{11}]^{2-}$ is achieved. Keeping the alkali metal in excess until the last increment of $Ru_3(CO)_{12}$ is added minimizes the formation of higher nuclearity clusters[3a] which result from the reaction of $Ru_3(CO)_{12}$ with the $[Ru_3(CO)_{11}]^{2-}$ generated. On the other hand $[Os_3(CO)_{11}]^{2-}$ can be prepared using a "one-pot" procedure[3b] because the $[Os_3(CO)_{11}]^{2-}$ formed has little tendency to form higher-nuclearity clusters with unreacted $Os_3(CO)_{12}$.

General Procedure

Equipment. All manipulations were performed under inert atmosphere conditions. Standard vacuum line and inert atmosphere techniques were employed.[5] Reactions were conducted using glassware with solvseal joints. Glassware with ground glass joints could not be used as these joints were prone to leak at low temperature or when solvent dissolved the vaccum grease.

Materials. Tetrahydrofuran (THF) is dried, distilled, and stored in the presence of sodium benzophenone ketyl. A lecture bottle of dimethyl ether was attached to the vacuum line via black vacuum rubber tubing.

■ **Caution.** *Since dimethyl ether is gas at room temperature, it is advisable to wear goggles and a face shield along with gloves and a lab coat when working with it as a solvent.*

In a glove box several grams of benzophenone and freshly cut sodium metal were added to a glass tube (40 cm long, 2 cm ID) topped with a Kontes valve and vacuum line adapter. This glass tube was attached to the vaccum line and evacuated. The tube was immersed in a − 78°C isopropanol/CO_2 slush bath. The lecture bottle valve on the $(CH_3)_2O$ tank was opened *slowly* and $(CH_3)_2O$ was condensed into the glass tube at − 78°C. The pressure of

$(CH_3)_2O$ was monitored via a manometer and never allowed to exceed 400 mm of pressure. Following condensation of 20 mL of $(CH_3)_2O$, the lecture bottle valve was closed and the tube, still immersed in the $-78°C$ bath, the contents were periodically swirled, and the tube quickly reinserted into the slush bath. After 30 min the solution appeared dark blue in color, indicating that the solvent was dry. The solvent may be stored as a liquid infinitely at $-78°C$ behind a safety shield.

It is extremely important that the solvents be very dry to achieve high yields of products. We have found that the sodium benzophenone ketyl is best for this purpose. In addition, the benzophenone (Fischer) is ground to a fine powder, dried in vacuo for at least 48 h, and stored in a dry box for use. $Ru_3(CO)_{12}$ is prepared by a published procedure.[4] $Os_3(CO)_{12}*$ is used as received. Metallic potassium (Mallinckrodt) in mineral oil is washed with hexane and stored in a controlled atmosphere box. All manipulations are performed either in an atmosphere of prepurified nitrogen or under a vacuum using standard techniques.[4]

A. [DIPOTASSIUM UNDECACARBONYL TRIRUTHENATE (2−)], $K_2[Ru_3(CO)_{11}]$

$$Ru_3(CO)_{12} + 2K[(C_6H_5)_2CO] \rightarrow K_2[Ru_3(CO)_{11}] + CO + 2(C_6H_5)_2CO$$

Procedure

In a controlled atmosphere box, $Ru_3(CO)_{12}$ (0.1663 g, 0.26 mmole) is added to a tip tube. Metallic potassium (0.024 g, 0.61 mmole) and benzophenone (0.105 g, 0.58 mmole) are added to a two-necked, 50-mL flask equipped with a glass-coated magnetic stir bar. The tip tube is connected to the side neck of the flask and the flask is then connected to an extractor and sealed with a Kontes 4-mm stopcock (Fig. 1), Stopcock A. The flask is attached to the vacuum line, evacuated by opening Stopcocks C and A and dry THF (3–4 mL) is condensed into it at $-78°C$. The slush bath is removed and the potassium–benzophenone mixture is allowed to warm to room temperature and stirred until the solid potassium is completely dissolved (2–5 h). The $Ru_3(CO)_{12}$ is then added portionwise (10–15 mg every 10–15 min) to the reaction mixture. This is best accomplished by cooling the solution in the flask to $-78°C$ to recondense any solvent which may wet the solid $Ru_3(CO)_{12}$. The $Ru_3(CO)_{12}$ is added carefully by tilting the tip-tube and then gently tapping the portion into the reaction solution, which is then warmed

---

* Strem Chemical Co.

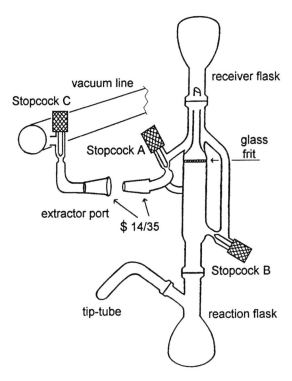

Figure 1. Diagram of a vacuum line extractor.

to room temperature and stirred. If $Ru_3(CO)_{12}$ becomes caked to the side of the tip-tube, solvent may be condensed into the tube and the dissolved $Ru_3(CO)_{12}$ then added to the reaction mixture. In addition, as the reaction proceeds, it is advisable to freeze the solution to $-196°C$ and pump away evolved CO two or three times. If the $Ru_3(CO)_{12}$ is not added in small quantities, the major product is $K_2[Ru_4(CO)_{13}]$.[2b] The reaction mixture is stirred at room temperature until solid $Ru_3(CO)_{12}$ is no longer visible at the bottom of the flask and the solution is the characteristic blue color of the potassium–benzophenone ketyl. Additional $Ru_3(CO)_{12}$ should not be added if the blue color of the potassium–benzophenone does not return after $Ru_3(CO)_{12}$ addition. A slight green discoloration is expected as the red-brown $[Ru_3(CO)_{11}]^{2-}$ is produced in the reaction. Carbon monoxide is formed in quantitative yield. After the reaction is complete, the THF is removed at room temperature by pumping it away through a removable U-trap maintained at $-196°C$. When the volume of the THF is approximately 1 mL, the flask is cooled to $-78°C$ and 5 mL of dry CH_2Cl_2 is condensed into the solution to induce precipitation of the product. The precipitate is filtered. The solvent in the filtrate is then recondensed into the upper portion of the extractor by opening stopcock B and cooling the upper portion of the extractor. This is accomplished as follows: A spatula, wrapped tightly with glass wool which itself is secured by copper wire, was intermittently dipped into liquid nitrogen and wiped across the upper portion of the extractor. The cooling of the glass brought about condensation of the solvent. In this manner, the product is washed several times with CH_2Cl_2, and dried under vacuum for 1 h. The yield is 0.160 g [90% based on $Ru_3(CO)_{12}$].

The sodium salt, $Na_2[Ru_3(CO)_{11}]$, and the lithium salt, $Li_2[Ru_3(CO)_{11}]$ (in solution), can be prepared by employing the same procedures as above with the exception that the sodium salt is precipitated from a concentrated CH_2Cl_2 solution with hexane.

I.R. Spectrum ($v(CO)$, *THF*). 2012 (m), 1978 (m, sh), 1955 (s), 1942 (s), 1914 (s, sh), 1883 (s, sh), 1715 (w, br), 1640 (m), and 1600 (vw) cm^{-1}.

Anal. Calcd. for: C, 19.16; K, 11.34; Ru, 43.97. Found: C, 19.35; K, 11.09; Ru, 43.75%.

Properties

The compound $K_2[Ru_3(CO)_{11}]$ is a red-brown powder. It is sensitive to air and moisture but can be stored indefinitely in an inert atmosphere or in a vaccum. It is soluble in ethers, such as THF or glyme, and insoluble in CH_2Cl_2 and alkanes. This potassium salt can be converted to the $[PPN]^+$, $[Ph_4P]^+$, and $[Ph_4As]^+$ salts by metathesis reactions with $[PPN]X$, $[Ph_4P]X$, and $[Ph_4As]X$ (X = halide) in THF-CH_2Cl_2. $K[HRu_3(CO)_{11}]$ can be generated by the reaction of $K_2[Ru_3(CO)_{11}]$ with one equivalent of HCl.

B. [DIPOTASSIUM UNDECACARBONYL TRIOSMATE (2−)], $K_2[Os_3(CO)_{11}]$

$$Os_3(CO)_{12} + 2K[(C_6H_5)_2CO] \rightarrow K_2[Os_3(CO)_{11}] + CO + 2(C_6H_5)_2CO$$

Procedure

Metallic potassium (0.0067 g, 0.17 mmole), benzophenone (0.0303 g, 0.17 mmole), and $Os_3(CO)_{12}$ (0.0778 g, 0.086 mmole) are added to a one-necked, 50-mL flask in a controlled atmosphere box. The flask contains a glass-coated magnetic stir bar and it is connected to an extractor with a one-necked, 50-mL receiver flask. The flask is sealed with a Kontes 4-mm vacuum adapter. The extraction apparatus was attached to an extractor port and evacuated via Stopcocks A and C on a vacuum line. Dry THF (3–4 mL) is condensed into the flask at − 78°C. The reaction is warmed to room temperature and stirred for 3 h. During this time the potassium is visibly consumed and a clear orange solution remains. Evolution of carbon monoxide gas is quantitative. When the reaction is complete, the THF is removed at room temperture by pumping it away through a removable U-trap maintained at − 196°C. Approximately 1 mL of Me_2O and 3–4 mL of dry CH_2Cl_2 are condensed into the flask at − 196°C, resulting in an orange solution. The Me_2O is then pumped away at − 78°C under a dynamic vacuum, causing the product to precipitate as a yellow-orange powder. This precipitate is filtered as previously described, washed with several portions of fresh CH_2Cl_2, and dried under vacuum for 1 h. The yield is 0.065 g [79% based on $Os_3(CO)_{12}$]. The compound $Na_2[Os_3(CO)_{11}]$ can be prepared in a similar manner.

I.R. Spectrum (v(CO), THF). 2040 (m), 1970 (s), 1950 (vs), 1890 (m, sh), 1665 (vw), and 1625 (m) cm^{-1}.

Anal. Calcd. for: C, 13.80; K, 8.17; Os, 59.62. Found: C, 13.60; K, 8.02; Os, 59.27%.

Properties

The compound $K_2[Os(CO)_{11}]$ is a bright yellow-orange powder. It is sensitive to air and moisture but can be stored indefinitely in an inert atmosphere or in a vacuum. It is soluble in ethers, such as THF or glyme, and insoluble in CH_2Cl_2 and alkanes. This potassium salt can be converted to the $[PPN]^+$, $[Ph_4P]^+$, and $[Ph_4As]^+$ salts by the reaction with [PPN]X, $[Ph_4P]X$, and $[Ph_4As]X$ (X = halide) in THF-CH_2Cl_2, $K[HOs_3(CO)_{11}]$ can be generated by the reaction of $K_2[Ru_3(CO)_{11}]$ with one equivalent of HCl.

References

1 (a) A. J. Deeming, *Adv. in Organomet. Chem.,* **26**, 1 (1986). (b) B. F. G. Johnson and J. Lewis, *Adv. in Inorg. and Radiochem.,* **24**, 225 (1981).
2. (a) M. J. Went, M. J. Sailor, P. L. Bogdan, C. P. Brock, and D. F. Shriver, *J. Am. Chem. Soc.,* **109**, 6023 (1987). (b) M. J. Sailor and D. F. Shriver, *Organometallics,* **4**, 1476 (1985).
3. (a) C. C. Nagel, J. C. Bricker, D. G. Alway, and S. G. Shore, *J. Organomet. Chem.,* **219**, C9 (1981). (b) A. A. Bhattacharyya, C. C. Nagel, and S. G. Shore, *Organometallics,* **2**, 1187 (1983).

4. C. R. Eady, P. F. Jackson, B. F. G. Johnson, J. Lewis, M. C. Malatesta, M. McPartlin, and W. J. H. Nelson, *J. Chem. Soc. Dalton Trans.*, **1980**, 383.
5. D. F. Shriver and M. A. Drezdzon, *The Manipulation of Air-Sensitive Compounds*, 2nd ed., John Wiley & Sons, New York, 1986.

43. [PPN]$_2$[Ru$_3$(CO)$_{11}$] AND [PPN]$_2$[Os$_3$(CO)$_{11}$], μ-NITRIDO-BIS(TRIPHENYLPHOSPHORUS)(1+) UNDECACARBONYLTRIRUTHENATE(2−) AND UNDECACARBONYLTRIOSMATE(2−)

Submitted by G. B. KARET,* E. J. VOSS,* M. J. SAILOR,* and D. F. SHRIVER*
Checked by E. ROSENBERG† and D. S. KOLWAITE†

The synthetically useful dianions [M$_3$(CO)$_{11}$]$^{2-}$ were first isolated by Shore and co-workers as the Ca^{2+} (M = Ru) and the K$^+$ (M = Os) salts by the reduction of M$_3$(CO)$_{12}$ using alkali metal benzophenone solutions in THF.[1] [Ru$_3$(CO)$_{11}$]$^{2-}$ reacts with Ru$_3$(CO)$_{12}$ to form the higher nuclearity clusters [Ru$_4$(CO)$_{13}$]$^{2-}$ and [Ru$_6$(CO)$_{18}$]$^{2-}$ but the triruthenium anion can be obtained in high purity by slowly adding triruthenium dodecacarbonyl to an excess of reducing solution using vacuum-line techniques.[2] Vacuum-line syntheses of both dianions have been described in detail.[1]

Described here are alternative syntheses of these dianions as the μ-nitrido-bis(triphenylphosphorus)(1 +){(Ph$_3$P)$_2$N$^+$ or PPN$^+$} salts, which are attained on a comparatively large scale in high yields using convenient Schlenk-like techniques. In the procedure for the synthesis of [PPN]$_2$-[Ru$_3$(CO)$_{11}$], undesirable condensation products are avoided by first forming [PPN][Ru$_3$Cl(CO)$_{11}$] by Cl$^-$ substitution.[4−6] This intermediate is then reduced.

Procedure

■ **Caution.** *Due to the toxicity of metal carbonyl compounds, all reactions must be carried out in an efficient fume hood.*

All manipulations performed under a dry N$_2$ atmosphere using standard Schlenk, syringe, and glovebox techniques.[7] Solvents are distilled under N$_2$

* Department of Chemistry, Northwestern University, Evanston, IL 60208-3113.
† Department of Chemistry, The University of Montana, Missoula, MN 59812.

from appropriate drying agents before use: Tetrahydrofuran and diethyl ether from $Na[Ph_2CO]$, dichloromethane from P_4O_{10}, and methanol from I_2 and Mg turnings.[8] Commercial [PPN]Cl* is dried in a 110°C oven for at least 12 h prior to use. $Ru_3(CO)_{12}$ and $Os_3(CO)_{12}$ are prepared according to published procedures.[9,10] Benzophenone is obtained commercially and used without further purification. Sodium beads are washed with pentane, vacuum dried, and stored under N_2. All glassware must be dried in a 110°C oven before use.

A. BIS[μ-NITRIDO-BIS(TRIPHENYLPHOSPHORUS)(1+)] UNDECACARBONYLTRIRUTHENATE(2−), [PPN]₂[Ru₃(CO)₁₁]

$$Ru_3(CO)_{12} + [PPN]Cl \rightarrow [PPN][Ru_3(CO)_{11}Cl] + CO$$

$$[PPN][Ru_3(CO)_{11}Cl] + [PPN]Cl + 2NaPh_2CO \rightarrow$$

$$[PPN]_2[Ru_3(CO)_{11}] + 2NaCl + 2Ph_2CO$$

Benzophenone (2.0 g, 11 mmol) and sodium beads (0.6 g, 26 mmol) are loaded into a 100-mL Schlenk flask equipped with a Teflon coated stir bar in a glove box. Under an N_2 atmosphere degassed THF, 60 mL, is added, and the solution is vigorously stirred for at least 1 h. The resulting solution of sodium benzophenone is deep blue or purple, and it must be protected from air by an N_2 atmosphere.

A 300-mL Schlenk flask is charged with 2.0 g (3.1 mmol) $Ru_3(CO)_{12}$, 4.0 g (7.0 mmol) [PPN]Cl, and a stir bar in air. The flask is evacuated and refilled with N_2, and 30 mL dry and degassed THF is added. The flask is immediately sealed. The reaction mixture is stirred for 1 h, while the orange $Ru_3(CO)_{12}$ dissolves. The resulting red-brown solution of $[PPN][Ru_3(\mu\text{-}Cl)(CO)_{11}]$ contains some $[PPN][Ru_3(\mu\text{-}Cl)(CO)_{10}]$ caused by loss of CO.[6] An IR spectrum taken at this point reveals $v(CO)$: 2098(w), 2059(s), 2025(vs), 2010(vs), 1991(s), 1966(s, sh), 1961(vs), 1879(w), 1827(s) cm^{-1}.

The $[PPN][Ru_3Cl(CO)_{11}]$ is stirred under N_2 and reduced by dropwise addition of the sodium benzophenone solution from a syringe until a dark green tint persists, indicating completion of reduction. Normally ca. 30 mL of reducing agent is required. Methanol, 2 mL, is added with stirring, and the solution is filtered in the Schlenk line under N_2 to remove NaCl, and the volume of the filtrate is reduced to ca. 20 mL under vacuum. Diethyl ether, 60 mL, is added dropwise with stirring. The resulting red-brown

---

* μ-Nitrido-bis(triphenylphosphorus) chloride is available from Alfa Ventron.

microcrystals are isolated by filtration on a medium porosity frit, washed with 3×5 mL methanol and 3×5 mL diethyl ether, and then dried in vacuo. The yield of crude $[PPN]_2[Ru_3(CO)_{11}]$ isolated at this point is 4.9 g [94% based on $Ru_3(CO)_{12}$].

The crude product may be recrystallized from CH_2Cl_2. $[PPN]_2$-$[Ru_3(CO)_{11}]$, 1.1 g, is dissolved in 15 mL CH_2Cl_2, filtered, and then immediately crystallized by the dropwise addition of 40 mL of diethyl ether, to yield red crystals, 650 mg, 59%. It is important to minimize the time of exposure of $[Ru_3(CO)_{11}]^{2-}$ to CH_2Cl_2 because this cluster is slowly decomposed upon standing in CH_2Cl_2. If the product has an IR band at 1760 cm^{-1}, $[PPN]$-$[Ru(CO)_{13}]$ is present, but it can be removed by washing the product with several small portions of degassed CH_3OH under a N_2 atmosphere using a 100-mL Schlenk filter.

Properties

The compound $[PPN]_2[Ru_3(CO)_{11}]$ is a red or orange-brown air and moisture-sensitive solid. It is very soluble in CH_3CN, less less soluble in acetone and CH_2Cl_2, slightly soluble in THF and CH_3OH, and insoluble in diethyl ether and pentane. IR, CH_3CN, $v(CO)$, cm^{-1} 2030(w), 1959(s), 1941(vs), 1878(m), 1680(w).

Anal. Calcd. %: C, 59.04; H, 3.58; N, 1.67; Ru, 17.96. Found %: C, 58.24; H, 3.56, N, 1.64; Ru, 18.10. A band at 1760 cm^{-1} indicates contamination by $Ru_4(CO)_{13}{}^{2-}$, which can be partially removed by washing the product with degassed methyl alcohol. $[PPN]_2[Ru_3(CO)_{11}]$ is a useful precursor for the synthesis of $[PPN]_2[Ru_3(CO)_9CCO]$.[4]

B. μ-NITRIDO-BIS(TRIPHENYLPHOSPHORUS)(1+) UNDECACARBONYLTRIOSMATE(2−), $[PPN]_2[Os_3(CO)_{11}]$

$$Os_3(CO)_{12} + 2[PPN]Cl + 2NaPh_2CO \rightarrow$$
$$2Ph_2CO + 2NaCl + [PPN]_2[Os_3(CO)_{11}] + CO$$

Glassware is treated before use with a solution of 0.5 mL Glassclad 6c in 10 mL of pentane.* The Glassclad solution is swirled around in the flask to coat its interior wall and poured out. The flask is then flushed with nitrogen to remove the hydrocarbon solvent and then cured in an oven at 125°C for

---

* United Chemical Technologies, 2731 Bartram Road, Bristol, PA 19007.

at least 1 h. This treatment minimizes the formation of $[HOs_3(CO)_{11}]^-$, which is formed by adsorbed water and surface OH groups on untreated glassware.

A sodium benzophenone solution is made exactly as for the synthesis of $[PPN]_2[Ru_3(CO)_{11}]$. A 100-mL Schlenk flask is charged with 0.515 g (0.0568 mmol) $Os_3(CO)_{12}$, 0.665 g (1.16 mmol) PPNCl, and a stir bar in air. The flask is evacuated and refilled with N_2 and then THF, 5 mL, is added. Under an N_2 flush, the suspension of yellow $Os_3(CO)_{12}$ and white PPNCl is treated with the sodium benzophenone solution until all of the $Os_3(CO)_{12}$ dissolves and a blue color, indicative of excess reducing agent, persists for more than 10 min. Because the reduction of $Os_3(CO)_{12}$ occurs slowly, the reducing agent is added in increments of 1 mL or less. Near the end of the reaction the suspension is red-orange when no excess reducing agent is present. Approximately 20 mL of reducing agent is required to reach the end point. The reaction mixture is filtered in a nitrogen atmosphere to remove NaCl and PPNCl. The volume of the reaction mixture is then reduced under vacuum to ca. 10 mL. An orange microcrystalline solid is precipitated by the addition of 60 mL diethyl ether. This solid is isolated on a medium-porosity frit, washed with 2 × 10 mL of dried and degassed diethyl ether, and dried under vacuum. Yield: 1.105 g (0.565 mmol, 99%) of crude material.

The product is pure enough for many purposes and is best used without further purification. It contains some [PPN]Cl, NaCl, and benzophenone as impurities. Further manipulation often results in the formation of $[PPN][HOs_3(CO)_{11}]$. At the risk of protonation of the product, NaCl and excess PPNCl may be removed by recrystallization from CH_2Cl_2 and diethyl ether, followed by washings of the solid with i-PrOH and diethyl ether. For very pure $[PPN]_2[Os_3(CO)_{11}]$, the vacuum-line synthesis of this compound is recommended.[3]

Properties

$[PPN]_2[Os_3(CO)_{11}]$ is a orange, air- and moisture-sensitive solid, which is soluble in CH_2Cl_2 and insoluble in diethyl ether, 2-propanol, and pentane. It is considerably more moisture-sensitive in solution than in the solid state. IR $v(CO)$ 2048(w), 1967(s), 1943(vs), 1877(m), 1637(w,br).

Anal. Calcd. %: H, 3.09; C, 50.97; Os, 29.17. Found: H, 3.11; C, 52.11; Os, 26.50. A common impurity is $[PPN][HOs_3(CO)_{11}]$, which has a strong infrared band at 1993 cm^{-1}. $[PPN]_2[Os_3(CO)_{11}]$ is useful for the synthesis of $[PPN]_2[Os_3(CO)_9CCO]$.[11]

References

1. C. C. Nagel, J. C. Bricker, D. G. Alway, and S. G. Shore, *J. Organomet. Chem.*, **219**, C9 (1981).
2. A. A. Bhattacharya, C. C. Nagel, and S. G. Shore, *Organometallics*, **2**, 1187 (1983).
3. T. J. Coffy, D. A. McCarthy, and S. G. Shore, *Inorg. Synth.*, **32** (1998).
4. M. J. Sailor, C. P. Brock, and D. F. Shriver, *J. Am. Chem. Soc.*, **109**, 6015 (1987).
5. S. -H. Han, G. L. Geoffroy, B. D. Dombek, and A. L. Rheingold, *Inorg. Chem.*, **27**, 4355 (1988).
6. G. Lavigne and H. D. Kaesz, *J. Am. Chem. Soc.*, **106**, 4647 (1984).
7. D. F. Shriver and M. A. Drezdzon, *The Manipulation of Air-Sensitive Compounds*, 2nd ed., Wiley, New York, 1986.
8. A. J. Gordon and R. A. Ford, *The Chemist's Companion*, Wiley, New York, 1972.
9. M. I. Bruce, C. M. Jensen, and N. L. Jones, *Inorg. Synth.*, **28**, 216 (1987).
10. S. R. Drake and P. A. Loveday, *Inorg. Synth.*, **28**, 230 (1987).
11. M. J. Went, M. J. Sailor, P. L. Bogdan, C. P. Brock, and D. F. Shriver, *J. Am. Chem. Soc.*, **109**, 6023 (1987).

44. PLATINUM–RUTHENIUM CARBONYL CLUSTER COMPLEXES

Submitted by R. D. ADAMS,* T. S. BARNARD,* J. E. CORTOPASSI,*
W. WU,* and Z. LI*
Checked by J. R. SHAPLEY† and KWANGYEAL LEE†

Mixed-metal cluster complexes can exhibit modified and enhanced reactivities due to the effects of cooperativity or synergism between the metal atoms.[1–4] It has recently been shown that the pentacarbonyl complexes of the iron subgroup readily react with $Pt(cod)_2$,[5] cod = 1,5-cyclooctadiene, under mild conditions to yield the hexanuclear cluster complexes $Pt_3Fe_3(CO)_{15}$,[6] $Pt_2Ru_4(CO)_{18}$,[7] and $Pt_2Os_4(CO)_{18}$.[8] The latter two complexes can be viewed as dimers of the unit $PtM_2(CO)_9$ and can be split upon reaction with 1,2-ethandiyldiphenylphosphine (dppe) and CO to yield two equivalents of the trinuclear species $PtRu_2(CO)_8(dppe)$[7] and $PtOs_2(CO)_{10}$, respectively.[9]

The $Pt_2Ru_4(CO)_{18}$ cluster reacts with H_2 to form $Pt_3Ru_6(CO)_{21}(\mu_3\text{-H})$-$(\mu\text{-H})_3$ in which the platinum and ruthenium atoms are arranged in triangular layers of the pure elements.[10] This complex can be converted to $Pt_3Ru_6(CO)_{20}(\mu_3\text{-C}_2Ph_2)(\mu\text{-H})_2$ by reaction with diphenylacetylene.[10] The latter complex was found to be an active catalyst for the hydrogenation

*Department of Chemistry and Biochemistry, University of South Carolina, Columbia, SC 29208.
†Department of Chemistry, University of Illinois, Urbana, IL 61801.

of diphenylacetylene to Z-stilbene.[4] The syntheses of the complexes $Pt_2Ru_4(CO)_{18}$ and $Pt_3Ru_6(CO)_{21}(\mu_3\text{-H})(\mu\text{-H})_3$ are described below.

General Procedure

■ **Caution.** *Due to the high toxicity of carbon monoxide and the high flammability of hydrogen, the reactions must be carried out in a well-ventilated hood.*

Reactions are performed under an atmosphere of dry nitrogen unless specified otherwise. Reagent-grade solvents are dried over 4 Å molecular sieves and deoxygenated by purging with nitrogen prior to use. The compound $Pt(cod)_2{}^5$ is prepared by the published procedure. $Ru(CO)_5$ is prepared by an adaptation of a previously reported procedure as described below.[11]

A. IN SITU RUTHENIUM PENTACARBONYL AND OCTADECACARBONYLDIPLATINUMTETRARUTHENIUM, $Pt_2Ru_4(CO)_{18}$

$$4Ru(CO)_5 + 2Pt(cod)_2 \rightarrow Pt_2Ru_4(CO)_{18} + 2CO + 4cod$$

■ **Caution.** *Photolysis by a 1000-W mercury UV lamb emits potentially blinding radiation. The apparatus should be placed in a hood whose shield is completely covered in reflective aluminum foil.*

A 1-L, three-necked, round-bottomed flask with ground glass joints is equipped with a Teflon-covered magnetic stirring bar, a nitrogen-inlet adapter, a condenser (in the center joint), and a rubber septum. Hexane (600 mL) is added to the flask. Then $Ru_3(CO)_{12}$* (250 mg, 0.391 mmol) is added to the flask and the solution is stirred until all of the solid is dissolved. The solution is then irradiated using a medium-pressure mercury UV lamp (1000 W, Cooper Industries, Vicksburg, MS, Model HC) in the presence of a slow purge of carbon monoxide at 25°C. The lamp source is placed ca. 10 in. from the flask and should be cooled by a slow stream of air from a compressed air line. *Note*: The use of a lamp with less power (e.g., a 250-W high-pressure mercury lamp) lengthens the time for the formation of $Ru(CO)_5$ considerably. Reaction is complete when the color of the solution has changed from orange to colorless (ca. 1.5 h) and the characteristic IR absorptions of $Ru_3(CO)_{12}$ [$\nu_{(CO)}$ 2061 (vs), 2031 (s), and 2012 (m)] disappear and the absorptions of

* Purchased from Strem Chemicals, Inc., Newburyport, MA.

Ru(CO)$_5$ appear [$v_{(CO)}$ 2038 (s), 2003 (vs)]. The yield of Ru(CO)$_5$ is assumed to be quantitative.

The flask of the Ru(CO)$_5$ solution is cooled to 0°C, evacuated, and refilled with nitrogen 10 times to remove excess CO. Under nitrogen, Pt(cod)$_2$ (145.3 mg, 0.353 mmol) is dissolved in CH$_2$Cl$_2$ (25 mL). Pt(cod)$_2$ is not stable in CH$_2$Cl$_2$ for long periods. This solution should be prepared just prior to use. Add this solution quickly (in a few seconds) to the Ru(CO)$_5$ solution, and stir the mixture at 0°C for 30 min*. During this 30-min period the reaction flask is evacuated and refilled with nitrogen every 5 min to remove the liberated CO. After 30 min, an additional quantity of Pt(cod)$_2$ (97.0 mg, 0.236 mmol) is dissolved in CH$_2$Cl$_2$ (25 mL) and added to the reaction mixture. The mixture is stirred at 0°C for an additional 30 min.* Once again, the flask is evacuated and refilled with nitrogen periodically to remove the liberated CO. The reaction mixture is stirred for an additional 9 h at 25°C with no further CO removal during this period. The resulting purple solution is reduced in volume to ca. 100 mL by using a rotary evaporator, and the components are then separated by silica-gel column chromatography (25 × 500 mm). The first yellow band eluting with hexane is Ru$_3$(CO)$_{12}$, 101.9 mg, 41% [based on Ru(CO)$_5$]. The second band is purple and elutes with a CH$_2$Cl$_2$/hexane (1/4) solvent mixture. This is the product Pt$_2$Ru$_4$(CO)$_{18}$, 174.4 mg, 46% based on platinum.

The periodic removal of CO is used to obtain the best yields of the product. The authors routinely perform this reaction with one-half the required amount of Pt(cod)$_2$ to increase the yield based on platinum added. The amount of Ru$_3$(CO)$_{12}$ is increased by this method, but it is easily recovered and can be reused to synthesize Ru(CO)$_5$ in subsequent preparations.

Anal. Calcd. for C$_{18}$O$_{18}$Pt$_2$Ru$_4$: C, 16.64. Found: C, 16.90.

Properties

The complex forms purple air-stable crystals and has good solubility in both polar and nonpolar solvents (e.g., acetone, dichloromethane, tetrahydrofuran, benzene, and hexane). It has been characterized crystallographically.[7] The structure is an open, but folded, ladder-like array of six metal atoms with the two platinum atoms in the center. The IR spectrum (hexane) exhibits $v_{(CO)}$ bands at 2085 (m), 2062 (vs), 2035 (vs), 2016 (w). Pt$_2$Ru$_4$(CO)$_{18}$ has been found to be very useful for the synthesis of new platinum–ruthenium cluster complexes.[10,12,13]

---

* Note: The checkers found that the yield of Pt$_2$Ru$_4$(CO)$_{18}$ is increased to 61% when the reaction flask is wrapped in foil and the hood light is turned off.

B. HENEICOSACARBONYLTETRAHYDRIDOTRIPLATINUM-HEXARUTHENIUM, $Pt_3Ru_6(CO)_{21}(\mu_3\text{-}H)(\mu\text{-}H)_3$

$$3Pt_2Ru_4(CO)_{18} + 4H_2 \rightarrow 2Pt_3Ru_6(CO)_{21}(\mu_3\text{-}H)(\mu\text{-}H)_3 + 12CO$$

A dry, 300-mL, three-necked flask with ground glass joints is equipped with a Teflon-covered magnetic stirring bar, a nitrogen-inlet adapter, a condenser (in the center joint) connected to an oil bubbler, and a rubber septum. A Variac-controlled heating mantle is placed under the flask. Heptane, 140 mL, is added to the flask, followed by $Pt_2Ru_4(CO)_{18}$ (25.0 mg, 0.019 mmol) which is dissolved by stirring. The solution is purged with hydrogen for 5 min and then heated to a slow reflux. *Note:* Vigorous refluxing or purging of the solution decreases the yield of $Pt_3Ru_6(CO)_{21}(\mu_3\text{-}H)(\mu\text{-}H)_3$ dramatically, and results in an increased yield of $Pt_3Ru_7(CO)_{22}(\mu_3\text{-}H)_2$. The heating and a slow hydrogen purge (ca. 3–4 bubbles per second) are continued for 15 min. The hydrogen purge is stopped and the solution is cooled to ca. 50°C by using a water bath. The solution is then filtered into a 250-mL round-bottomed flask. The solvent is removed by using a rotary evaporator and the solid is scraped from the sides of the flask with a spatula. This solid is then washed with 5-mL portions of cold pentane until the washings are colorless. The washings are collected in a separate round-bottomed flask. After washing, the remaining solid is the pure product $Pt_3Ru_6(CO)_{21}(\mu_3\text{-}H)(\mu\text{-}H)_3$, 15.8 mg. If desired, the compounds in the washings can be separated by TLC in air (silica gel, 7/3 hexane/CH_2Cl_2 as the eluent) to yield 1.4 mg of a combination of $Ru_3(CO)_{12}$ and $Ru_4(CO)_{12}(\mu\text{-}H)_4$ in the first band (yellow), 0.6 mg of unreacted $Pt_2Ru_4(CO)_{18}$ in the second band (purple), 1.4 mg of the product $Pt_3Ru_6(CO)_{21}(\mu_3\text{-}H)(\mu\text{-}H)_3$ in the third band (brown), and 0.9 mg of $Pt_3Ru_7(CO)_{22}(\mu_3\text{-}H)_2{}^{12}$ in the fourth band (green). The combined amounts of $Pt_3Ru_6(CO)_{21}(\mu_3\text{-}H)(\mu\text{-}H)_3$ gives 17.2 mg (76% yield).

Anal. Calcd. for $C_{21}H_4O_{21}Pt_3Ru_6$: C, 14.14; H, 0.23. Found: C, 14.49; H, 0.26.

Properties

The product is a black microcrystalline solid at 25°C. It is readily soluble in dichloromethane, sparingly soluble in hexane and benzene, and decomposes in coordinating solvents such as tetrahydrofuran, acetonitrile, and acetone. It is stable in air at $-20°C$ as a solid for periods of several weeks. Its structure has been determined crystallographically.[10] The metal atoms are arranged into a face-shared bioctahedral cluster. The central triangular layer contains the three platinum atoms. The outer triangles contain the ruthenium atoms. Three of the hydride ligands bridge the edges of one of the ruthenium triangles, while the fourth hydride is a triply bridging ligand on the face of the other ruthenium triangle. In CH_2Cl_2 solvent, the IR spectrum shows absorptions at 2081 (w, sh), 2066 (vs), 2052 (m, sh), 2026 (w) cm^{-1}. The ^1H NMR spectrum at $-88°C$ shows resonances at $\delta = -15.84$ (s, 3H) and -19.26 (s, 1H) in acetone-d_6. Due to a dynamical averaging process, the hydride resonances are broadened at higher temperatures and are not observed at 25°C. $Pt_3Ru_6(CO)_{21}(\mu_3\text{-}H)(\mu\text{-}H)_3$ has been used for the synthesis of new higher nuclearity platinum–ruthenium mixed–metal cluster complexes.[14,15]

References

1. R. D. Adams, Ed. in *Comprehensive Organometallic Chemistry II*, E. W. Abel, F. G. A. Stone, and G. Wilkinson, Ed.-in-Chief, Pergamon, Oxford, 1995, Vol. 10.
2. R. D. Adams, *Polyhedron*, **7**, 2251 (1988).
3. R. Giordano and E. Sappa, *J. Organomet. Chem.*, **448**, 157 (1993).
4. R. D. Adams, T. S. Barnard, Z. Li, W. Wu, and J. Yamamoto. *J. Am. Chem. Soc.*, **116**, 9103 (1994).
5. J. L. Spencer, *Inorg. Synth.*, **19**, 213 (1979).
6. R. D. Adams, I. Arafa, G. Chen, J.-C. Lii, and J.-G. Wang. *Organometallics*, **9**, 2350 (1990).
7. R. D. Adams, G. Chen, and W. Wu, *J. Cluster Sci.*, **4** , 119 (1993).
8. R. D. Adams, M. Pompeo, and W. Wu, *Inorg. Chem.*, **30**, 2425 (1991).
9. P. Sundberg, *J. Chem. Soc., Chem. Commun.*, **1987**, 1307.
10. R. D. Adams, T. S. Barnard, Z. Li, W. Wu, and J. Yamamoto, *Organometallics*, **13**, 2357 (1994).
11. R. Huq, A. J. Poë, and S. Chavala, *Inorg. Chim. Acta*, **38**, 121 (1980).
12. R. D. Adams, Z. Li, J.-C. Lii, and W. Wu, *Organometallics* **11**, 4001 (1992).
13. R. D. Adams and W. Wu, *Organometallics*, **12**, 1248 (1993).
14. R. D. Adams T. S. Barnard, and J. E. Cortopassi, *Organometallics*, **14**, 2232 (1995).
15. R. D. Adams, T. S. Barnard, J. E. Cortopassi, and L. Zhang, *Organometallics*, **15**, 2664 (1996).

45. TRI(μ-CARBONYL)NONACARBONYLTETRARHODIUM, $Rh_4(\mu\text{-}CO)_3(CO)_9$

Submitted by Ph. SERP, Ph. KALCK*, R. FEURER*, and R. MORANCHO*
Checked by KWANG YEOL LEE† and JOHN R. SHAPLEY†

$$RhCl_3 \cdot 3H_2O + 3CO \xrightarrow{CH_3OH/N_2}$$
$$H[RhCl_2(CO)_2] + HCl + (CH_3O)_2CO$$

* Ecole Nationale Supérieure de Chimie de Toulouse 118, rue de Narbonne, 31077 Toulouse cedex, France.
† Department of Chemistry, University of Illinois at Urbana-Champaign, Urbana, IL 61801.

$$H[RhCl_2(CO)_2] + NaCl \xrightarrow{CO} Na[RhCl_2(CO)_2] + HCl$$

$$4Na[RhCl_2(CO)_2] + 6CO + 2H_2O \xrightarrow{Na_2 \ citrate}$$

$$Rh_4(CO)_{12} + 2CO_2 + 4NaCl + 4HCl$$

The $Rh_4(CO)_{12}$ complex finds extensive application in homogeneous as well as in heterogeneous catalysis where it is a very useful precursor. This compound is also a convenient starting material for the synthesis of various rhodium clusters.[1]

Since the first synthesis in 1943 by Hieber and Lagally by treatment of $RhCl_3$ with CO at high temperatures and pressures,[2] several methods were described in the literature for the preparation of $Rh_4(CO)_{12}$. Two main routes can be distinguished. Firstly, high carbon monoxide pressure syntheses require metal powders acting as halogen acceptors and carbonylating agents to reduce $Rh_2Cl_2(CO)_4$ or $RhCl_3 \cdot 3H_2O$.[2-4] Second, procedures under an ambient CO atmosphere were reported starting from $Rh_2Cl_2(CO)_4$ and $NaHCO_3$ in *n*-hexane with small amounts of water.[5-8] Similarly $K_3[RhCl_6] \cdot nH_2O$ or $RhCl_3 \cdot 3H_2O$ can be reduced with CO in the presence of copper powder in water.[9,10] The preparation of $Rh_4(CO)_{12}$ in CO matrix from monoatomic rhodium vapor should be mentioned.[11] Recently, it has been shown that the surface-mediated carbonylation of $Rh_2Cl_2(CO)_4$ leads to a mixture of $Rh_4(CO)_{12}$ and $Rh_6(CO)_{16}$ from which $Rh_4(CO)_{12}$ can be extracted with a 47% yield.[12]

All these syntheses require long reaction times and tedious handling of carbon monoxide during crucial steps such as filtration. In our hands, the low-pressure procedures afforded yields of isolated $Rh_4(CO)_{12}$ in the range 55–60%, as previously reported by Cattermole and Osborne[10] rather than 90% as claimed by Chini and co-workers.[5-8] We have developed a straightforward and easy method to prepare $Rh_4(CO)_{12}$ with yields higher than 90%. It involves the in situ generation of the key species $[RhCl_2(CO)_2]^-$ which is reduced by CO in the presence of disodium citrate.

Procedure

■ **Caution.** *All operations with toxic CO gas must be carried out in a well-ventilated fume hood. For preparations on a large scale, the best way to eliminate carbon monoxide is to burn it in a small flame.*

A three-necked, 250-mL, round-bottomed flask, equipped with separate gas inlets for CO and for N_2 and a condenser having on the top a pressure-release bubbler, is charged with 25 mL methanol containing 0.500 g of

$RhCl_3 \cdot 3H_2O$.* The solution is heated at reflux with a CO purge for 15 h with vigorous stirring. The initially dark-red solution gradually becomes lemon yellow. The reaction progress needs to be checked by solution infrared spectroscopy; only two ν_{CO} bands are characteristic of the anion $[RhCl_2(CO)_2]^-$, 2070 and 1994 cm^{-1}. The solution is cooled to room temperature under a slow nitrogen stream. Then 0.175 g NaCl (2.99 mmole) is introduced and the magnetic stirring is maintained for 20 min under a CO stream. The reflux condensor is replaced by a dropping funnel with a pressure-equalizing tube. It is charged with an aqueous solution saturated with nitrogen containing 1.2 g of disodium citrate (100 mL of 0.4 M solution). The solution is slowly (requiring about 1 h) introduced from the funnel in the methanol mixture with vigorous stirring. The solution turns orange quickly, due to the formation of $Rh_4(CO)_{12}$.

The stirring is maintained for 5 h under a gentle CO bubbling at room temperature [we observed that for longer reaction times small amounts of $Rh_2Cl_2(CO)_4$ are formed in addition to $Rh_4(CO)_{12}$ whose yields decreases slightly]. Most of $Rh_4(CO)_{12}$ precipitates during this operation. The precipitate is filtered under a nitrogen atmosphere, then dried under vacuum. The complex is dissolved in 30 mL of CH_2Cl_2 and thus separated from solid NaCl. The CH_2Cl_2 solution is rapidly evaporated in vacuum to avoid formation of metallic rhodium. The complex can be used without any further purification. The yields of isolated product are higher than 90%. The complex is preferentially stored under a CO atmosphere.

Properties

The cluster $Rh_4(CO)_{12}$ is obtained as an orange powder and can be recrystallized in *n*-hexane at $-70°C$ to give dark-red microcrystals. This complex is characterized by its infrared spectrum in *n*-hexane solution showing ν_{CO} bands at 2074.5 (s), 2071 (s), 2062 (w,sh), 2044 (m), 2024 (w), 1886 (s) cm^{-1} (s = strong, m = medium, w = weak, sh = shoulder). Other properties and the reactivity of this compound were reviewed.[1,13]

References

1. P. Chini and B. T. Heaton, *Top. Curr. Chem.*, **71**, 1 (1977).
2. W. Hieber and H. Lagally, *Z. Anorg. Allg. Chem.*, **251**, 96 (1943).
3. B. L. Booth, M. J. Else, R. Fields, H. Goldwhite, and R. N. Haszeldine, *J. Organometal. Chem.*, **14**, 417 (1968).

* Available from Alfa Products, Ventron Corp. P.O. Box 299, Danvers MA 92923. In our case this salt was generously supplied by the Comptoir Lyon-Alemand-Louyot, 13 rue de Montmorency, 75139 Paris cedex 03, France.

4. S. H. Chaston and F. G. A. Stone, *J. Chem. Soc. (A)*, 500 (1969).
5. P. Chini and S. Martinengo, *Chem. Commun.*, 251 (1968).
6. P. Chini and S. Martinengo, *Inorg. Chim. Acta*, **3**, 315 (1969).
7. G. Giordano, S. Martinengo, D. Strumolo, and P. Chini, *Gazz. Chim. Ital.*, **105**, 613 (1975).
8. S. Martinengo, G. Giordano, and P. Chini, *Inorg. Synth.*, **20**, 209 (1980).
9. S. Martinengo, P. Chini, and G. Giordano, *J. Organometal. Chem.*, **27**, 389 (1971).
10. P. E. Cattermole and A. G. Osborne, *Inorg. Synth.*, **17**, 115 (1977).
11. L. A. Hanlan, and G. A. Ozin, *J. Am. Chem. Soc.*, **96**, 6324 (1974).
12. D. Roberto, R. Psaro, and R. Ugo, *Organometallics*, **12**, 2292 (1993).
13. R. P. Hughes in *Comprehensive Organometallic Chemistry*, Vol. 5, G. Wilkinson, F. G. A. Stone, and E. W. Abel, Eds., Pergamon Press, 1982, p. 317.

46. HIGH NUCLEARITY HYDRIDODECARUTHENIUM CLUSTERS

Submitted by MARIE P. CIFUENTES* and MARK G. HUMPHREY*
Checked by JOHN R. SHAPLEY* and KWANGYEOL LEE†

High nuclearity carbonyl clusters are of interest as models of metal crystallites with chemisorbed ligands. However, systematic exploration of their chemistry has been hampered by the lack of high-yielding syntheses. The synthesis of $[Os_{10}(\mu_6\text{-}C)(CO)_{24}]^{2-}$ in excellent yield has permitted a detailed examination of its chemistry; although reactive toward electrophiles, it is remarkably inert toward nucleophiles.[1] The related decaruthenium cluster $[Ru_{10}(\mu\text{-}H)(\mu_6\text{-}C)\text{-}(CO)_{24}]^-$ was previously available in 15% yield as one of at least six products from the residues obtained from pyrolyzing $Ru_3(CO)_{12}$ in mesitylene for 3–5 days,[2] a procedure which does not encourage systematic reactivity studies. We report below the synthesis of $[Ru_{10}(\mu\text{-}H)(\mu_6\text{-}C)\text{-}(CO)_{24}]^-$, as both the $[Ru_2(\mu\text{-}H)(\mu\text{-}NC_5H_4)_2(CO)_4(NC_5H_5)_2]^+$ or the metathesized $[PPh_4]^+$ salts, from a triruthenium precursor $Ru_3(\mu\text{-}H)\text{-}(\mu\text{-}NC_5H_4)(CO)_{10}$, itself available from $Ru_3(CO)_{12}$ in excellent yield; the decaruthenium cluster anion is thus available in an overall yield of at least 70% from $Ru_3(CO)_{12}$ and the synthetic procedures can be comfortably performed in 5 h laboratory work. The decaruthenium cluster is a useful starting material for the facile syntheses of nucleophile-substituted decametallic derivatives. For example, while phosphine substitution on the decaosmium cluster requires refluxing xylene and Os–Os cleavage by halogens for the reaction to proceed, stirring the decaruthenium cluster at room

* Department of Chemistry, Australian National University, Canberra, ACT 0200, Australia.
† Department of Chemistry, University of Illinois at Urbana-Champaign, Urbana, IL 61801.

temperature in acetone with one or more equivalents of phosphines affords the corresponding substituted cluster anions within minutes in excellent yield;[3] the synthesis of the mono(triphenylphosphine)-substituted cluster is described as an illustration.

A. DECACARBONYL(μ-HYDRIDO)(μ-PYRIDYL)-TRIRUTHENIUM[4]

$$Ru_3(CO)_{12} + NC_5H_5 \rightarrow Ru_3(\mu\text{-}H)(\mu\text{-}NC_5H_4)(CO)_{10} + 2CO$$

Procedure

■ **Caution.** *This reaction should be carried out in a well-ventilated hood, as carbon monoxide—a highly poisonous, colorless, and odorless gas—is evolved. Pyridine is toxic and manipulations involving its use should be carried out in an efficient fume hood.*

Dodecacarbonyltriruthenium, $Ru_3(CO)_{12}$, is commercially available (Strem, Aldrich) or may be prepared by a published procedure.[5]

A 250-mL, two-necked, round-bottomed flask is fitted with a reflux condenser connected to a nitrogen bubbler. The flask is charged with 636 mg $Ru_3(CO)_{12}$ (0.995 mmol) and cyclohexane (100 mL). Pyridine (94 mg, 1.19 mmol, 1.2 equivalents) is added and the mixture deoxygenated before heating to reflux with stirring for around 40 min (although reaction time may vary with scale). The reaction must be monitored carefully by solution IR spectroscopy to ensure reaction does not proceed to the bis(pyridyl) cluster complex, $Ru_3(\mu\text{-}H)_2(\mu\text{-}NC_5H_4)_2(CO)_8$ (easily identified by a signal in the IR spectrum at 2080w), which can be difficult to separate. The mixture is allowed to cool and the solvent is evaporated in vacuo. The orange residue is dissolved in a minimum amount of dichloromethane and purified by thin-layer chromatography, (TLC; for example, Merck, GF_{254} silica gel) using petroleum ether (60–80°C boiling point range) as eluent, to remove any trace amounts of unreacted $Ru_3(CO)_{12}$. The product is obtained as a yellow microcrystalline powder by evaporation from dichloromethane. Yield: 554 mg (0.836 mmol, 84%).

Anal. Calcd. for $C_{15}H_5NO_{10}Ru_3$: C, 27.20; H, 0.76; N, 2.11%. Found: C, 27.68; H, 0.65; N, 2.15%.

Properties

The compound $Ru_3(\mu\text{-}H)(\mu\text{-}NC_5H_4)(CO)_{10}$ is an orange crystalline solid, air stable at room temperature for long periods. It is readily soluble in organic solvents such as hexane, benzene, and dichloromethane, solutions of which are moderately stable in air for some hours. The product is readily identified by its IR spectrum, which contains the following CO absorptions (cyclohexane): 2100(m), 2062(vs), 2051(vs), 2024(m), 2015(vs), 2010(m), 1999(m), and 1985(w) cm^{-1}, and by the proton NMR spectrum which contains a characteristic hydride signal at -14.44 (s, 1H, RuH; CDCl$_3$).

B. TETRACARBONYL(μ-HYDRIDO)BIS(PYRIDINE)BIS-(μ-PYRIDYL)DIRUTHENIUM TETRACOSACARBONYL-(μ_6-CARBIDO)(μ-HYDRIDO)DECARUTHENATE[6]

$$Ru_3(\mu\text{-}H)(\mu\text{-}NC_5H_4)(CO)_{10} \rightarrow$$

$$[Ru_2(\mu\text{-}H)(\mu\text{-}NC_5H_4)_2(CO)_4(NC_5H_5)_2][Ru_{10}(\mu\text{-}H)(\mu_6\text{-}C)(CO)_{24}]$$

Procedure

■ **Caution.** *This reaction should be carried out in a well-ventilated hood, as carbon monoxide—a highly poisonous, colorless, and odorless gas—is evolved.*

A 100 mL, two-necked, round-bottomed flask is fitted with a reflux condenser connected to a nitrogen bubbler. The flask is charged with Ru$_3$(μ-H)-(μ-NC$_5$H$_4$)(CO)$_{10}$ (238 mg, 0.359 mmol) and chlorobenzene (25 mL) and the mixture is deoxygenated. The thoroughly stirred solution is heated at reflux temperature under an atmosphere of dry nitrogen until no starting material was observed by IR (disappearance of the band at 2100 m cm^{-1}; this takes around 40 min, but reaction time may vary with scale). The resulting black solution was filtered in air through filter-aid to remove a small amount of insoluble black material, washing through with chlorobenzene, and the combined filtrate and washings reduced to dryness in vacuo. This residue is triturated with petroleum ether (60–80°C boiling point range) until the washings are clear (usually 2×20 mL) to remove any unreacted starting material, leaving a black microcrystalline solid [Ru$_2$(μ-H)(μ-NC$_5$H$_4$)$_2$(CO)$_4$(NC$_5$H$_5$)$_2$][Ru$_{10}$(μ-H)(μ_6-C)(CO)$_{24}$] [**1**, 178 mg, 85%, calculated yield based on Ru$_3$(μ-H)(μ-NC$_5$H$_4$)(CO)$_{10}$], which is spectroscopically clean and sufficiently pure for subsequent chemistry as in **C**.

Anal. Calcd. for C$_{49}$H$_{20}$N$_4$O$_{28}$Ru$_{12}$: C, 25.31; H, 0.87; N, 2.41. Found: C, 25.03; H, 1.01; N, 1.78.

If desired, the complex can be recrystallized by slow evaporation of an acetone solution to give green-black needles of **1** (137 mg, 0.0589 mmol, 66%).

Anal. Found: C, 25.82; H, 0.78; N, 1.95.

Properties

Compound **1** is stable as a solid under inert atmosphere for periods of days only, slowly decomposing to an insoluble material. It is readily soluble in organic solvents such as dichloromethane, methanol, and ethanol; these solutions can be handled in air for a short time. The IR spectrum of **1** contains the following CO absorptions (CH_2Cl_2): 2052(s), 2009(s), 1997(m, sh) cm^{-1}. The ^1H NMR spectrum (d_6-acetone) shows characteristic singlets due to the metal-bound hydrides at -10.74 (Ru_2H) and -13.55 ($Ru_{10}H$) ppm, as well as signals due to the aromatic protons of the cation (δ 8.61–7.09). Further spectroscopic data and the X-ray crystal structure have been reported.[6] Stirring **1** and two equivalents of [PPN]Cl [bis(triphenylphosphine)iminium chloride] in acetone for 20 min provides a convenient and quantitative route into the corresponding dianion $[Ru_{10}(\mu_6 - C)(CO)_{24}]^{2-}$.[7]

C. TETRAPHENYLPHOSPHONIUM TETRACOSACARBONYL-(μ_6-CARBIDO)(μ-HYDRIDO)DECARUTHENATE[6]

$$Ru_3(\mu\text{-}H)(\mu\text{-}NC_5H_4)(CO)_{10} + [PPh_4][BF_4] \rightarrow$$
$$[PPh_4][Ru_{10}(\mu\text{-}H)(\mu_6\text{-}C)(CO)_{24}]$$

Procedure

■ **Caution.** *This reaction should be carried out in a well-ventilated hood, as carbon monoxide—a highly poisonous, colorless, and odorless gas—is evolved.*

A 100-mL, two-necked, round-bottomed flask is fitted with a reflux condenser connected to a nitrogen bubbler. The flask is charged with $Ru_3(\mu\text{-}H)$-$(\mu\text{-}NC_5H_4)(CO)_{10}$ (50 mg, 0.075 mmol), $[PPh_4][BF_4]$ (96 mg, 0.23 mmol), and chlorobenzene (10 mL) and the mixture is deoxygenated. The thoroughly stirred solution is heated at reflux temperature under an atmosphere of dry nitrogen until no starting material was observed by IR (disappearance of the band at 2100 m cm^{-1}; this takes around 40 min). The resulting black solution was filtered in air through filter-aid to remove a small amount of insoluble black material, washing through with chlorobenzene, and the combined filtrate and washings reduced to dryness in vacuo. This residue is taken up in a minimum amount of acetone and subjected to preparative thin-layer chromatography. Elution with 1:1 acetone in petroleum ether (60–80°C boiling point range) affords two main products. The first brown band ($R_f = 0.6$) contains $[PPh_4][Ru_{10}(\mu\text{-H})(\mu_6\text{-C})(CO)_{24}]$ with trace amounts of the hexaruthenium anion $[HRu_6(CO)_{18}]^-$. Crystallization from $CH_2Cl_2/$ butanol affords $[PPh_4][Ru_{10}(\mu\text{-H})(\mu_6\text{-C})(CO)_{24}]$ as a black microcrystalline solid (**2**, 21 mg, 45%). The second brown band ($R_f = 0.2$) can be similarly crystallized to afford $[PPh_4]_2[Ru_{10}(\mu_6\text{-C})(CO)_{24}]$ (1 mg, 0.0004 mmol, 8%). If metathesis of the cation (replacement of the diruthenium cation present in **B** by $[PPh_4]^+$) has been only partially successful, a minor brown band is observed as the first band, $R_f \sim 0.65$ (i.e., very close to the metathesized product). In this case, care should be taken with the TLC to ensure separation. The purity of the desired product can be confirmed by IR (absence of the band at 1999 cm^{-1} corresponding to the diruthenium cation) and ^{1}H NMR (absence of the diruthenium cation hydride signal at -10.74 ppm).

Anal. Calcd. for $C_{49}H_{21}O_{24}Ru_{10}$: C, 28.92; H, 1.04. Found: C, 28.23; H, 0.99.

Properties

Compound **2** has similar stability and solubility to **1**. The IR spectrum of **2** contains the following CO absorptions (CH_2Cl_2): 2052(s), 2008(s) cm^{-1}. The ^{1}H NMR spectrum (d_6-acetone) shows the singlet due to the deca-ruthenium-bound hydride at -13.55 ppm, as well as signals due to the cation (δ 8.05–7.74).

D. TETRACARBONYL (μ-HYDRIDO)BIS(PYRIDINE)BIS-(μ-PYRIDYL)DIRUTHENIUM TRICOSACARBONYL-(μ_6-CARBIDO)(μ-HYDRIDO)(TRIPHENYLPHOSPHINE)-DECARUTHENATE

$[Ru_2(\mu\text{-H})(\mu\text{-NC}_5H_4)_2(CO)_4(NC_5H_5)_2][Ru_{10}(\mu\text{-H})(\mu_6\text{-C})(CO)_{24}] + PPh_3 \rightarrow$

$[Ru_2(\mu\text{-H})(\mu\text{-NC}_5H_4)_2(CO)_4(NC_5H_5)_2][Ru_{10}(\mu\text{-H})(\mu_6\text{-C})(CO)_{23}(PPh_3)] + CO$

Procedure

■ **Caution.** *This reaction should be carried out in a well-ventilated hood, as carbon monoxide—a highly poisonous, colorless, and odorless gas—is evolved.*

A 100-mL, two-necked, round-bottomed flask is fitted with a reflux condenser connected to a nitrogen bubbler. The flask is charged with $[Ru_2(\mu\text{-}H)(\mu\text{-}NC_5H_4)_2(CO)_4(NC_5H_5)_2][Ru_{10}(\mu\text{-}H)(\mu_6\text{-}C)(CO)_{24}]$ (63 mg, 0.027 mmol) and acetone (15 mL) and the mixture is deoxygenated. Triphenylphosphine (7.0 mg, 0.027 mmol) is then added, under a gentle stream of nitrogen to the stirred solution, resulting in an immediate change in the IR spectrum. The mixture is taken to dryness in vacuo and crystallized from acetone/*n*-butanol at $-20°C$ to give black microcrystalline $[Ru_2\text{-}(\mu\text{-}H)(\mu\text{-}NC_5H_4)_2(CO)_4(NC_5H_5)_2][Ru_{10}(\mu\text{-}H)(\mu_6\text{-}C)(CO)_{23}(PPh_3)]$ (**3**, 45 mg, 0.016 mmol, 60%). The ^1H NMR spectrum at this stage indicates the presence of residual *n*-butanol in an otherwise pure compound.

Anal. Calcd. for $C_{66}H_{35}N_4O_{27}PRu_{12} \cdot C_4H_9OH$: C, 31.92; H, 1.72; N, 2.13. Found: C, 31.86; H, 1.68; N, 1.91.

An analytically pure sample may be obtained by careful recrystallization of the crude material by slow evaporation of a dichloromethane solution at room temperature.

Anal. Calcd. for $C_{66}H_{35}N_4O_{27}PRu_{12}$: C, 30.97; H, 1.38; N, 2.19. Found: C, 31.10; H, 1.32; N, 2.18.

Properties

Compound **3** is a relatively air-stable, green-black, crystalline solid. It is soluble in organic solvents such as dichloromethane, methanol, ethanol, and acetone. The IR spectrum of **3** contains the following CO absorptions (CH_2Cl_2): 2077(w), 2047(s), 2016(m), and 2000(m). The ^1H NMR spectrum $(CDCl_3)$ contains a doublet corresponding to the decaruthenium-bound hydride at -11.68 ppm [$J(HP)$ 7 Hz] and a singlet due to the diruthenium-bound hydride at -10.93 ppm. The ^{31}P NMR spectrum (d_6-acetone, 121 MHz) contains a characteristic signal at 44.4 ppm due to the decaruthenium-bound triphenylphosphine moiety. Other spectroscopic data and the X-ray crystal structure have been reported.[3]

E. TETRAPHENYLPHOSPHONIUMTRICOSACARBONYL-(μ_6-CARBIDO)(μ-HYDRIDO)(TRIPHENYLPHOSPHINE)-DECARUTHENATE)

$[PPh_4][Ru_{10}(\mu\text{-}H)(\mu_6\text{-}C)(CO)_{24}] + PPh_3 \rightarrow$

$$[PPh_4[Ru_{10}(\mu\text{-}H)(\mu_6\text{-}C)(CO)_{23}(PPh_3)] + CO$$

Procedure

- **Caution.** *This reaction should be carried out in a well-ventilated hood, as carbon monoxide—a highly poisonous, colorless, and odorless gas—is evolved.*

A 100-mL, two-necked, round-bottomed flask is fitted with a reflux condenser connected to a nitrogen bubbler. The flask is charged with $[PPh_4]$ $[Ru_{10}(\mu\text{-H})(\mu_6\text{-C})(CO)_{24}]$ (10 mg, 0.0049 mmol) and acetone (10 mL) and the mixture is deoxygenated. Triphenylphosphine (1.3 mg, 0.0049 mmol) is then added, under a gentle stream of nitrogen, to the stirred solution, resulting in an immediate change in the IR spectrum. The mixture is taken to dryness, in vacuo and the residue taken up in a minimum amount of acetone. Preparative thin-layer chromatography, using 1:1 acetone in petroleum ether (60–80°C boiling point range), affords two main green bands. Crystallization of the contents of these bands from $CH_2Cl_2/^iPrOH$ at $-20°C$ affords green-black microcrystalline $[PPh_4][Ru_{10}(\mu\text{-H})(\mu_6\text{-C})(CO)_{23}(PPh_3)]$ (**4,** 9.0 mg, 0.0043 mmol, 82%) and $[PPh_4][Ru_{10}(\mu\text{-H})(\mu_6\text{-C})(CO)_{22}(PPh_3)_2]$ (1 mg, 0.0004 mmol, 8%).

Anal. Calcd. for $C_{66}H_{36}O_{23}P_2Ru_{10}$: C, 34.93; H, 1.60. Found: C, 34.20; H, 1.65.

Properties

Compound **4** has stability and solubility properties, and ^1H and ^{31}P NMR spectra similar to **3**. The IR spectrum of **4** contains the following CO absorptions (CH_2Cl_2): 2075(w), 2046(s), 2015(s), and 1999(m). Spectroscopic data have been reported fully.[7]

References

1. M. P. Cifuentes and M. G. Humphrey in *Comprehensive Organometallic Chemistry II*, Vol. 7, E. W. Abel, F. G. A. Stone, and G. Wilkinson, Eds., Elsevier, Amsterdam, 1995, pp. 907–960.
2. P. J. Bailey, B. F. G. Johnson, J. Lewis, M. McPartlin, and H. R. Powell, *J. Organomet. Chem.*, **377**, C17 (1989).
3. M. P. Cifuentes, M. G. Humphrey, B. W. Skelton, and A. H. White, *Organometallics*, **14**, 1536 (1995).
4. M. I. Bruce, M. G. Humphrey, M. R. Snow, E. R. T. Tiekink, and R. C. Wallis, *J. Organomet. Chem.*, **314**, 311 (1986).
5. M. I. Bruce, C. M. Jensen, and N. L. Jones, *Inorg. Synth.*, **26**, 259 (1989).
6. M. P. Cifuentes, M. G. Humphrey, B. W. Skelton, and A. H. White, *Organometallics*, **12**, 4272 (1993).
7. M. P. Cifuentes, M. G. Humphrey, B. W. Skelton, and A. H. White, *J. Organomet. Chem.*, **507**, 163 (1996).

Chapter Five

MAIN GROUP AND TRANSITION METAL HYDRIDES

47. DICHLORODIHYDRO(N,N,N',N',-TETRAETHYL-1,2-ETHANEDIAMINE-N,N')SILICON

Submitted by STEVEN D. KLOOS* and PHILIP BOUDJOUK*
Checked by LI GUO[†] and CLAIRE A. TESSIER[†]

Dichlorosilane is a useful reagent in silicon chemistry. It may be treated with Grignard reagents,[1] alkyl lithium compounds,[2] or alkenes and alkynes in the presence of a catalyst to form organosilanes.[3] However, the volatility (bp = 8.6°C) and flammability of SiH_2Cl_2, as well as its high reactivity toward protic reagents leading to the release of HCl, are significant barriers to broader use of the compound.

As part of a study to elucidate the mechanisms of reactions of complexed halosilanes, we discovered that N,N,N',N'-tetraethylethylenediamine (teeda) reacts with trichlorosilane in an unexpected fashion to give [$SiH_2Cl_2 \cdot$ teeda].

* Center for Main Group Chemistry, Department of Chemistry, North Dakota State University, Fargo, ND 58105-5516.
† Department of Chemistry, University of Akron, Akron, OH 44325-3601.

TABLE I. Reaction of HSiCl₃ with teeda in CH₂Cl₂/Pentane

HSiCl₃		teeda		HSiCl₃ : teeda	CH₂Cl₂	Pentane	Isolated Yields
mL	mmol	mL	mmol	Molar Ratio	(mL)	(mL)	(%)
2.37	23.4	5.0	23.4	1:1	15	16	17
0.805	8.00	1.7	8.0	1:1	5	10	26
4.73	46.9	5.0	23.4	2:1	15	25	50
28.4	281	20.0	93.8	3:1	50	100	86
37.9	375	20.0	93.8	4:1	50	100	98
17.0	169	9.0	42.2	4:1	20	20	90
9.50	93.8	3.5	16.4	5.7:1	10	30	94

This complex functions as a safer and more convenient source of H_2SiCl_2. General features of the synthesis are short reaction times, high yields, simple apparatus, and a routine workup that produces very pure product. The major advantage, of course, is the elimination of the need to handle dichlorosilane. A synthesis of SiH_2Ph_2 from $[SiH_2Cl_2 \cdot teeda]$ is included to show the utility of this new reagent.

Table I summarizes a variety of reaction conditions we explored in an effort to optimize conditions. We found that high trichlorosilane: teeda ratios (> 3:1) give high yields and that methylene chloride is greatly preferred as a solvent over pentane. The latter gives the product as a yellow, sticky solid. The volumes of pentane* in the table refer to the amounts used to precipiate $[SiH_2Cl_2 \cdot teeda]$.

Procedure

■ **Caution.** *HSiCl₃ and SiCl₄ react rapidly with atmospheric water to release HCl, which is corrosive. All syringe transfers should be carried out in a well-ventilated hood.*

Glass syringes (preflushed with nitrogen) are used for all transfers. Trichlorosilane[†] is freshly distilled and degassed by freeze-pump-thaw techniques before use. Teeda[†] is distilled from CaH_2 after stirring at room temperature

---

* The checkers found hexane could be a replacement for pentane.
[†] Purchased from Aldrich Chemical Co., Milwaukee, WI 53233.

over CaH_2 for at least 8 h. CH_2Cl_2 is freshly distilled from P_2O_5. Pentane is washed with H_2SO_4 before distillation from CaH_2. Tetrahydrofuran (THF) and diethyl ether are distilled from sodium benzophenone immediately prior to use.

A dry, nitrogen-filled 250-mL, one-necked, round-bottomed flask is equipped with a septum that is securely fastened with a wire band and sealed with Parafilm. In the flask is placed 20 mL CH_2Cl_2 and 17.0 mL $HSiCl_3$ (169 mmol), and a syringe needle in the septum or in the nitrogen line with a safety oil bubbler is used to release positive pressure. Teeda (9.0 mL, 42.2 mmol) is added to this mixture, resulting in a temperature rise. Following cooling (ca. 10 min after the teeda addition), 20 mL pentane is added, causing the precipitation of teeda $\cdot H_2SiCl_2$. This mixture is stored at $-20°C$ for 16 h,* and a liquid, a mixture of dichloromethane, pentane, trichlorosilane, and silicon tetrachloride, separates from the white solid and is removed by syringe. The flask is allowed to warm to room temperature, at which point more liquid is released from the solid and removed by syringe. The septum is replaced with a gas inlet. Residual volatiles are removed by vacuum, giving 10.40 g (90.2%) of product, a white solid; mp 101–103°C.

Anal. Calcd for $C_{10}H_{26}Cl_2N_2Si$: C, 43.94; H, 9.59; N, 10.25. Found: C, 44.05; H, 9.66; N, 10.34.

Properties

The compound $[SiH_2Cl_2 \cdot teeda]$ is a white solid that is stable under N_2 at room temperature. It is soluble in CH_2Cl_2 and 1,2-dichloroethane, slightly soluble in CH_3CN and THF, and virtually insoluble in pentane or hexane. Solutions of $[SiH_2Cl_2 \cdot teeda]$ stored at room temperature begin to yellow within 1 h, but solutions stored at temperatures below $-20°C$ are stable for several weeks. Upon exposure to air, $[SiH_2Cl_2 \cdot teeda]$ is hydrolyzed to a mixture of polymeric siloxane and $[teeda \cdot xH][Cl]_x$ (x = 1, 2). ^{1}H and ^{13}C NMR spectra may be recorded at room temperature, but [29]Si are best done below $-20°C$. Characteristic NMR data are: ^{1}H NMR (20°C, CD_3CN): δ 4.99 (s, 2H, SiH), δ 2.91 (br q, 8H, J = 7.2 Hz, CH_2CH_3), δ 2.78 (br s, 4H, CH_2CH_2), δ 1.13 (t, 12H, J = 7.2 Hz, CH_3); (20°C, $CDCl_3$) δ 5.14 (s, 2H, SiH), δ 2.86 (q, 8H, J = 7.3 Hz, CH_2CH_3), δ 2.68 (br s, 4H, CH_2CH_2), δ 1.11 (t, 12H, J = 7.3 Hz, CH_3); ^{13}C NMR (20°C, CD_3CN): 48.5 (br, CH_2CH_3), 47.57 (CH_2CH_2), 10.0 (br, CH_2CH_3); ^{29}Si NMR ($-30°C$,

---

* The checkers found that cooling at $-78°C$ for 7 h was equally effective.

CD$_3$CN) δ $-$ 120.2 (d, J_{SiH} = 404 Hz). Within 1 h of preparing NMR samples at room temperature, ^{1}H NMR show new peaks at 11.5 and 6.1 ppm. The IR spectrum, which must be recorded in an inert atmosphere, has a strong, characteristic band at 2178 cm^{-1} (v(Si–H), KBr). As [SiH$_2$Cl$_2$·teeda] is hydrolyzed by atmospheric water, this band moves to higher wavenumbers (up to 2190 cm^{-1}) and a new broad peak at ca. 2300–2800 cm^{-1} appears. The structure of [SiH$_2$Cl$_2$·teeda] was confirmed by X-ray crystallography.[4]

Synthesis of SiH$_2$Ph$_2$. In a drybox, [SiH$_2$Cl$_2$·teeda] (3.00 g, 11.0 mmol) is placed in a 250-mL, two-necked, round-bottomed flask. The flask is then taken from the dry box and fitted with a stir bar, condenser, and septum. CH$_2$Cl$_2$ (10 mL) and THF (30 mL) are added, followed by addition of PhMgCl (2.0 M, 13.2 mL, 26.3 mmol) over 5 min. These operations were carried out under an atmosphere of N$_2$. The solution is stirred overnight, and the solvents are removed on a rotary evaporator. Diethylether (30 mL) is added to dissolve the product followed by HCl (80 mL, 0.2 M) to destroy unreacted starting materials and form water soluble teeda·2HCl. The combined liquids are extracted with ether, and the extract is distilled to yield 1.51 g (75%) of Ph$_2$SiH$_2$, bp = 128°C/12 torr. NMR analysis shows: ^{1}H NMR (CDCl$_3$): δ 7.75 (m, 2H, phenyl), δ 7.52 (m, 3H, phenyl), δ 5.10 (s, 2H, SiH); ^{29}Si NMR (CDCl$_3$): δ $-$ 33.2.

Acknowledgments

Financial support from Dow Corning, the Air Force Office of Scientific Research through Grant No. 91-0197, and the National Science Foundation through Grants EHR-918770 and OSR 9452892 is gratefully acknowledged.

References

1. Y. van den Winkel, B. L. M. van Baar, F. Bickelhaupt, W. Kulik, C. Sierakowski, and G. Maier, *Chem. Ber.*, **124**, 185 (1991).
2. W. Watanabe, T. Ohkawa, T. Muraoka, and Y. Nagai, *Chem. Lett.*, **9**, 1321 (1981).
3. R. A. Benkeser, E. C. Mozdzen, W. C. Muench, R. T. Roche, and M. P. Siklosi, *J. Org. Chem.*, **44**, 1370 (1979).
4. S. D. Kloos, Ph.D. Thesis, North Dakota State University, 1995.

48. TRICARBONYL(HYDRIDO)[1,2-BIS(DIPHENYL-PHOSPHINO)ETHANE]MANGANESE AS PRECURSOR TO LABILE SITE DERIVATIVES

Submitted by MILTON ORCHIN,* SANDOSH K. MANDAL,*
and JULIAN FELDMAN*
Checked by WAY-ZEN LEE[†] and DONALD J. DARENSBOURG[†]

Unlike many, if not most, transition metal hydrides which are relatively unstable and sensitive to air and moisture, the manganese hydride described here is completely stable in air. It serves as the starting material for the synthesis of a variety of derivatives in which the hydrogen atom is replaced by other atoms (or groups of atoms) directly bonded to manganese. Earlier methods of preparation involved the displacement of CO by bis(diphenylphosphino)ethane from the difficult to prepare and hard to handle $HMn(CO)_5$[1] as well as from the decarboxylation of a postulated intermediate $(diphos)Mn(CO)_3COOH$.[2] Our preparation is a modification of a previously described procedure[3] in which 1-propanol was used as reagent and as solvent. In addition to a much shorter reaction time, this procedure results in a pure, almost colorless product uncontaminated by an unknown dimer which frequently accompanies the hydride.[4]

A. TRICARBONYL(HYDRIDO)[1,2-BIS(DIPHENYLPHOSPHINO)-ETHANE]MANGANESE

$$Mn_2(CO)_{10} \;+\; 2\underset{\text{dppe}}{(C_6H_5)_2PCH_2CH_2P(C_6H_5)_2} \;+\; CH_3(CH_2)_3CH_2OH \longrightarrow$$

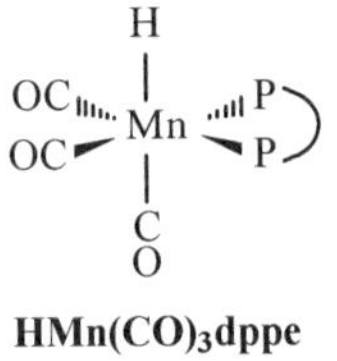

$$+ \; 4CO \; + \; CH_3(CH_2)_3CHO$$
(and aldol products)

HMn(CO)₃dppe

Procedure

■ **Caution.** *Because carbon monoxide is evolved during the reaction and because part of the workup involves the use of benzene, a suspected carcinogen, the reaction should be conducted in a well-ventilated hood.*

* Department of Chemistry, University of Cincinnati, OH 45221-0172.
† Department of Chemistry, Texas A&M University, College Station, TX 77843.

To a 250-mL, one-necked, round-bottomed (rb) flask containing a stirring bar are added dimanganese decacarbonyl* (1.949 g, 5.000 mmol), 1,2-bis-(diphenylphosphino)ethane (dppe)[†] (recrystallized once from benzene) (4.003 g, 10.050 mmol), and 75 mL of 1-pentanol.[‡]

The flask is attached to a small condenser and placed in an oil bath resting on a hotplate-stirrer. A stirring bar is placed in the oil bath and a thermometer is suspended in it. The flask is adjusted in the bath so that both the stirring bar in the bath and in the flask are able to rotate. Cooling water is passed through the jacket of the condenser and the oil-bath temperature is raised rapidly to about 150°C so as to reflux the pentanol (bp 137°C). As the bath temperature reaches about 120°C, the suspended solids dissolve to form a yellow solution which then first turns a deep red as carbon monoxide is being evolved, and then yellow as the heating continues (about 10–20 min total). Refluxing with stirring is continued for a total of 1 h. After cooling, the stirring bar is removed and washed with a little benzene while being held over the flask. The flask is transferred to a rotovap and the solvent is removed. The slightly pink, solid residue is dissolved in 100 mL of benzene and the solution is stirred for 15 min with 1 g of powdered silica gel (Merck, Silica gel-60). The pale yellow solution is filtered into a 250-mL rb flask and the flask is placed on a rotovap. After removal of the solvent, the resinous residue is dissolved in 120 mL of hot 1-propanol. On *slow* cooling to room temperature, nearly colorless needles gradually precipitate. After standing overnight at room temperature, the flask is placed in the refrigerator for about 2 h, after which the crystals are collected by filtration. The crystals are washed twice with about 10 mL of hexane and dried in a vacuum oven at 60°C. The yield of this first crop is 3.737 g (69.5%), mp 142–145°C. A small additional quantity of the hydride is present in the mother liquor, but it is difficult to separate this from impurities, mostly free dppe.

Properties

The IR spectrum in benzene shows three strong carbonyl bands at 1997.2, 1923.5, and 1903.5 cm^{-1} associated with the three *facial* coordinated carbon monoxides. In CH_2Cl_2 solution, the two lower-frequency bands appear as one strong broad band at 1911 cm^{-1}, which is about the average of the two

---

* Prepared from methylcyclopentadienylmanganesetricarbonyl (a commerical antiknock compound with the acronym MMT) according to a published procedure.[5] Also available from Strem Chemical, Newburyport, MA 01950.

† Pressure Chemicals, 3419 Smallman, St. Pittsburgh, PA 15201.

‡ Fisher Scientific Co.

lower-frequency bands observed in benzene. The mass spectrum (CI) gives a parent peak, m/z, of 538; calculated for $C_{29}H_{25}MnO_3P_2$, 538.

No precautions to exclude air are necessary as the product is completely stable in air. The mass spectrum of the product showed that no complexed phosphine oxides are present. The tricarbonyl hydride produced by this simple procedure is the starting material for the preparation of the ionic tetracarbonyl salt described below in the next procedure as well as for preparation of the covalent tosylate obtained by treatment of the hydride with *p*-toluenesulfonic acid.

B. TETRACARBONYL[1,2-BIS(DIPHENYLPHOSPHINO)-ETHANE]MANGANESE(I) TETRAFLUOROBORATE(1−)

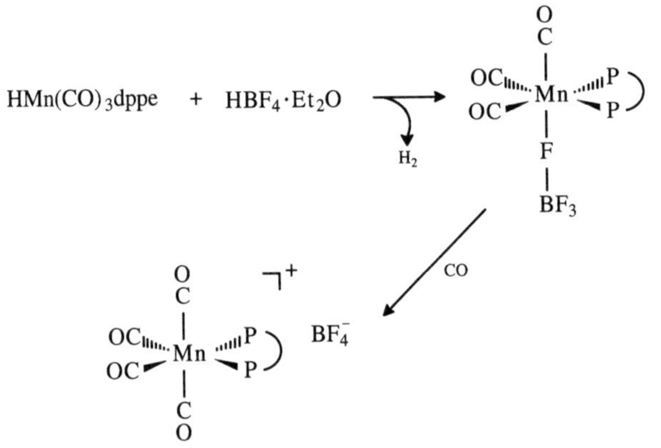

$$HMn(CO)_3dppe \quad + \quad HBF_4 \cdot Et_2O$$

This tetracarbonylmanganese salt provides a convenient entry into compounds possessing a carbon atom bonded directly to the manganese atom. Thus, for example, nucleophiles such as alkoxides attack a coordinated carbon monoxide to give a alkoxycarbonyl (ester) complex. In a method reported earlier, the starting material, tricarbonyl(bromol)[1,2-bis(diphenylphosphino)ethane]manganese, is the less readily available.[6] The conversion to the tetrafluoroborate salt from this bromide involves the reaction with silver tetrafluoroborate. The procedure described herein, first reported elsewhere[7] involves the hydride as the starting material as well as the use of HBF_4 in place of the expensive silver salt.

Procedure

■ **Caution.** *This preparation should be carried out in a well-ventilated hood because carbon monoxide is used in the reaction and dihydrogen is evolved.*

Care must be exercised in handling tetrafluoroboric acid because of its corrosive character.

Tricarbonyl(hydrido)[1,2-bis(diphenylphosphino)ethane]manganese (1.076 g, 2 mmol) is dissolved in 100 mL of methylene chloride in a 125-mL Erlenmeyer flask containing a stirring bar. The solution is stirred vigorously by a magnetic stirrer while carbon monoxide is bubbled through the solution. While continuing to bubble carbon monoxide, there is added dropwise to this solution over a period of about 0.5 min, 0.5 mL (6 mmol) of 85% tetrafluoroboric acid-diethyl ether complex.* The solution turns deep yellow, due to the intermediate tetrafluoroborato complex while dihydrogen is being evolved. The color fades completely as the carbon monoxide is absorbed. According to the infrared monitor, after about 15 min the hydride is completely consumed. The solution is concentrated by rotoevaporation to a volume of about 5 mL and the pale yellow product is precipitated by addition of 50 mL of diethyl ether; yield 98%. The salt is recrystallized from methyl ethyl ketone–benzene (1:2) to give white crystals.

Properties

IR (cm^{-1}, CH$_2$Cl$_2$) ν_{CO} 2093 (s), 2026 (s, sh), 2014 (vs), 1997 (s, sh); in CH$_3$OH, 2094 (s), 2029 (s), 2012 (vs), 2002 (sh); in CH$_3$CN, 2094 (s), 2069 (w), 2028 (s), 2013 (vs). Calcd. for C$_{30}$H$_{24}$BF$_4$MnO$_4$P$_2$: C, 55.2; H, 3.7. Found: C, 54.6; H, 3.8 ^{1}H NMR (δ, CDCl$_3$) 7.6 (m, 20H); 3.36 (m, 4H). The salt is insoluble in hydrocarbon solvents and in ethers.

C. TRICARBONYL(p-TOLUENESULFONATO)-[1,2-BIS(DIPHENYLPHOSPHINO)ETHANE] MANGANESE

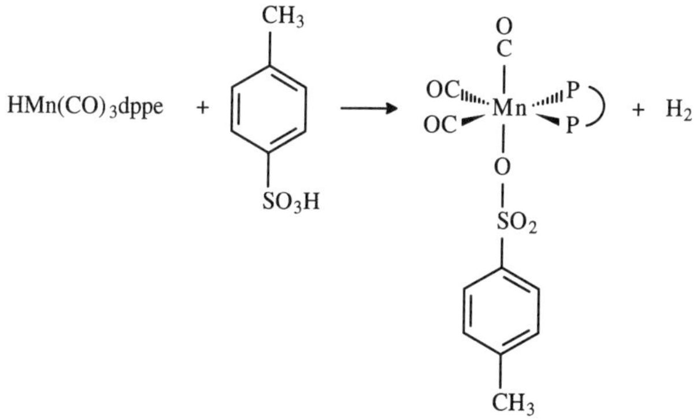

The synthesis of an organomanganese compound with a tosylate group directly bonded to the manganese makes available a useful intermediate because the tosylate, a good leaving group, can readily be replaced by other nucleophiles (alkoxides, azides, etc.), to give a variety of other compounds in which functional groups are directly bonded to the manganese atom. This procedure is essentially the same as that published earlier from this laboratory[8] except for the purification of the crude product. Its preparation from the corresponding hydride in one step makes this important intermediate readily available.

Procedure

To a 125-mL Erlenmeyer flask containing a stirring bar is added 1.078 g (2.000 mmol) of tricarbonyl(hydrido)[1,2-bis(diphenylphosphino)ethane] manganese in 60 mL of methylene chloride. To the stirred solution, 2.11 g (11.1 mmol) of solid *p*-toluenesulfonic acid hydrate is added in small increments. The theoretical quantity of dihydrogen is evolved in less than 10 min. The suspension is washed twice with equal volumes of water, then with dilute sodium bicarbonate solution, then with water again. The methylene chloride is evaporated to dryness and the product is crystallized from benzene: methylene chloride (2 : 1) to give yellow crystals in 77% yield.

Properties

Melting point with decomposition $> 240°C$, IR (cm^{-1} in CH_2Cl_2) v_{CO} 1920 s, 1965 s, 2033 vs. ^{31}P NMR (CD_2Cl_2): δ 71.5 (s). Calcd. for $C_{36}H_{31}MnO_6P_2S$: C, 61.0; H, 4.4. Found: C, 60.9; H, 4.5.

Acknowledgment

We thank the Ethyl Corporation for a generous quantity of methylcyclopentadienylmanganese tricarbonyl.

References

1. B. L. Booth and R. N. Haszeldine, *J. Chem. Soc., Inorg., Phys., Theoret.*, **157** (1996).
2. D. J. Darensbourg and J. A. Froelich, *Inorg. Chem.*, **17**(11), 3300 (1978).
3. B. D. Dombek, *Ann. N. Y. Acad. Sci.*, **415**, 176 (1983).
4. S. K. Mandal, J. Feldman, and M. Orchin, *J. Coord. Chem.*, **33**, 219 (1994).
5. R. B. King, J. C. Stokes, and T. F. Kosenowski, *J Organomet. Chem.*, **11**, 641 (1968).
6. S. K. Mandal, D. M. Ho, and M. Orchin, *Polyhedron*, **11**(16), 2055 (1992).
7. S. K. Mandal, J. Feldman, and M. Orchin, *J. Coord. Chem.*, **33**, 219 (1994).
8. S. K. Mandal, D. M. Ho, and M. Orchin, *Inorg. Chem.*, **30**, 2244 (1991).

49. PENTAHYDRIDOBIS(TRICYCLOHEXYLPHOSPHINE)-IRIDIUM(V) AND TRIHYDRIDOTRIS(TRIPHENYLPHOS-PHINE)IRIDIUM(III)

Submitted by SUSANNE BRINKMANN,* ROBERT H. MORRIS,* RAVINDRANATH RAMACHANDRAN,* and SUNG-HAN PARK* Checked by ROBERT H. CRABTREE,† BEN P. PATEL,† and BRADFORD J. PISTORIO†

Polyhydrido phosphine complexes of iridium are versatile starting materials used widely in the preparation of dihydride or dihydrogen complexes.[1] Various preparative methods for these iridium polyhydrides have been established[2-8] with commercially available appropriate iridium precursors. Pentahydridebis(tricyclohexylphosphine)iridium, IrH_5L_2 (**1**) (L = PCy_3) has been obtained in a multistep synthesis involving intermediates, $[Ir_2Cl_2(COD)_2]$,[9] $[Ir(C_5H_5N)_2(COD)](PF_6)$,[10] $[Ir(C_5H_5N)(PCy_3)(COD)]$-$(PF_6)$,[11] and $[Ir(PCy_3)_2(\eta^2\text{-}H_2)_2(H)_2](PF_6)$.[8] This synthesis of **1** is tedious, involving several steps with relatively low yield and also has the problem of the separation of $[NEt_3H](PF_6)$ in the final step. An alternative preparation of iridium pentahydrides with P^iPr_3 ligands,[12] or PPh_2Me ligands,[6] involves only two steps starting from $IrCl_3 \cdot 3H_2O$ (Eqs. 1–4).

$$IrCl_3 \cdot 3H_2O + 3P^iPr_3 \xrightarrow{\text{HCl/EtOH}} [HP^iPr_3][IrCl_4(P^iPr_3)_2] + H_2O \quad (1)$$

$$[IrCl_4(P^iPr_3)_2][HP^iPr_3] \xrightarrow[\text{THF}]{\text{excess LiAlH}_4} IrH_5(P^iPr_3)_2 \quad (2)$$

$$IrCl_3 \cdot 3H_2O + 3PPh_2Me \xrightarrow{\text{HCl/EtOH}} IrCl_3(PPh_2Me)_3 + H_2O \quad (3)$$

$$IrCl_3(PPh_2Me)_3 \xrightarrow[\text{THF}]{\text{excess LiAlH}_4} IrH_5(PPh_2Me)_2 \quad (4)$$

In an attempt to synthesize iridium pentahydrides by similar routes with other phosphine ligands (PPh_3 or PCy_3), we obtained different products. Under the conditions of reaction (1) a rose-colored complex identified as $[IrHCl_2(PCy_3)_2]$[13] (**2**) was obtained for PCy_3. For PPh_3, a pale yellow powder was obtained as an impure product consisting of $IrHCl_2(PPh_3)_3$ (**3**)

---

* Department of Chemistry, University of Toronto, 80 St. George Street, Toronto, Ontario, M5S 3H6, Canada.
† Department of Chemistry, Yale University, New Haven, CT 06520.

as the major species.[3,5] Complexes (2) and (3), upon reaction with NaOR (R = Me, Et) in THF under hydrogen, give the title polyhydrides, $IrH_5(PCy_3)_2$ (1) and $IrH_3(PPh_3)_3$ (4), in good yields (75–85%). This new method for making these two types of polyhydrides has several advantages: the use of sodium alkoxides instead of $LiAlH_4$, the good yield, and the production of a single isomer for (4) under mild conditions. Other methods give isomeric mixtures of 4 (i.e., *mer*-[Ir] and *fac*-[Ir]).[3,4,7] This easy access to (1) or (4) should facilitate further study of these complexes, which contain very reactive metal bound hydrides.

General Techniques

All preparations are carried out under an atmosphere of dry argon using conventional Schlenk techniques unless mentioned otherwise. Tetrahydrofuran (THF), diethyl ether (Et_2O), and *n*-hexane are dried over and distilled from sodium benzophenone ketyl. Ethanol (EtOH) is distilled from magnesium ethoxide. Distilled water should be degassed by several freeze-pump-thaw cycles under argon before use. Triphenylphosphine, tricyclohexylphosphine, hydrochloric acid solution, and sodium methoxide are purchased from Aldrich Chemical Company, Inc. Iridium trichloride hydrate is obtained from Johnson-Matthey Co. Sodium ethoxide is generated in the reaction of sodium metal with water-free ethanol under argon and dried to white powder before use.

NMR spectra are obtained on a Unity XL-400, operating at 400.00 MHz for ^1H, 161.98 MHz for ^{31}P, or on a Gemini-300 operating at 300.00 MHz for ^1H, 121.45 MHz for ^{31}P. ^{31}P NMR chemical shifts are measured relative to H_3PO_4 as external reference. ^1H NMR chemical shifts are measured relative to deuterated solvent peaks and referenced indirectly to tetramethylsilane. IR spectra are recorded on a Nicolet 550 Magna-IR spectrometer. Fast-atom bombardment mass spectrometry (FAB MS) is carried out with a VG 70-250S instrument using a 3-nitrobenzylalcohol (NBA) matrix. All FAB MS samples are dissolved in acetone and placed in the matrix under a blanket of nitrogen.

■ **Caution.** *Hydrogen gas can form explosive mixtures with air.*

A. HYDRIDODICHLOROBIS(TRICYCLOHEXYLPHOSPHINE)-IRIDIUM(III)

$$IrCl_3 \cdot 3H_2O + 2PCy_3 \xrightarrow{HCl/EtOH} [IrHCl_2(PCy_3)_2]$$

A two-necked, 100-mL, round-bottomed flask containing $IrCl_3 \cdot 3H_2O$ (1.00 g, 2.84 mmoles) is equipped with a magnetic stir bar and a reflux condenser and then flushed with argon. By use of syringes, dioxygen-free ethanol (20 mL) and then concentrated HCl (1.8 mL) are added with stirring to yield a dark brown solution. The solution is stirred and refluxed for 6 h, during which time it turns greenish brown. The solution is then cooled to room temperature and PCy_3 (1.6 g, 5.70 mmoles) is added quickly, causing the solution to turn deep green instantly. After 1 h of reflux, the solution turns red and a red solid begins to precipitate out. Refluxing is then continued for a further period of 15 h. After cooling to room temperature, the red solid is filtered in air from a slightly yellow mother liquor, washed with ice-cold ethanol (10 mL) twice, and dried under vacuum for 6 h to yield a pink solid (1.68 g) (72%).

Properties

$[IrHCl_2(PCy_3)_2]$ is a pink solid which is stable under nitrogen and can also be handled in the air for short periods of time. It is moderately soluble in dichloromethane and chloroform, sparingly soluble in acetone, alcohols, or THF. Spectroscopic data for $[IrHCl_2(PCy_3)_2]$: NMR ($CDCl_3$, δ). $^{31}P\{^1H\}$: 21.3 (s). ^1H: -47.9 (t, 1H, $^2J_{PH} = 11.3$ Hz, IrH), 0.8–2.3 [m, 66H, $P(C_6H_{11})_2$]. The chemical shift of the hydride is in agreement with the literature value.[14]

Anal. Calcd. for $C_{36}H_{67}Cl_2IrP_2$: C, 52.4; H, 8.19. Found: C, 52.28; H, 8.35.

B. PENTAHYDRIDOBIS(TRICYCLOHEXYLPHOSPHINE)-IRIDIUM(V)

$$[IrHCl_2(PCy_3)_2] \xrightarrow[\text{THF}]{\text{NaOMe/H}_2} IrH_5(PCy_3)_2$$

$IrHCl_2(PCy_3)_2$ (1.2 g, 1.4 mmoles) and an excess of NaOMe (0.5 g, 9.3 mmoles) are added to a 150-mL, two-necked, round-bottomed flask equipped with a magnetic stir bar and a gas inlet for introducing argon, hydrogen, or vacuum. Dry, deoxygenated THF (50 mL) is added against a flow of argon to produce a suspension. The contents are then purged with hydrogen and stirred under an atmosphere of dihydrogen gas for 5 h, after which time the pink color of the suspension fades to a pink-white. Stirring under hydrogen is continued for another 24 h, during which time the yellow-white color of the suspension slowly turns to grey (the reaction time might have to be adjusted according to the color of the suspension). Solvent is removed by vacuum to yield a grey-white solid. It is then stirred with 100 mL of degassed water for 1 h under argon, and then collected on a Schlenk frit under argon and washed with a further 40 mL water. After drying the solid in vacuum for 4 h, it is then stirred with hexanes (40 mL) for 1 h under argon. After filtration under argon, the white solid is washed with ether (40 mL) until the filtrate is clear. Then the product is dried in vacuum. Yield: 0.93 g (85% based on $[IrHCl_2(PCy_3)_2]$).

Properties

$IrH_5(PCy_3)_2$ is a white solid which is stable in the solid state in air at room temperature, but slowly decomposes in solution. It is insoluble in ether and hexanes, sparingly soluble in dichloromethane, chloroform, and toluene, and moderately soluble in benzene. Spectroscopic data for $IrH_5(PCy_3)_2$: NMR. $^{31}P\{^1H\}(C_6H_6/C_6D_6, \delta)$: 31.6 (s). ^1H NMR (C_6D_6, δ): -10.5 (t, $^2J_{PH} = 12.0$ Hz, IrH), 1.0–2.2 [m, 66H, P$(C_6H_{11})_2$]. IR (Nujol): $\nu(IrH)$ 1945 cm^{-1}.

Anal. Calcd. for $C_{36}H_{71}IrP_2$: C, 56.96; H, 9.43. Found: C, 56.80; H, 9.64.

This complex has been used for the synthesis of unusual complexes containing thiol ligands,[1a] dihydrogen ligands,[1c] or ligands with short proton–hydride distances.[1d]

C. CRUDE HYDRIDODICHLOROTRIS-(TRIPHENYLPHOSPHINE)IRIDIUM(III)

$$IrCl_3 \cdot 3H_2O \xrightarrow[\text{reflux in EtOH}]{\text{(i) HCl (ii) 3PPh}_3} fac\text{- or } mer\text{-}IrHCl_2(PPh_3)_3$$

$IrCl_3 \cdot 3H_2O$ (1.0 g, 2.84 mmoles) is suspended in EtOH (15 mL) in a Schlenk flask containing a stirring bar under argon. To this is slowly added concentrated hydrochloric acid (1.8 mL) while the suspension is stirred. The flask is then fitted with an argon-filled condensor. Heating at reflux for 5 h produces a dark greenish yellow solution. This is cooled to room temperature and then 3 equiv. of triphenylphosphine (1.49 g, 5.68 mmoles) are added. The solution is again refluxed for a further 12 h to produce a pale yellow precipitate. After cooling the solution to 0°C, the product is filtered in air and washed with two portions of cold ethanol and dried in vacuo. The product is a pale yellow powder containing a mixture of a major $[IrHCl_2(PPh_3)_3]$ and minor species. Yield 2.3 g [77% based on $IrHCl_2(PPh_3)_3$]. This is used without further purifications for the preparation of *fac*-$IrH_3(PPh_3)_3$ as indicated below.

Properties

Spectroscopic data for $IrHCl_2(PPh_3)_3$: NMR (CD_2Cl_2, δ). $^{31}P\{^1H\}$: -8.9 (d, $J_{PP} = 13.3$ Hz), -29.1 (t, $J_{PP} = 13.3$ Hz); ^1H: -13.88 (dt, 1H, $^2J_{PH(cis)} = 16.4$ Hz, $^2J_{PH(trans)} = 162$ Hz, Ir-H), 6.8–8.0 {m, 45H, $P(C_6H_5)_3$}. NMR $(CDCl_3, \delta)$ for the minor species. $^{31}P\{^1H\}$: 6.7 (d, J = 13.5 Hz), -1.8 (d, J = 16.4 Hz), -6.8 (t, J = 16.4 Hz); ^1H: -19.16 (quartet, J = 14.1 Hz). MS(FAB): m/z, calc. for $C_{54}H_{46}{}^{35}Cl_2{}^{193}IrP_3$, 1050; observed, 1049 (M^+-H), 1015 (M^+-Cl), 979 (M^+-2Cl). This data is consistent with either the *fac* or *mer* stereochemistry. Further experiments are needed to establish the structure. Crude $IrHCl_2(PPh_3)_3$ is unstable in both the solid state and solution and should be used immediately after its preparation.

D. *fac*-TRIHYDRIDOTRIS(TRIPHENYLPHOSPHINE)-IRIDIUM(III)

$$fac\text{- or } mer\text{-}IrHCl_2(PPh_3)_3 \xrightarrow[\text{THF}]{\text{NaOEt/H}_2} fac\text{-}IrH_3(PPh_3)_3$$

The crude $IrHCl_2(PPh_3)_3$ (1.0 g, ca. 0.95 mmol) and a large excess of dried sodium ethoxide (0.8 g, 12 mmol) are suspended in THF (15 mL) in a 50-mL Schlenk flask containing a stirring bar under dihydrogen gas. After 30 min the stirred solution becomes orange yellow and over a period of 24 h slowly turns pale yellow. All the solvent is then removed under vacuum to dryness. A sufficient amount (ca. 20 mL) of degassed distilled water is added to the residue to dissolve sodium salts. The solution is stirred for 30 min and filtered under argon. The product is washed with water (ca. 5 mL portions) twice under argon and then with cold ethanol (ca. 1 mL) and dried in vacuo. Finally, the pale yellow powder is washed with ether several times (3 × 5 mL) until the filtrate is clear. This is redissolved in CH_2Cl_2 (ca. 35 mL) and filtered through Celite to give a colorless solution. The solvent is evaporated in vacuo to give *fac*-$IrH_3(PPh_3)_3$ as a white powder (0.84 g, 67%).

Properties

fac-$IrH_3(PPh_3)_3$ is stable under argon and can be handled in air for short periods of time. It is light sensitive[15] and slowly decomposes under argon. It is moderately soluble in chlorinated solvents, benzene, and toluene, but insoluble in alcohols, ether, or hexanes. NMR $(CDCl_3, \delta)$ for *fac*-$IrH_3(PPh_3)_3$. $^{31}P\{^1H\}$: 30.05 (s); ^1H: -12.25 (AA'A″ XX'X″, Ir-H), 6.8–7.8 [m, 45H, $P(C_6H_5)_3$]. The hydride resonance of the 500 MHz ^1H NMR spectrum (toluene-d_8, 25°C) has been simulated approximately[15] by use of the parameters $J_{PH(trans)} = +120$ Hz, $J_{PH(cis)} = -18$ Hz, $J_{HH(cis)} = -3.5$ Hz,

$J_{PP(cis)} = -3.2$ Hz. T_1(min)(hydride): 0.144 s ($-60°$C, 300 MHz) MS(FAB): m/z, calc. for $C_{54}H_{48}{}^{193}IrP_3$, 982; observed, 982 (M$^+$), 979 (M$^+$–3H).

Anal. Calcd. for $C_{54}H_{48}IrP_3 \cdot H_2O$: C, 64.85; H, 5.04. Found: C, 64.76; H, 4.88. This complex is a starting material to make complexes with novel hydride–proton interactions.[16] The *mer* isomer has been used to make heterobimetallic clusters.[17]

References

1. (a) P. G. Jessop and R. H. Morris, *Inorg. Chem.*, **32**, 2236 (1993); (b) R. H. Morris and P. G. Jessop, *Coord. Chem. Rev.*, **121**, 155 (1992) and references therein; (c) R. H. Crabtree, M. Lavin, and L. Bonneviot, *J. Am. Chem. Soc.*, **108**, 4032 (1986); (d) W. Xu, A. J. Lough, and R. H. Morris, *Inorg. Chem.*, **35**, 1549 (1996) and references therein.
2. G. G. Hlatky and R. H. Crabtree, *Coord. Chem. Rev.*, **65**, 1 (1985) and references therein.
3. (a) J. J. Levison and S. D. Robinson, *J. Chem. Soc., A*, **1970**, 2947; (b) N. Ahmad, S. D. Robinson, and M. F. Uttley, *J. Chem. Soc., Dalton Trans.*, **1972**, 843; (c) G. L. Geoffroy and R. Pierantozzi, *J. Am. Chem. Soc.*, **98**, 8054 (1976).
4. (a) L. Malatesta, M. Angoletta, A. Araneo, and F. Canziani, *Angew. Chem.*, **73**, 273 (1961); (b) M. Angoletta, *Gazz. Chim. Ital.*, **92**, 811 (1962).
5. (a) L. Vaska, *J. Am. Chem. Soc.*, **83**, 756 (1961); (b) R. G. Hayter, *J. Am. Chem. Soc.*, **83**, 1259 (1961).
6. R. Bau, C. J. Schwerdtfeger, L. Garlaschelli, and T. F. Koetzle, *J. Chem. Soc., Dalton Trans.*, **1993**, 3359.
7. J. Chatt, R. S. Coffey, and B. L. Shaw, *J. Chem. Soc.*, **1965**, 7391.
8. R. H. Crabtree, H. Felkin, and G. E. Morris, *J. Organomet. Chem.*, **141**, 205 (1977).
9. J. L. Herde, J. C. Lambert, and C. V. Senoff, *Inorg. Synthesis*, **15**, 18 (1975).
10. B. Denise and G. Pannetier, *J. Organomet. Chem.*, **63**, 423 (1973).
11. R. Crabtree and G. E. Morris, *J. Organomet. Chem.*, **135**, 395 (1977).
12. M. G. Clerici, S. Di Gioacchino, F. Maspero, E. Perrotti, and A. Zanobi, *J. Organomet. Chem.*, **84**, 379 (1975).
13. B. R. James and M. Preece, *Inorg. Chim. Acta*, **34**, L219 (1979).
14. D. G. Gusev, V. I. Bakhmutov, V. V. Grushin, and M. E. Volpin, *Inorg. Chim. Acta*, **177**, 115 (1990).
15. G. L. Geoffroy and R. Pierantozzi, *J. Am. Chem. Soc.*, **98**, 8054 (1976).
16. S. H. Park, A. J. Lough, and R. H. Morris, *Inorg. Chem.*, **35**, 3001 (1996).
17. A. Albinati, C. Anklin, P. Janser, H. Lehner, D. Matt, P. S. Pregosin, and L. M. Venanzi, *Inorg. Chem.*, **28**, 1105 (1989).

Chapter Six

TITANIUM(III) CHLORIDE*

50. AN ACTIVE FORM OF TITANIUM(III) CHLORIDE

$$Me_3SiSiMe_3 + 2TiCl_4 \rightarrow 2Me_3SiCl + 2TiCl_3$$

Submitted by ANN R. HERMES[†] and GREGORY S. GIROLAMI[†]
Checked by RICHARD A. ANDERSEN[‡]

The reduction of $TiCl_4$ with hexamethyldisilane does not afford titanium(II) chloride as reported by Narula and Sharma.[1] Instead, the method affords an active form of titanium(III) chloride that is useful for the preparation of other titanium(III) species.

Procedure

A 50-mL, three-necked, round-bottomed flask is equipped with a dropping funnel, reflux condenser, and a magnetic stir bar. The glassware should be oven dried, assembled, and purged with dry nitrogen gas. The flask is charged with 1.6 mL (2.8 g, 15 mmol) of titanium tetrachloride and the dropping funnel is charged with 3.0 mL (2.15 g, 15 mmol) of hexamethyldisilane. The flask is cooled in an ice bath and the hexamethyldisilane is added dropwise over 5 min. The stirred solution turns orange. The ice bath is removed and the reaction mixture is heated to 115°C for 4 h. The dark brown solid is

* A correction to *Inorganic Syntheses*, **24**, 181 (1986).
† School of Chemical Sciences, The University of Illinois at Urbana-Champaign, Urbana, IL 61801.
‡ Department of Chemistry, The University of California at Berkeley, Berkeley, CA 94720.

collected by filtration, washed twice with 10 mL of dichloromethane, and dried under vacuum. Yield: 2.0 g (86%).[§]

Anal. Calcd. for $TiCl_3$: Ti, 31.0; Cl, 68.9. Found: Ti, 28.3; Cl, 62.5; C, 2.1.

Properties

The microanalytical data suggest that the product consists of $TiCl_3$ contaminated with about 9 wt % of organic impurities (probably containing Me_3Si- or Me_3SiO- groups). The β-form of $TiCl_3$ is brown,[2,3] and it is likely that this is the material formed in the synthesis. The product is air sensitive and insoluble in hydrocarbons and chlorocarbons. The identification of the product as titanium(III) chloride rather than titanium(II) chloride is supported by the 1:2.98 Ti:Cl ratio and by its reactivity. The product dissolves readily in hot tetrahydrofuran to generate blue solutions from which $TiCl_3(thf)_3$ can be crystallized. By comparison, the purple form of $TiCl_3$ dissolves slowly (22 h) in refluxing tetrahydrofuran.[4] In addition, the checker finds that the brown form dissolves in hot acetonitrile to generate blue solutions, which after concentration and cooling afford $TiCl_3(MeCN)_3$.[5] [It is notable that $TiCl_2(MeCN)_2$ is a black insoluble solid.[6]] The checker also finds that addition of 3 equivalents of $LiN(SiMe_3)_2$ to the brown form of $TiCl_3$ gives a blue solution; taking the solution to dryness and crystallizing the resulting solid from pentane affords $Ti[N(SiMe_3)_2]_3$.[7] In contrast, this amide complex cannot be made from the purple form of $TiCl_3$.

The reactivity studies show that the reduction of $TiCl_4$ with $Me_3SiSiMe_3$ gives a form of $TiCl_3$ that is more reactive than the purple form of $TiCl_3$ available commercially.

References

1. S. P. Narula and H. K. Sharma, *Inorg. Synth.*, **24**, 181 (1986).
2. G. Natta, P. Corradini, I. W. Bassi, and L. Porri, *Atti Accad. Nazl. Lincei, Rend. Classe Sci. Fis. Mat. Nat.*, **24**, 121 (1958).
3. G. Natta, P. Corradini, and G. Allegra, *Polym. Sci.*, **51**, 399 (1961).
4. L. E. Manzer, *Inorg. Synth.*, **21**, 135 (1982).
5. R. J. H. Clark, J. Lewis, D. S. Machin, and R. S. Nyholm, *J. Chem. Soc.*, **1963**, 379.
6. G. W. A. Fowles, T. E. Lester, and R. A. Walton, *J. Chem. Soc. A*, **1968**, 1081.
7. R. G. Copperthwaite, *Inorg. Synth.*, **18**, 112 (1978).

[§] Communication to the editor: Narula subsequently confirmed the above results and has isolated pure ($\sim$99%) $TiCl_3$ by performing the experiment at reduced pressure followed by repeated (10–15) washings of the product with 5–7 mL portions of warm (33°C) CCl_4/CH_3Cl under anhydrous conditions. *Anal.* Calcd. $TiCl_3$, Ti, 31.06; Cl, 68.93; Found: Ti, 30.67, Cl, 68.63%. A dark brown resinous mass ($\sim$0.1 g) was recovered from the filtrate after evaporation of solvent. ^{1}H NMR spectra identify (Me_3Si/Me_3SiO) groups in the impurity.

CONTRIBUTOR INDEX
Volume 32

SUBJECT INDEX

Note: The Subject Index for *Inorganic Syntheses, Vol. 32* is based on the Chemical Abstracts Service (CAS) Registry nomenclature. All non-cross referenced entries consist of the CAS Registry Name, the CAS Registry Number and the page reference. The inverted form of the CAS Registry Name (parent index heading) is used in the alphabetically ordered index. For the reader's convenience cross references from synonyms to CAS Names are included.

314

FORMULA INDEX
Volume 32

Note: The formulas in the *Inorganic Syntheses, Vol. 32* Formula Index follow the CAS Registry system.

323

CHEMICAL ABSTRACTS SERVICE
REGISTRY NUMBER INDEX